Probiotics and Prebiotics

Where Are We Going?

Edited by

Gerald W. Tannock

University of Otago, Dunedin
New Zealand

Copyright © 2002
Caister Academic Press
32 Hewitts Lane
Wymondham
Norfolk NR18 0JA
England

www.caister.com

British Library Cataloguing-in-Publication Data

A catalogue record for this book is available from the British
Library

ISBN: 0-9542464-1-1

*Printed and bound in Great Britain
by IBT Global, Barking, London IG11 0JP*

Contents

List of Contributors   vi

Preface   viii

Chapter 1
Probiotics and Prebiotics: Where Are We Going?   1
Gerald W. Tannock

Chapter 2
Fluorescence *In Situ* Hybridisation as a Tool in Intestinal   41
Bacteriology
Hermie J. M. Harmsen and Gjalt W. Welling

Chapter 3
From Composition to Functionality of the Intestinal Microflora   59
Sergey R. Konstantinov, Nora Fitzsimon, Elaine E. Vaughan,
and Antoon D. L. Akkermans

Chapter 4
Genus- and Species-Specific PCR Primers for the Detection and   85
Identification of Bifidobacteria
Takahiro Matsuki, Koichi Watanabe, and Ryuichiro Tanaka

Chapter 5
Prebiotic Oligosaccharides: Evaluation of Biological Activities   107
and Potential Future Developments
Robert A. Rastall and Glenn R. Gibson

Chapter 6
Prebiotics and Calcium Bioavailability   149
Kevin Cashman

Chapter 7
The Possible Role of Probiotic Therapy in Inflammatory   175
Bowel Disease
Michael Schultz and Heiko C. Rath

Chapter 8
Gut Microflora and Atopic Disease   239
Clare S. Murray and Ashley Woodcock

Chapter 9
Genomic Perspectives on Probiotics and the Gastrointestinal   263
Microflora
Olivia E. McAuliffe and Todd R. Klaenhammer

Chapter 10
Intestinal Microflora and Homeostasis of the Mucosal Immune   311
Response: Implications for Probiotics?
Stephanie Blum and Eduardo J. Schiffrin

Index   331

Journals of Related Interest

Current Issues in Intestinal Microbiology

Editor-in-Chief: *Gerald W. Tannock, University of Otago*
Editors: *Rod D. Berg, Louisiana State University Medical School; Antoon Akkermans, Wageningen Agricultural University; Ian Rowland, University of Ulster*

ISSN 1466-531X (print)
ISSN 1466-5328 (on-line)

CIIM is abstracted in Index Medicus, Medline, Pubmed, CAB Abstracts, CAB Health, EMBASE (Excerpta Medica), and BIOBASE (Current Awareness in Biological Sciences).

Current Issues in Intestinal Microbiology (CIIM) is a peer-reviewed, international journal. CIIM publishes high-quality papers, in English, in all areas of intestinal microbiology pertaining to human and other animal systems, including: microbial ecology, pathogenic organisms, probiotics and prebiotics. The journal is published in print and electronic forms. Two issues per year. Subscription information from: www.horizonpress.com

Journal of Molecular Microbiology and Biotechnology

Editor-in-Chief: *Professor Milton H. Saier, University of California at San Diego*
Senior Editors: *Professor Dr. Peter Dürre, Universität Ulm; Professor Masayori Inouye, Robert Wood Johnson Medical School; Professor Takeshi Mizuno, Nagoya University; Professor George P.C. Salmond, University of Cambridge; Professor Peter Reeves, University of Sydney*

ISSN 1464-1801

The Journal of Molecular Microbiology and Biotechnology (JMMB) is abstracted in Index Medicus, Medline, Pubmed, BIOSIS, Chemical Abstracts, Cambridge Scientific Abstracts, CAB Abstracts, CAB Health, EMBASE (Excerpta Medica), and BIOBASE (Current Awareness in Biological Sciences), Science Citation Index Expanded, Research Alert, Biotechnology Citation Index, and Journal Citation Reports Science.

JMMB publishes high-quality primary research papers in all areas of molecular microbiology and molecular biotechnology. It also features written symposia on unified topics, timely reviews, minireviews, and a section devoted to correspondence and comments. JMMB provides a medium for the amalgamation of the disciplines of molecular microbiology and biotechnology in the genomics era. Subscription information from: www.horizonpress.com

Molecular Biology Today

Editors: Samy Ashkar, Boston MA 02115; Asim K. Bej, Univ. Alabama at Birmingham; Andrea Cabibbo, Milano; Keith Firman, Portsmouth

MBT is abstracted in BIOSIS, CAB Abstracts, CAB Health, EMBASE (Excerpta Medica), and BIOBASE (Current Awareness in Biological Sciences). Subscription information from: www.caister.com

Current Issues in Molecular Biology

Editors: Liam Good, Karolinska Institutet; Ram H. Datar, University of Southern California; Geoff Symonds, Johnson & Johnson Research Laboratories; J. David McDonald, Wichita State University

CIMB is abstracted in Index Medicus, Medline, Pubmed, BIOSIS, CAB Abstracts, CAB Health, Cambridge Scientific Abstracts, EMBASE (Excerpta Medica), and BIOBASE (Current Awareness in Biological Sciences). Subscription information from: www.caister.com

Contributors

Antoon D. L. Akkermans
Laboratory of Microbiology
Wageningen University
Hesselink van Suchtelenweg 4
6703 CT Wageningen
The Netherlands
Email:
antoon.akkermans@algemeen.micr.wag-
ur.nl

Stephanie Blum
Nestle Research Centre
Vers-chez-les-Blanc
1000 Lausanne 26
Switzerland
Email:
stephanie.blum-sperisen@rdls.nestle.com

Kevin Cashman
Dept. Food Science, Food Technology,
and Nutrition, and Dept. Medicine
University College Cork
Cork
Ireland
Email: k.cashman@ucc.ie

Nora Fitzsimons
Dairy Products Research Centre
Teagasc, Moorepark
Fermoy
Co. Cork
Ireland
Email: nfitzsimons@moorepark.teagasc.ie

Glenn R. Gibson
Unit of Food Microbial Sciences
School of Food Biosciences
University of Reading
Whiteknights
PO Box 226
Reading RG6 6AP
United Kingdom
Email: g.r.gibson@reading.ac.uk

Hermie J. M. Harmsen
Department of Medical Microbiology
University of Groningen
Hanzeplein 1
9713 GZ Groningen
The Netherlands

Todd R. Klaenhammer
Department of Food Science
Campus Box 7624
North Carolina State University
Raleigh, NC 27695-7624
USA
Email: klaenhammer@ncsu.edu

Sergey R. Konstantinov
Laboratory of Microbiology
Wageningen University
Hesselink van Suchtelenweg 4
6703 CT Wageningen
The Netherlands
Email:
sergey.konstantinov@algemeen.micr.wag-
ur.nl

Takahiro Matsuki
Yakult Central Institute for Microbiological
Research
1796 Yaho, Kunitachi
Tokyo 186-8650
Japan
Email: takahiro-matsuki@yakult.co.jp

Olivia E. McAuliffe
Department of Food Science
Campus Box 7624
North Carolina State University
Raleigh, NC 27695-7624
USA
Email: oemcauli@unity.ncsu.edu

Clare S. Murray
North West Lung Centre
Wythenshawe Hospital
Manchester M23 9LT
United Kingdom

Heiko C. Rath
Klinik und Poliklinik für Innere Medizin I
Klinikum der Universität
93042 Regensburg
Germany
Email:
Heiko.Rath@klinik.uni-regensburg.de

Robert A. Rastall
Unit of Food Microbial Sciences
School of Food Biosciences
University of Reading
Whiteknights
PO Box 226
Reading RG6 6AP
United Kingdom
Email:
R. A. Rastall@reading.ac.uk

Eduardo Schiffrin
Nestle Research Centre
Vers-chez-les-Blanc
1000 Lausanne 26
Switzerland

Michael Schultz
Klinik und Poliklinik für Innere Medizin I
Klinikum der Universität
93042 Regensburg
Germany
Email:
michael.schultz@klinik.uni-regensburg.de

Ryuichiro Tanaka
Yakult Central Institute for Microbiological
Research
1796 Yaho, Kunitachi
Tokyo 186-8650
Japan

Gerald W. Tannock
Department of Microbiology
University of Otago
PO Box 56
Dunedin
New Zealand
Email:
gerald.tannock@stonebow.otago.ac.nz

Koichi Watanabe
Yakult Central Institute for Microbiological
Research
1796 Yaho, Kunitachi
Tokyo 186-8650
Japan

Elaine E. Vaughan
Laboratory of Microbiology
Wageningen University
Hesselink van Suchtelenweg 4
6703 CT Wageningen
The Netherlands
Email:
elaine.vaughan@algemeen.micr.wag-ur.nl

Gjalt W. Welling
Department of Medical Microbiology
University of Groningen
Hanzeplein 1
9713 GZ Groningen
The Netherlands
Email: g.w.welling@med.rug.nl

Ashley Woodcock
North West Lung Centre
Wythenshawe Hospital
Manchester M23 9LT
United Kingdom
Email: awoodcock@fs1.with.man.ac.uk

Preface

My concluding remarks published three years ago in "*Probiotics: a critical review*" contained three predictions.

1. New technologies would lead to better analysis of complex intestinal communities.
2. Improved, controlled, trials of probiotics would be carried out.
3. Greater collaboration between research disciplines would lead to greater gains in knowledge.

These predictions have proved to be accurate at least as far as research concerning the composition of the gut microflora, and the impact of probiotic consumption on it, is concerned. This can be confirmed by a quick search of the gut microflora/ probiotic literature published since 1999.

Although I had envisioned a five year period would be necessary to await further significant developments, such has been the progress in the field of gut microflora research, and the popularity of *Probiotics: a critical review*, that the publication of a new book on the topic seemed necessary.

The chapters of this book contain state-of-the-art commentaries on aspects of probiotics, prebiotics and the gut microflora that will lead us into the next period of exciting research. Readers will note a particular emphasis on the possible application of probiotics and prebiotics to specific health problems of humans in this book. In my view, this will be a rewarding aspect of gut microflora research during the next few years. Good research requires good technologies, and the acknowledged experts in the field of microflora analysis cover this aspect in some of the chapters. In the penultimate chapter we learn how genomic research will aid in our understanding of the gut microflora.

I am indebted to the contributing authors for their timely and informative chapters that together provide an authoritative view of the "state-of-play". Thank you very much.

G. W. Tannock

From: *Probiotics and Prebiotics: Where Are We Going?*
Edited by: Gerald W. Tannock

Chapter 1

Probiotics and Prebiotics: Where Are We Going?

Gerald W. Tannock

Abstract

Perspectives concerning the composition of the human gut microflora have changed drastically since the concept of probiotics, and even prebiotics, was introduced. Culture-independent, nucleic acid-based, methods of analysis have provided results that demonstrate the complexity of the gut microflora and the quantitative and qualitative dominance of previously little known bacterial species. The need for invasive sampling techniques has also become apparent because reliance on the analysis of the faecal microflora may fail to accurately reflect the true state of affairs in the proximal colon. Probiotic administration transiently alters the gut microflora by donating bacterial cells to the ecosystem. It may be possible to use carefully formulated products in the alleviation of inflammatory bowel diseases or allergic diseases if the safety of such probiotics can be guaranteed. Prebiotics may be useful in the study of the regulation of the gut ecosystem as well as in attempts to rectify "abnormal" microfloras. The molecular foundations of the gut microflora-host relationship should be pursued using functional genomics, and clarification of the impact of the gut microflora on host nutrition and host physiology is necessary. To recognise the potential advantages of probiotic and prebiotic use, we need to understand how bacterial cells function in the gut ecosystem, and how bacterial functions impact on the host.

Metchnikoff Had It Easy

Elie Metchnikoff (1845-1916), a Nobel prize winner for his pioneering observations and descriptions of phagocytosis, was interested in the ageing process. While modern research on this topic concentrates on the maintenance of non-mutated DNA sequences, Metchnikoff focussed on the gut microflora as a source of intoxication from within. According to Metchnikoff, the bacterial community residing in the large bowel of humans was a source of substances toxic to the nervous and vascular systems of the host. These toxic substances, absorbed from the bowel and circulating in the bloodstream, contributed to the ageing process. Gut bacteria were thus identified as the causative agents of "autointoxication". The offending bacteria were capable of degrading proteins (putrefaction), releasing ammonia, amines and indole that, in appropriate concentrations, were toxic to human tissues. Metchnikoff inferred that low concentrations of toxic bacterial products could escape detoxification by the liver and enter the systemic circulation (Metchnikoff 1907; 1908). His solution for the prevention of autointoxication was radical: surgical removal of the large bowel. A less frightening and more popular remedy, however, was to attempt to replace or diminish the number of putrefactive bacteria in the intestine by enriching the gut microflora with bacterial populations that fermented carbohydrates and had little proteolytic activity. Oral administration of cultures of fermentative bacteria would, it was proposed, "implant" the "beneficial" bacteria in the intestinal tract. Lactic acid-producing bacteria were favoured as fermentative bacteria to use for this purpose since it had been observed that the natural fermentation of milk by these microbes prevented the growth of non-acid-tolerant bacteria, including proteolytic species (Metchnikoff 1907). If a lactic fermentation prevented the putrefaction of milk, would it not have the same effect in the digestive tract if appropriate bacteria were used? Eastern Europeans, some of whom were apparently long-lived, consumed fermented dairy products as part of their diet. Taken as proof of efficacy, fermented milk products were introduced to Western Europe as health-related foods: the birth of probiotics ("live microbial feed supplement which beneficially affects the host animal by improving its intestinal microbial balance" [Fuller 1989]).

In those days, it was easy to believe that the composition of the gut microflora could be so easily modifed by the consumption of bacteria in food. Even until the 1960s, descriptions of the gut microflora were quite simple. Clostridia, lactobacilli, enterococci and *Escherichia coli* were commonly considered to be the numerically predominant bacteria in faecal samples, reflecting the relative ease of culture of these bacteria (Rettger *et al.* 1935). Rumen microbiologists, thanks to the innovative techniques derived by Hungate, demonstrated the importance of obligately anaerobic bacteria and other microbes in the rumen fermentation, and hence the reliance of the ruminant host on microbial metabolic products for nutritional well-being (Hungate 1966). Holdeman and Moore modified the roll tube techniques of Hungate for use in investigations of anaerobic infections of humans (Holdeman and Moore 1972). The derivation of selective media for anaerobes commonly implicated in anaerobic infections, together with the use of anaerobic gloveboxes, accelerated the acquisition of knowledge concerning the reservoir of these anaerobes: the gut microflora (Aranki and Freter 1972; Summanen *et al.,* 1993). Major studies were carried out in the USA during the 1970s in relation to the influence of diet on the compostion of the gut microflora and the occurrence of colon cancer. These studies gathered a wealth of information concerning the composition of the intestinal microflora of humans. Some bacterial populations were detected at levels of 10^{10} per gram (wet weight) of faeces and it was estimated, by statistical extrapolation from the number of species that had actually been cultured, that something like 400 bacterial species might reside in the human gut (Moore and Holdeman 1974). It was considered, however, that 30-40 species constituted 99% of the faecal microflora for any given individual (Drasar and Hill 1974a).

Perspectives change as knowledge increases.

Uncertain Times

An ecosystem containing hundreds of bacterial species for study should be paradise for a bacteriologist. Yet analysis of this complex bacterial community is fraught with difficulties. Until the 1990s, analysis of the composition of the gut microflora relied on the use of traditional bacteriological methods of culture, microscopy and identification (O'Sullivan 1999). Selective bacteriological culture media were essential for accurate analysis of the microflora because they enabled enumeration of specific bacterial populations to be made (Summanen *et al.,* 1993). Unfortunately, few culture media used in the analysis of the microflora are absolutely selective and misinterpretations by the novice are easily made. Not all of the species comprising a population may be able to proliferate with equal ease on the selective medium. This introduces bias to the results.

Even in the 1970s, researchers had observed that the total microscopic count of bacterial cells in human faecal smears was always higher than the total viable count (CFU, colony-forming units) obtained by culture on a non-selective agar medium. It was claimed, however, that good bacteriological methods would permit the culture of 88% of the total microscopic count. But this comparison was obtained by using total microscopic 'clump' counts (aggregates of bacterial cells) rather than by counting individual bacterial cells in smears (Moore and Holdeman 1974). While valid from the point of view that 'colony forming units' on agar plates have not necessarily arisen from a single bacterial cell, the comparisons gave a false sense of confidence with regard to analytical results at that time. Total bacteria microscopic counts, utilising the 4',6-diamidino-2-phenylindole (DAPI) stain and computer imaging, has revealed average total bacterial cell counts in human faeces approaching 1×10^{11} per gram (wet weight) (Tannock *et al.*, 2000). State-of the-art bacteriological methodologies still only permit about 40% of this bacterial community to be cultivated on non-selective agar medium in the laboratory (Tannock *et al.*, 2000). Thus a large proportion of the bacterial cells seen in microscope smears have never been investigated. Although some of these cells may be non-viable, it is likely that many are viable but non-cultivable due to their fastidious requirements for anaerobiosis or, more likely, due to the complex nutritional interactions that can occur between the inhabitants of bacterial communities. These nutritional complexities may be difficult, if not impossible, to achieve in laboratory culture media.

Non-cultivable bacteria are not the only problem in studying the human gut microflora. Most investigations of the gut microflora of humans have been studies of the bacterial content of faeces. Does the study of faecal samples provide accurate information concerning the gut microflora? Based on Savage's (Savage 1983) studies of the murine gastrointestinal tract, the answer is "no". One could not deduce that significant epithelial associations occur in the digestive tract of mice (Table 1) from the bacteriological examination of murine faeces (Savage 1977). Yet, unlike mice, the gut microflora of humans is essentially confined to the ileum and colon, and it has not been clearly demonstrated by culture or microscopy that extensive association of bacteria with mucosal surfaces, such as those demonstrated in mice, occurs in the human host. Bacteria associated with biopsies collected from various regions of the human colon can be detected by cultural (Alander *et al.*, 1999)

Table 1. Mucosal associations in the murine gastrointestinal tract

Gastrointestinal Region	Type of Association
Forestomach	Lactobacilli adhere directly to the surface of the stratified, keratinised, squamous epithelium lining this part of the murine stomach. A layer of *Lactobacillus* cells covers the epithelium.
Ileum	Segmented, filamentous bacteria (related to the genus *Clostridium*) form attachment sites in epithelial cells on villous surfaces. One end of the filament is attached to an epithelial cell. The bacteria are often densely associated with the mucosal surface in the vicinity of Peyer's patches.
Proximal colon	The mucus layer on the mucosal surface, particularly between the rugae, is densely colonised by a mixture of fusiform- and rod-shaped bacteria. Very motile, helical bacterial cells predominate in the crypts from which mucus flows.

For further details see Savage 1977 and 1983.

and molecular methods but bacterial numbers are variable (8.6×10^4-1.1×10^6 cells per biopsy; [Zoetendal 2001]).

While Moore *et al.* (1978) demonstrated that the composition of the faecal microflora reflected that of the distal colon, we know that the microflora is more metabolically active in the proximal colon compared to the distal colon (Cummings and Macfarlane 1991). This is a reflection of the relative availability of fermentable substrates in the two sites and is of significance with respect to the analysis of the effects of non-digestible food additives (oligosaccharides, resistant starch) on the intestinal ecosystem. Biochemical changes occurring in the proximal colon may not be detectable in the distal colon or, therefore, in the faeces (Crittenden 1999). Recently, it has been reported that the composition of the microflora of the proximal large bowel differs from that of the faeces (Marteau *et al.,* 2001). Somewhat surprisingly, facultative anaerobes were more numerous in the caecal contents compared to the faeces where obligate anaerobes predominated. Perhaps, the facultatively anaerobic bacteria detected in faeces represent remnants of much larger populations inhabiting the proximal colon?

Sampling methods for gut microflora research
will become more sophisticated and more invasive.

5

Table 2. Cultivated bacterial species within the *Bacteroides-Prevotella* group

Porphyromonas salivosa	*Bacteroides caccae*
Porphyromonas catoniae	*Bacteroides ovatus*
Porphyromonas gingivalis	*Bacteroides thetaiotaomicron*
Porphyromonas cangingivalis	*Bacteroides forsythus*
Porphyromonas endodontalis	*Bacteroides distasonis*
Rikenella microfusus	*Bacteroides merdae*
Cytophaga fermentans	*Prevotella heparinolytica*
Bacteroides putredinis	*Prevotella pallens*
Bacteroides splanchnicus	*Prevotella veroralis*
Bacteroides vulgatus	*Prevotella denticola*
Bacteroides fragilis	*Prevotella oulora*
Bacteroides stercoris	*Prevotella oris*
Bacteroides uniformis	*Prevotella oralis*
Bacteroides eggerthii	*Prevotella ruminicola*

Table 3. Cultivated bacterial species within the *Clostridium coccoides* group

Clostridium polysaccharolyticum	*Eubacterium xylanophilum*
Clostridium herbivorans	*Eubacterium ventriosum*
Clostridium populeti	*Eubacterium eligens*
Clostridium coccoides	*Eubacterium formicigenerans*
Clostridium nexile	*Eubacterium contortum*
Clostridium oroticum	*Eubacterium rectale*
Clostridium clostridiiforme	*Eubacterium hadrum*
Clostridium celerecrescens	*Eubacterium halii*
Clostridium xylanolyticum	*Eubacterium ramulus*
Clostridium symbiosum	*Lachnospira pectinoschiza*
Clostridium aminovalericum	*Ruminococcus obeum*
Clostridium aminophilum	*Ruminococcus hansenii*
Coprococcus eutactus	*Ruminococcus productus*
Butyrivibrio crossotus	*Ruminococcus torques*
Butyrivibrio fibrisolvens	*Ruminococcus gnavus*
	Roseburia cecicola

Table 4. Cultivated bacterial species within the *Clostridium leptum* group

Fusobacterium prausnitzii	*Ruminococcus bromii*
Eubacterium siraeum	*Eubacterium plautii*
Clostridium sporosphaeroides	*Clostridium viride*
Clostridium leptum	*Eubacterium desmolans*
Ruminococcus flavefaciens	*Termitobacter aceticus*

Family Trees and Even Newer Perspectives

Carl Woese's molecular phylogenetic studies of microorganisms revolutionised our understanding of biological diversity and evolution (Woese 1987). The phylogenetic framework provided by the comparison of 16S ribosomal RNA (rRNA) gene sequences provides a conceptual approach to microbial identification and taxonomy. 16S ribosomal RNA gene sequences (16S rDNA) contain regions conserved across all bacterial species interspersed with regions (V1-V9) in which the nucleotide base sequences are variable among bacterial types (Stackebrandt and Goebel 1994). Sometimes, the variable region sequences are highly species-specific. Comparison of 16S rDNA similarities can therefore be used in the identification of bacterial species and consequently, in the analysis of bacterial communities (Raskin *et al.,* 1997). Universal primers can be used in polymerase chain reactions (PCR) to amplify 16S rDNA from bacterial cells in natural samples. The amplified 16S rDNA sequences are cloned, screened (some sequences will have been cloned more than once), and sequenced. Alignment of the sequences with those stored in databanks permits the recognition of which species were represented in the habitat, and detects those that cannot be cultivated by conventional bacteriological techniques. In a study of this type reported by the Dore group (Suau *et al.,* 1999), three bacterial divisions represented 95% of the faecal microflora of a human subject: *Bacteroides-Prevotella*, *Clostridium coccoides* group, and *Clostridium leptum* group. The diversity within these divisions is, however, vast (Tables 2-4) and a full inventory of the inhabitants of the human gut awaits the results of further nucleic acid-based studies.

The Welling group at the University of Groningen, The Netherlands (Franks *et al.,* 1998) has pioneered fluorescent in situ hybridisation (FISH) as a means of investigating the intestinal microflora. They have derived oligonucleotide probes, labelled with fluorescent dyes, that target 16S rRNA sequences. Although an array of hundreds of probes could be used to enumerate bacterial species in faecal samples, due to practical considerations, a small collection of probes that recognise large phylogenetic groups of bacteria have been used. It has been estimated that these probes currently detect about ninety per cent of the members of the faecal microflora (Harmsen and Welling, Chapter 2). Technical difficulties can influence the accuracy of the results. The oligonucleotide probes must reach their target sequence, which is inside the bacterial cell, by passing through the cell wall. This is more easily achieved with some bacterial species than with others (Welling *et al.*, 1997). The method is best for enumeration of the numerically predominant members of the microflora, the lowest level of detection being 10^6 cells per gram (Welling *et al.*, 1997).

DNA-RNA hybridisations provide a means of determining the proportions in which specific bacterial groups occur within the microflora. In this method, bacterial RNA is extracted from samples and dot-blotted to membranes. The membranes are probed with a collection of radioactively labelled oligonucleotides, each specific for a particular bacterial group. A universal probe (currently Bact338) that hybridises to a conserved rRNA sequence in the majority of bacterial cells is used as a reference against which the hybridisation results of the other probes are compared. This provides a means of calculating the proportions that the various populations form in the total bacterial community. In work carried out in France, six oligonucleotide probes detected, on average, 70% of the 16S rRNA hybridised by the universal probe in faecal extracts from 27 human subjects. *Bacteroides-Prevotella-Porphyromonas* accounted for 37% of the faecal microflora, *Clostridium coccoides* group for 16% and the *Clostridium leptum* group for 14%. Bifidobacteria, enterobacteria and *Lactobacillus-Streptococcus-Enterococcus* each comprised only about 1% of the microflora (Sghir *et al.*, 2000).

Oligonucleotide probes can only be derived if the microbial members of the ecosystem are known. Phylogenetic analysis of the community, such as that carried out by Suau *et al.* (1999), or perusal of the 16S rDNA sequences derived from cultured bacterial species, are thus prerequisite to successful use of FISH or DNA-RNA hybridisations. Another molecular approach to monitoring the composition of complex communities overcomes this limitation. PCR coupled with temperature gradient gel electrophoresis (TGGE) has been shown by Zoetendal *et al.* (1998) to provide an excellent method for comparative monitoring of the faecal microflora. Since this pioneering work, TGGE has been superseded by denaturing gradient gel electrophoresis (DGGE).

In PCR/DGGE, fragments of the 16S rRNA gene are amplified by PCR. One of the primers has a GC-rich 5' end (GC clamp). 16S fragments are amplified from the bacterial community in the sample and denaturing gradient gel electrophoresis separates the 16S molecular species within the resulting mixture. The double-stranded 16S fragments migrate through a polyacrylamide gel containing a gradient of urea and formamide until they are partially denatured by the chemical conditions. The fragments do not completely denature because of the GC clamp, and migration is radically slowed when partial denaturation occurs. Because of the variation in the 16S sequences of different bacterial species, chemical stability is also different; therefore different 16S 'species' can be separated by this electrophoretic method. A profile of 16S sequences amplified from the sample is thus obtained. DNA fragments can be cut from the gel and re-amplified for sequencing and hence identification of the bacterial species or group from which the fragment was derived (Muyzer and Smalla 1998; Zoetendal *et al.,* 1998). A sequence from the bacterial species or phylogenetic division identified in this way might be used in further studies as the target for FISH enumeration, or DNA-RNA hybridisation, of the specific bacterial population. PCR, wonderful molecular biological tool though it is, poses some problems: while culture bias is removed, another bias is introduced because polymerase chain reactions are known to amplify DNA sequences from mixed populations with differing efficiency (Reysenbach *et al.,* 1992). Chimeric sequences can be derived during PCR where there is a mixture of template DNAs in the reaction mix (Kopczynski *et al.,* 1994), and there can be heterogeneity with regard to 16S rRNA sequences within species, and even within a single bacterial cell (Nubel *et al.,* 1996). Nevertheless, the depth and breadth of gut microflora research has been considerably improved through the use of nucleic acid-based methodologies.

> *Nucleic acid-based methods of analysis are now*
> *essential tools in gut microflora research.*

Getting the Right Perspective on Lactobacilli and Bifidobacteria

Members of the genus *Bifidobacterium* are non-sporing, non-motile, gram-positive, rod-shaped bacteria. Some species, under particular conditions of culture, may show cells with a branching morphology. They are obligately anaerobic bacteria, producing major amounts of acetic and lactic acid from the fermentation of glucose (Scardovi 1986). They represent no more than 10% of the faecal microflora of adult humans when analysis is by culture, and 1-3% when analysis is by molecular methodologies (Langendijk *et al.,*

1995; Franks *et al.,* 1998; Sghir *et al.,* 2000). Japanese authors have reported that the numbers of bifidobacteria in faeces decline in elderly subjects (Mitsuoka 1992). There is considerable commercial interest in boosting bifidobacterial numbers in the gut through the consumption of yoghurts containing bifidobacteria ('bifidus' yoghurts) or by the addition to foods of 'prebiotics' (non-digestible carbohydrates which pass to the colon where they become fermentable substrates for some members of the microflora, including bifidobacteria). This commercial activity is aimed at maintaining, or even improving, the health of the consumer because bifidobacteria are putatively beneficial to the intestinal ecosystem (Gibson *et al.,* 1995; Gibson and Roberfroid 1995; Gibson 1998). The population size of bifidobacteria required to produce a 'healthy' effect has not, however, been elucidated.

Interest in bifidobacteria is derived mainly from the observation of Tissier that these bacteria were commonly present in the faeces of infants (Tissier 1905). A more varied microflora was evident in the faeces of cow's milk formula-fed infants. Since paediatricians recognised, and still recognise, that infants receiving mothers milk have less intestinal upsets than formula-fed children, the inference that bifidobacteria were associated with health was made. Modern formula feeds for infants resemble much more closely human milk. Probably for this reason, it is now more difficult to recognise different microfloras associated with breast-fed or formula-fed infants on the basis of bifidobacterial numbers. Recent investigations using FISH, however, show that bifidobacteria are much more numerous than in adult faeces comprising up to 75% of the total faecal microflora of formula-fed infants and up to 91% of the microflora of breast-fed infants (Harmsen *et al.,* 2000).

Members of the genus *Lactobacillus* are non-sporing, gram-positive, rod-shaped bacteria. Intestinal isolates grow best under anaerobic conditions, but many can be cultivated microaerophilically. Lactobacilli produce a major amount of lactic acid when they ferment glucose (Kandler and Weiss 1986) and are common components of 'probiotics' (Goldin and Gorbach 1992). Despite the intense interest in lactobacilli as probiotics, these bacteria make up no more than one per cent of the total bacterial community of human faeces . Lactobacilli may be more numerous in the proximal colon (caecum) compared to the faeces (Marteau *et al.,* 2001), although this was not encountered by earlier investigators (Gorbach 1967). Finegold *et al.* (1983) detected lactobacilli in 73% of 62 healthy Americans when a single faecal sample from each was examined. Many of the *Lactobacillus* species detected in human faeces are utilised in the preparation of fermented food products (Tannock *et al.,* 2000; Walter *et al.,* 2001) and it seems likely that these are allochthonous strains. Possibly relatively few *Lactobacillus* species are

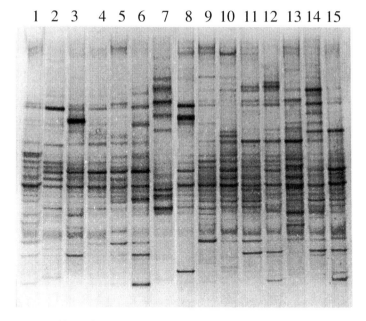

Figure 1. Faecal bacterial community profiles generated by PCR/DGGE. Lanes 1-15 show the profiles obtained from each of 15 healthy adult New Zealanders (Stebbings, Munro, and Tannock, Unpublished data).

autochthonous to the human digestive tract. *Lactobacillus ruminis* and *Lactobacillus salivarius* appear noteworthy in this respect (Tannock *et al.,* 2000).

Lactobacilli comprise a tiny portion, numerically, of the faecal microflora. Their population density in the proximal large bowel requires further investigation. Bifidobacteria form a major portion of the infant faecal microflora.

Everything Changes, Yet Everything Stays the Same

In general, ecosystems are recognised to have a large degree of stability over time (Alexander 1971).This is because, when the environment changes, homeostatic reactions come into play to restore the relationships that pre-existed among the populations forming the community. PCR/DGGE analysis of the faecal microflora has provided two important observations. Firstly, each human had a unique bacterial community in their faeces because the PCR/DGGE profile of each human was distinct from the profiles of other humans (see Figure 1 for examples). The uniqueness of human gut microfloras is supported by the results of analyses using other nucleic acid-

Table 5. Variation in the composition of the faecal microflora of 15 healthy humans

Bacterial group	% of total faecal microflora (n = 15)[1]
Bacteroides-Prevotella	1.5-96.3
Bifidobacterium	0.4-44.6
Atopobium-Eggerthella-Colinsella	0.1-23.0
Eubacterium rectale-Clostridium coccoides	6.2-69.0
Clostridium leptum	0.5-38.5

[1]Measurements made by FISH.
Stebbings, Harmsen, Welling, Munro, and Tannock, 2001, unpublished data.

based methods (Table 5). Second, homeostasis was marked because the same distinctive profile was generated from all samples obtained from each individual. DNA fragments that had migration distance in common could be discerned in profiles from different humans, though, so it is tempting to think that these represent the "true (or core) faecal microflora" of humans These bacteria that are harboured in common obviously require further investigation.

Homeostasis makes modification of the composition of the gut microflora a difficult prospect because of the tendency of ecosystems to resist alterations. This is true of the gut ecosystem where detailed studies have shown that consumption of probiotic bacteria results in their transient detection in the faeces. Once consumption ceases, the probiotic bacteria are no longer detected (Spanhaak *et al.,* 1998; Dunne *et al.,* 1999; Tannock *et al.,* 2000; Satokari *et al.,* 2001). These studies showed that introducing lactobacilli or bifidobacteria at one end of the digestive tract resulted in their appearance at the other end. It is not known whether probiotic bacteria metabolise during transit through the human gut. Experiments to determine the metabolic activity of allochthonous bacteria in the gut could be carried out, perhaps using fluorescent green protein technology (Drouault *et al.,* 1999). Colonisation of the gut need not be prerequisite for the delivery of bioactive substances (cytokines, immunogens) to the gut environment, but metabolic activity would presumably be useful (Mercenier 1999; Steidler *et al.,* 2000).

The strain composition of bacterial populations in faeces can be altered transiently by administering a probiotic.

Differentiating Abnormal From Normal

Metchnikoff argued that the products of bacterial metabolism of proteins in the intestine provided a source of substances (amines, ammonia, urea) potentially toxic to the nervous and vascular systems and that the absorption of these substances from the intestine into the blood circulation contributed to the ageing process. He argued, in effect, that one could differentiate between "good" (lactic acid-producing bacteria) and "bad" (putrefactive) bacteria in the gut. This concept of good and bad bacteria is still mentioned today even though it is tenuous: both fermentation and proteolysis are characteristics of bacterial cells that live in an anaerobic environment and from which they obtain carbon and energy from organic compounds. There are few bacterial species associated with the human body that have not been implicated in at least one disease process, rare and opportunistic though these occurrences may be (Rautio *et al.,* 1999; Brouqui and Raoult 2001; Farina *et al.,* 2001; Woo *et al.,* 2002). Metchnikoff's view of the gut microflora was in a way, however, the start of the search for "abnormal microfloras".

The frequency of cancer of the colon varies geographically. The observed differences in occurrence of the disease are not due solely to genetic (ethnic) differences because migrants from a "low" incidence country settling in a "high" incidence country show a changed (increased) susceptibility to the disease once they assume the cultural habits of their adopted country. Comparison of statistics relating to cultural factors between low and high incidence countries suggest that dietary habits, especially fat and protein consumption, are linked to the development of the disease (Drasar and Hill 1974b). Dietary components and host secretions can be metabolised by members of the gut microflora, so it has been proposed that chemicals produced in the large bowel by bacterial metabolic activities may induce cancer of the colon. In the 1970's, dietary differences were believed to alter the composition of the gut microflora and two large studies were carried out in the USA to investigate this proposition (Finegold and Sutter 1978; Moore *et al.,* 1978). The studies did not shed light on the relationship between colon cancer and the microflora but the idea that certain diseases may be associated with a microflora of "altered" (abnormal) composition remained a tantalising concept for microbial ecologists.

Crohn's disease is a serious inflammatory disease of the gastrointestinal tract. It mainly affects the ileum and colon, but may occur in any region of the gut. Crohn's disease usually causes diarrhoea, abdominal pain, often fever, and sometimes rectal bleeding. Loss of appetite and subsequent weight loss may occur. While Crohn's disease afflicts people of all ages, it is mainly diagnosed in humans less than 30 years of age (Chadwick and Chen 1999).

Ulcerative colitis is also a disease that is mainly first diagnosed in young adults. It is characterised by episodes of diarrhoea with the passage of blood per rectum. Malaise, anorexia and weight loss vary greatly in degree. Most cases follow a chronic relapsing course with exacerbations and remissions. Ulcerative colitis involves the colon as a diffuse disease with distal predominance. The rectum is virtually always involved, and additional portions of colon may be involved extending proximally from the rectum in a continuous pattern. Unlike Crohn's disease in which all regions of the intestine may be affected, and in which there can be normal healthy bowel in between patches of diseased bowel, ulcerative colitis involves the mucosa of the colon in a continuous manner (Chadwick and Chen 1999). Crohn's disease and ulcerative colitis come under the umbrella term "inflammatory bowel diseases" (IBD).

Inflammatory bowel diseases appear to result from an inappropriate inflammatory response to luminal pathogen or an abnormal luminal constituent, autoimmunity, or an abnormal immune response to a normal luminal constituent such as members of the microflora, or dietary antigens. An abnormal response towards the microflora is currently the favoured explanation because decreasing bacterial numbers in the intestine can lead to clinical improvement and decreased inflammation. Members of the intestinal microflora provide a constant antigenic stimulus for the host's immune system. Normally, immunological tolerance towards the microflora prevents continuous intestinal inflammation (Duchmann *et al.*, 1995).While the aetiologies of Crohn's disease and ulcerative colitis remain unknown, there is compelling evidence from animal models that immunological tolerance is lost in genetically susceptible hosts which develop chronic immune-mediated colitis. This concept is supported by rodent models in which targeted deletion (gene knockout) or over-expression (transgenic) of a variety of genes that regulate immune or mucosal barrier function lead to an overly aggressive cellular immune response confined to the intestine (Elson *et al.*, 1995). In addition, transfer of subsets of normal T cells into immunodeficient mice demonstrates that different T cell populations induce or prevent colitis (Morrissey *et al.*, 1993). In most of these models, T-helper-1 (Th1) CD4+ lymphocytes secreting high levels of interferon-γ (IFN-γ) mediate disease (Berg *et al.*, 1996), while oral tolerance and protection is determined by T_{R1} (a new class of regulatory T cells) and T-helper-3 (Th3) regulatory cells producing interleukin-10 (IL-10) and transforming growth factor β (TGFβ) (Groux *et al.,* 1997). Results from experiments with germfree rodents, in which intestinal inflammation is absent or only mildly expressed, have indicated that bacteria are indispensable contributors to the pathogenesis of chronic, immune-mediated, intestinal inflammation (Sartor 1997). Germfree IL-10-deficient mice rapidly develop colitis when they are colonised with bacteria from healthy specific-pathogen-free animals (Rath *et al.,* 1996; Sellon *et al.,* 1998). IL-10, produced predominantly by macrophages and regulatory T lymphocytes (T_{R1} and T-helper-2 [TH2 cells]), suppresses TH1 responses by inhibiting production of interleukin-12 (IL-12) (D'Andrea *et al.,* 1993). IL-10-deficient mice colonised with a conventional microflora develop lethal enterocolitis but the disease is attenuated and confined to the colon when IL-10-deficient mice are colonised with bacteria from specific-pathogen-free animals, and is absent when the mice are maintained germfree (Sellon *et al.,* 1998). Colitis is evident by one week following bacterial colonisation. The colitis is severe with transmural disease and a 50% mortality by 5 weeks after colonisation of ex-germfree mice (Sellon *et al.,* 1998). Madsen and colleagues have demonstrated that ciprofloxacin or a combination of metronidazole/neomycin can prevent and treat colitis in IL-10-deficient mice, further documenting the role of intestinal bacteria in causing disease in this model (Madsen *et al.,* 2000).

To investigate the influence of HLA-B27 on inflammatory disorders, human HLA-B27 and β2-microglobulin transgenic rats were derived by micro-injection of both genes into fertilised one-cell eggs. These rats spontaneously develop colitis, gastritis, arthritis, dermatitis, orchitis, epididymitis, carditis, alopecia and nail changes. Non-bloody, sometimes episodic, diarrhoea is the earliest clinical sign of disease beginning at about 2 months of age. Almost 100% of animals are diseased by 6 months of age. The colon is moderately thickened with a granular mucosa, but without grossly detectable mucosal ulcerations or adhesions. Histological findings in the inflamed caecum and colon are infiltration into the mucosa of inflammatory cells (predominantly mononuclear cells), reduction in the number of goblet cells, diffuse crypt hyperplasia, rare crypt abscesses, and early ulcers limited to the lamina propria. Arthritis occurs in more than 70% of the transgenic animals, with a later onset at 3 months of age. It occurs mostly in the tarsal joints of the hind limbs and persists for several days to months, sometimes in a pattern of exacerbation and remission. Histological manifestations are accumulation of neutrophils in the joint space, hyperplasia, chronic synovial inflammation, and bony erosions (Sartor *et al.,* 1996). Germfree HLA-B27/β2 microglobulin transgenic rats do not exhibit colitis or arthritis, but develop these inflammatory conditions after colonisation by an intestinal microflora (conventionalisation) or defined mixtures of intestinal bacteria (Rath *et al.,* 1999). Treatment with metronidazole attenuates colitis and arthritis, suggesting a particularly important role for anaerobic bacteria (Rath *et al.,* 1995; 1998).

Observations from both animal models of colitis and humans with IBD therefore suggest that chronic intestinal inflammation is caused by unrestrained immune responses to the gut microflora and/or their products. In experimental animal models, T-helper 1 (Th1) CD4+ lymphocytes secreting high levels of IFN-γ mediate disease (Simpson *et al.,* 1997). The bacteria critical in stimulating this response in IBD models are not known, but nucleic acid-based methods of analysis will provide a means to pinpoint the members of the bacterial community associated with early and late phases of colitis in the well-characterized experimental animal models. This will add a more accurate bacteriological dimension to the immunologically sophisticated models used in IBD research. It may be, however, that the specific composition of the gut microflora is not important in IBD. Rather, it may be functional aspects of the gut microflora that are critical: the production of bacterial products that stimulate the production of IFN-γ on an immune dysfunctional background may be all that matters.

Inflammatory bowel diseases provide the best hope of detecting an "abnormal microflora" that may be rectified by the use of novel probiotics or prebiotics.

Biological Freudianism

'Biological Freudianism' was a term used by Dubos and colleagues to denote the lasting effects of early environmental influences on the physiology of animals (Dubos *et al.,* 1966). They reported, as an example, a lasting influence that depended on the composition of the gut microflora of mice. A specific-pathogen-free colony of Swiss mice (the NCS colony) was derived at the Rockefeller Institute in New York by sterile caesarean delivery of the initial breeding stock of animals. The mice were maintained under conditions of high sanitation and, unlike the mice from which they were derived (the SS colony), did not harbour gram-negative bacteria such as *Escherichia coli, Proteus vulgaris,* or *Pseudomonas* species as members of the intestinal microflora. The two types of mice were found to differ markedly in their response to the parenteral administration of endotoxin. NCS mice were tolerant to doses that were lethal when administered to SS mice. Exposure to SS mice during early life, or prior injection with small doses of heat-killed gram-negative bacteria that therefore gave exposure to minor amounts of endotoxin, increased the susceptibility of the NCS mice. Their sensitivity to endotoxin now matched that of the SS mice. Thus it was concluded that the SS microflora containing enterobacteria (bacteria that are normally numerous in the murine intestinal tract early in life) sensitised the animals so that, as adults, they were highly susceptible to endotoxin.

Comparison of the immunological characteristics of germfree (raised in the absence of microbes) and conventional (raised in the presence of microbes) animals has demonstrated that the presence of the normal microflora stimulates the reticuloendothelial tissues of the animal, particularly those tissues associated with the intestinal tract. Peyer's patches and mesenteric lymph nodes are more developed, the lamina propria of the intestinal wall contains more neutrophils and lymphocytes, and the serum contains a higher concentration of immunoglobulins in conventional animals compared to germfree (Gordon and Pesti 1971). Other striking differences between germfree and conventional animals concern intestinal intra-epithelial lymphocytes that in conventional mice are about equally divided into $\alpha\beta+$ and $\gamma\delta+$ subpopulations. In germfree mice, $\gamma\delta+$ lymphocytes are predominant in the intestinal epithelium. The $\alpha\beta+$ subpopulation quickly increases in size when ex-germfree mice acquire an intestinal microflora (conventionalisation) (Kawaguchi *et al.*, 1993; Umesaki *et al.*, 1993).

The influence of particular bacterial groups on the development of the immune system during early life has not been investigated systematically. Studies with germfree mice, monoassociated with bacterial species, or injected with heat-killed bacteria have been reported. These studies, reviewed by McCracken and Gaskins (1999), are open to several criticisms, and firm conclusions about the significance of specific bacteria in 'programming' the immune system is lacking.

The influence of the gut microflora on the development of the immune system needs to be studied intensively. A practical application of such studies would be to understand the increasing incidence of atopic disorders such as asthma of children in affluent countries (Burr *et al.*, 1989; Burney *et al.*, 1990; Bjorksten 1997). Atopic disorders are characterised by dominant T helper 2 (Th2) mechanisms and the production of immunoglobulin E (IgE) to common environmental antigens. Asthma is an atopic disorder characterised by activation and recruitment of eosinophils to the lung resulting in chronic swelling and inflammation of the airways. The reasons for the increase in atopic disorders in developed countries are unknown, but several factors in addition to hereditary predisposition that may be of significance, all concerning microbial exposure in childhood, have recently been identified in retrospective studies.

(1) A steady decline in the extent to which the human population is exposed to major human diseases such as tuberculosis, measles, and influenza (Romagnani 1997). These infections characteristically induce a T helper 1 (Th1) type immune response that produces an immunological environment rich in interferon-gamma. This, in turn, is viewed as a potent suppresser of Th 2 activity. The less frequent exposure of children to these infections, at an age when allergen sensitisation is occurring, has been speculated to increase the risk of developing atopy (Bjorksten 1997; Martinez and Holt 1999).

(2) Immunisation with whole-cell pertussis vaccine has been reported as a predictor of atopic disorders (Farooqi and Hopkin 1998). *Bordetella pertussis* is well known for its Th2-promoting potential and has been used experimentally as an adjuvant to stimulate antibody production. *Bordetella pertussis* is a strong promoter of IgE synthesis with an effect on primary and secondary antibody responses to heterologous antigens (Pauwels *et al.,* 1983).

(3) Treatment with oral antibiotics during the first two years of life has been identified as a predictor of subsequent atopic disease (Farooqi and Hopkin 1998; Wickens *et al.,* 1999). Three categories of broad-spectrum antibiotics were identified in this role in a study conducted in the United Kingdom: penicillins, cephalosoprins and macrolides. The mechanism by which antibiotics influence the programming and development of the immunological system most likely involves alterations to the collection of bacterial species inhabiting the intestinal tract. Treatment of young children with broad spectrum, oral antibiotics might produce perturbations in the composition of the intestinal microflora such that bacteria important in promoting Th1 mechanisms are depleted at a crucial age. This could result in Th2 dominance over Th1 immune responses to environmental antigens and an increased incidence of atopic disorders (Farooqi and Hopkin 1998).

The aetiology of atopic disorders is doubtless complex, but it is important that their molecular origins be determined in order to develop effective countermeasures. One such countermeasure may be the administration of lactobacilli or bifidobacteria to children if these bacterial types can be shown to programme the immune system towards a Th1 response as has been demonstrated with avirulent *Mycobacterium bovis* (Erb *et al.*, 1998). Promising ameliorating results have been obtained in the management of atopic eczema through the administration of lactobacilli to children, and in influencing the Th1 versus Th2 response in mice (Isolauri *et al.*, 1993; Murosaki *et al.*, 1998; Isolauri *et al.*, 2000). If these observations prove to be repeatable, targeted uses for probiotics can be envisaged.

Perhaps the immune system can be educated
by manipulation of the gut microflora?

Prebiotics May be Dietary Regulators of the Gut Ecosystem

Molecular typing methods such as ribotyping and pulsed field gel electrophoresis of DNA digests prepared from bacterial isolates have permitted the analysis of the gut microflora at the level of bacterial strains. Examination of bifidobacterial populations in monthly faecal samples collected over a 12-month period from two healthy humans has shown that there can be marked variation in the complexity and stability in the composition of these bacterial populations between human subjects (McCartney *et al.*, 1996). In this study, one subject harboured a relatively simple (five strains of bifidobacteria detected during the 12-month period; Figure 2A) and stable collection of bifidobacteria, but the other subject harboured 32 strains, some of which appeared, disappeared and sometimes reappeared during the course of the study (Figure 2B). The collection of strains detected in each subject was unique to the individual, in that a strain common to both individuals was not detected. This study was later extended to a further 10 healthy humans. Two faecal samples were obtained from each subject. About half of the subjects harboured a relatively simple bifidobacterial population and the others harboured a more complex collection of these bacteria. Unique collections of bifidobacterial strains that persisted throughout the study were detected in each subject (Kimura *et al.*, 1997).

The observations made during these studies posed an important question. What regulates the types and behaviour of bifidobacterial strains present in

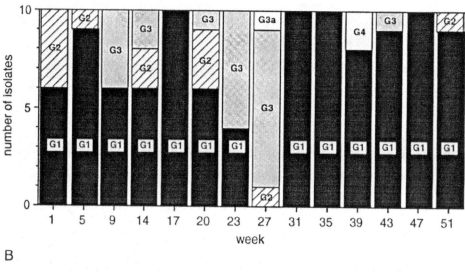

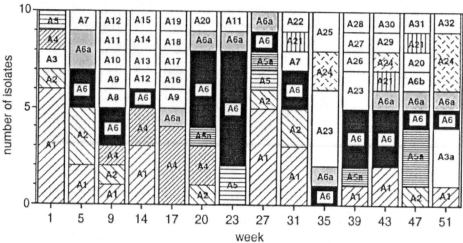

Figure 2. *Bifidobacterium* strains detected in faecal samples collected from two human subjects (A and B) over a 12-month period. Bifidobacterial strains from subject A were designated G1 to G4. Strains from subject B were designated A1 to A32 (Reproduced from McCartney *et al.*, 1996, with permission).

the microflora of a particular human? Perhaps human genetic factors are involved, because differences in host physiology might directly or indirectly influence the types of bacteria colonising the bowel. Dietary components might also be important. Most of the nutrients in the diet are absorbed in the small bowel, so that really only materials that cannot be readily digested by human processes pass to the large bowel. This is mostly plant structural material, but about 10% of dietary protein apparently reaches the colon as well. Some foods derived from milk or plants contain oligosaccharides (galacto- or fructo-oligosaccharides) and polysaccharides (inulin, resistant starch) that are not digested or absorbed in the small bowel. On average, it has been estimated that North Americans daily ingest 2.6 grams of inulin and 2.5 grams of fructo-oligosaccharides, about 70% of which comes from wheat and 25% from onions (Moshfegh *et al.,* 1999). As well as occurring naturally in foods, these non-digestible oligosaccharides can be derived by enzyme-catalysed industrial processes from various carbohydrate sources. They have become relatively common as low-calorie additives in foods during the past decade, and are considered to confer health benefits on the consumer – in which case they are referred to as "prebiotics" (Crittenden 1999). Non-hydrolysed oligosaccharides and polysaccharides pass to the large bowel where some members of the gut microflora are able to utilise them as fermentable sources of energy and carbon (Gibson and Roberfroid 1995).

It appears that bifidobacteria have the greatest capability to ferment the oligosaccharides (Crittenden 1999). It would be very interesting to investigate the preferential utilisation of different oligosaccharides by bifidobacterial species and strains. Perhaps the results of such experiments would explain the predominance of certain bifidobacteria in the gut of particular human hosts? Could the composition of the bifidobacterial population of an individual be modified by the administration of a specific oligosaccharide? What would be the consequence of this modification to the human host?

It is generally accepted that bifidobacterial numbers are increased in the faeces of humans who have ingested fructo-oligosaccharides. The average size of the increase is quite small, however, and the biological significance of these increases for the general human population or in specific disease states does not appear to be known (Tuohy *et al.,* 2001).

Prebiotics may be useful tools in the investigation of the regulation of the composition of bifidobacterial populations in the large bowel.

You Are What You Eat

The contribution of the gut microflora to the nutrition of the host is best documented in the case of ruminants where the microbial inhabitants of the rumen-reticulum ferment plant structural material that cannot be digested by the host (Hungate 1966). The fermentation products and microbial cell constituents nourish the host. The activities of the gut microflora may, however, impinge on the effective digestion and absorption of nutrients and may provide an interesting area of research. Perhaps most of the attention, in this respect, will be focussed on the nutrition of monogastric farm animals, such as pigs and poultry, where efficient feed conversion makes the difference between profit and loss for the farmer. A bacterial factor that could affect feed conversion is the production of bile salt hydrolase. Comparison of the characteristics of lactobacillus-free and lactobacillus-colonised mice revealed that lactobacilli were responsible for most of the bile salt hydrolase activity in the gut of mice (Tannock *et al.,* 1989). Bile salt hydrolases catalyse the hydrolysis of conjugated fatty acids, which results in the production of a free amino acid (taurine or glycine) and an unconjugated bile acid molecule. Conjugated bile acids enter the small bowel in bile and are important in the emulsification, digestion and absorption of dietary lipid that occurs in the jejunum of mammals and the distal jejunum-ileum of poultry (Renner 1965). Unconjugated bile acids are much less efficient in these roles (Drasar and Barrow 1985). The concentrations of unconjugated bile acids in the small bowel of mice colonised by lactobacilli are higher than those of lactobacillus-free mice (Tannock *et al.,* 1994). Therefore bile salt hydrolases produced by lactobacilli are active in catalysing the deconjugation of bile acids in the small bowel environment. The growth rates of lactobacillus-free and lactobacillus-colonised mice are the same, therefore the loss of conjugated bile salt molecules mediated by bile salt hydrolase does not impair adequate lipid digestion and absorption in well-nourished laboratory mice (Bateup *et al.,* 1995). Yet what might happen if the bile salt hydrolase activity in the poultry gut or pig gut was reduced by the removal of lactobacilli from the digestive tract (Feighner and Dashkevicz 1987)? In both of these animal species, lactobacilli are numerous in the small bowel because, as in the case of mice, the bacteria colonise proximal regions of the digestive tract: crop and pars oesophagea respectively (Fuller and Brooker 1974; Fuller *et al.,* 1978). Poultry utilise dietary lipid to a greater degree than other vertebrates and a proportion of the energy in commercial feeds is derived from fat. (Blem

Table 6. Comparison of selected biochemical properties of the intestinal tracts of germfree and conventional animals

Property	Conventional[1]	Germfree[1]
Bile acid metabolism	Deconjugation, dehydrogenation, and dehydroxylation	Absence of deconjugation dehydrogenation, and dehydroxylation
Bilirubin metabolism	Deconjugation and reduction	Little deconjugation and absence of reduction
Cholesterol	Reduction to coprostanol	Absence of coprostanol
β-aspartylglycine	Absent	Present
Intestinal gases	Hydrogen, methane, and carbon dioxide	Absence of hydrogen and methane. Less carbon dioxide
Short chain fatty acids	Large amounts of several acids	Small amounts of a few acids
Tryptic activity	Little activity	High activity
Urease	Present	Absent
β-glucuronidase (pH 6.5)	Present	Absent
Extent of degradation of mucins	More	Less
Serum cholesterol concentration	Lower	Higher

[1]Conventional = raised in association with a normal microflora; germfree = raised in the absence of demonstrable microbes.

1976). In the absence of bile salt hydrolase-producing bacteria, even greater growth gains might be achieved.

The nutritional requirements of intensively farmed livestock will provide potentially fertile research topics in which the impact of the gut microflora on animal physiology can be studied.

Bacteria Can Speak to Animals

Comparison of the characteristics of germfree animals (raised in the absence of microbial associates) and their conventional counterparts (presence of microbial associates) has demonstrated that the gut microflora as a whole has major impacts on the biochemistry, physiology, immunology, disease resistance, and nutrition of the animal host (Table 6). The impact of specific bacteria, lactobacilli, on the biochemistry of the murine gut milieu has been investigated (Table 7), but it is only recently that the molecular mechanisms associated with gut microflora-host relationships have begun to be studied (Hooper *et al.,* 2001). Made possible by the development of molecular biological tools, this research coincides with the complete genome sequencing of selected animal species and offers unprecedented opportunities to identify the molecular foundations of gut microflora-host relationships.

The intimate contact between the bacterial community of the gut and the animal host is a notable feature of this relationship. Numerous bacterial cells are confined within a relatively small, defined space and are separated from the sterile tissues of the host by an epithelium composed of a single layer of enterocytes. In the ileum of mice, and some other animal species, filamentous segmented bacteria related to the clostridia (Snel *et al.,* 1994) attach by one end to enterocytes, particularly in the vicinity of Peyer's patches (Klaasen *et al.,* 1992). The method of attachment involves the insertion of the base of the filament into a murine enterocyte. Actin rearrangements within the enterocyte form a socket by which the filament becomes permanently attached to the mucosal surface (Jepson *et al.,* 1993). The mechanism by which the bacteria cause rearrangement of the enterocyte cytoskeleton is not known, but likely involves the passage of effector molecules from the bacterial cell to the enterocyte: much as occurs when membrane ruffling of enterocytes is induced by *Salmonella typhimurium* (Galan 1996).

The presence of filamentous segmented bacteria in the mouse gut results in the increased expression of a murine gene involved in the fucosylation of the asialo GM1 glycolipid associated with enterocytes (Umesaki *et al.,* 1995). Curiously, a similar phenomenon has also been observed when ex-germfree mice were colonised with *Bacteroides thetaiotaomicron* (Bry *et al.,* 1996). This bacterial species utilises L-fucose, salvaged from intestinal glycoconjugates, as an energy source. The induction of fucosylation of glycoconjugates in the bowel of mice was shown to be dependent on the presence of a critical concentration of *Bacteroides* cells (10^6-10^7 CFU/ml) and on the ability of the bacteria to utilise L-fucose. A mutant strain, unable

Table 7. The impact of lactobacilli on the biochemical characteristics of mice

Characteristic	Reference
Bile salt hydrolase activity in intestinal contents is due mainly to presence of lactobacilli	Tannock *et al.* 1989
More unconjugated bile acids in intestinal contents when lactobacilli are present	Tannock *et al.* 1994
Azoreductase activity in large bowel contents reduced when lactobacilli present	McConnell and Tannock 1991
β-glucuronidase activity reduced in large bowel contents of male mice when lactobacilli are present	McConnell and Tannock 1993b
Enzyme activities associated with duodenal enterocytes unaffected by presence of lactobacilli	McConnell and Tannock 1993a
Serum cholesterol concentrations unaffected by presence of lactobacilli in the gastrointestinal tract	Tannock and McConnell 1994

to utilise L-fucose, was less efficient at inducing fucosylation. The linkage between L-fucose utilisation and the signalling system that induces fucosylation of intestinal glycoconjugates is mediated by bacterial protein FucR (Hooper *et al.*, 1999; Hooper *et al.*, 2000). It is proposed that the bacteroides, by influencing host biochemistry, ensure that L-fucose is constantly available as an energy source in a highly competitive environment. *Bacteroides thetaiotaomicron* affects the expression of other genes in the mucosal cells of the ileum of ex-germfree mice. Using high density oligonucleotide-based DNA microarrays for the analysis of the host transcriptional responses, it was shown that these bacteria regulated the expression of a broad range of mouse genes (Table 8) that participate in diverse and fundamental physiological functions (Hooper *et al.*, 2001). Defining the mechanisms by which the bacteria communicate with host cells will surely prove a fascinating area of research.

It will be impossible to understand the impact of probiotics
and prebiotics on the consumer without an understanding
of the molecular foundations of gut microflora-host relationships.

Table 8. Examples of murine genes regulated by colonisation of the gut by *Bacteroides thetaiotaomicron*

Murine gene	Change in gene expression relative to germfree[1]	Function
Sodium/glucose cotransporter (SGLT-1)	2.6 (0.9)[2]	Glucose uptake
Colipase	6.6 (1.9)	Lipd metabolism
Liver fatty acid-binding protein (L-FABP)	4.4 (1.4)	Lipid metabolism
Metallothionein	-5.4 (0.7)	Copper/Zinc sequestration
Polymeric immunoglobulin receptor (pIgR)	2.6 (0.7)	Transepitehlial IgA transport
Decay-accelerating factor (DAF)	5.7 (1.5)	Complement inactivation
Small proline-rich protein 2a (sprr2a)	205 (64)	Crosslinking protein (fortification of intestinal barrier)
Glutathione S-transferase (GST)	-2.1 (0.1)	Glutathione conjugation to electrophiles
Multidrug resistance protein (Mdr1a)	-3.8 (1.0)	Export of glutathione conjugates from epithelium
Lactase-phlorizin hydrolase	-4.1 (0.6)	Lactose hydrolysis
Adenosine deaminase (ADA)	2.6 (0.5)	Adenosine inactivation
Angiogenin-3	9.1 (1.8)	Angiogenesis factor

[1]Real-time quantitative PCR values using ileal RNA samples.
[2]Mean of triplicate measurements (standard deviation).
After (Hooper *et al.* 2001).

Incognito

Perhaps one of the greatest enigmas concerning the gut microflora is the mechanism by which large numbers of bacterial cells can persist in quite intimate association with the bodies of animals without inducing a marked inflammatory response on the part of the host. Antigens from gut bacteria are "seen" by the host because antibodies that react with them are present in the sera of healthy humans (Gillespie *et al.*, 1950; Cohen and Norins 1966; Evans *et al.*, 1966; Hoiby and Hertz 1979; Kimura *et al.*, 1997). IgM antibodes reactive with the cells of lactobacilli and bifidobacteria have been detected in the serum of the majority of humans tested (Kimura *et al.*, 1997). The detection of this class of immunoglobulin suggests exposure of the host to small amounts of antigen, and that this exposure was relatively recent (Berg 1983).

Some autochthonous members of the gut microflora do not elicit as great an immune response from the host as do allochthonous bacteria. For example, Berg and Savage (Berg and Savage 1972) injected conventional mice intraperitoneally with heat-killed cells of *E. coli* or *Bacteroides* species of murine origin. They compared the immune response to these strains with that resulting from injection with strains from humans. Four days after inoculation, the immune response to the murine strains was less than that to the human strains. Similarly, ex-germfree mice monoassociated with murine strains of *E. coli*, lactobacilli or bacteroides), did not show a detectable, systemic immune response but responded to strains of human origin. It is likely, therefore, that some autochthonous bacterial strains resemble their host antigenically. The immunological environment of the host may provide a selective force for the evolution of a special relationship. Immunological selection of genetic variants less antigenically "foreign" to the host may favour the establishment of certain bacterial strains as members of the gut microflora. Some credence to this hypothesis is provided by the observation that *Vibrio cholerae* cells of differing antigenicity were selected while the bacteria were inhabiting the gut of gnotobiotic mice (Sack and Miller 1969; Miller *et al.*, 1972). More recently, *Bacteroides fragilis* has been demonstrated to be able to modulate its surface antigenicity by producing at least 8 distinct capsular polysaccharides, and to be able to regulate their expression by the reversible inversion of DNA segments containing the promoters for their expression. This phase variation with respect to capsular polysaccharide may help the bacteroides maintain their residence in the gut in the face of constant immune pressure from the host (Krinos *et al.*, 2001).

Interestingly, induction of secretory IgA (sIgA) responses to pathogens requires co-stimulation by antigen-specific T cells, whereas induction of sIgA responses against non-pathogenic bacteria are T cell-independent in mice (Macpherson *et al.*, 2000). This independence may allow the host to respond to the changing microbial status of the gut as the diet introduces new, allochthonous, but non-pathogenic bacterial types. The allochthonous bacteria can be recognised and neutralised without eliciting a deleterious immune response. Non-pathogenic bacteria may also directly influence the intestinal epithelium to limit immune activation. An avirulent *Salmonella* strain, for example, has been demonstrated to inhibit ubiquitination and degradation of IκB, thereby blocking NF-κb-directed transactivation of genes encoding inflammatory mediators (Neish *et al.*, 2000).

The results of this research raise a number of questions concerning the relationship between probiotic bacteria and the immune system of the consumer. Probiotic bacteria are allochthonous and, upon initial administration, will presumably be "non-self" to the immune system. Is the resulting immune response of the type directed at overt pathogens, or like that directed at autochthonous bacteria? If the same probiotic bacteria continue to be ingested day after day, will a state of oral tolerance be induced? Does this mean that the "immunomodulating" effect of a probiotic product can be lost over time?

The impact of the consumption of a probiotic on the human immune system needs to be studied temporally in order to improve knowledge of gut microflora-immune system relationships.

This is Where We Are Going

Probiotics and prebiotics have become part of the lexicon of food technologists. Foods and dietary supplements that conform to these categories are retailed throughout the world, including via the World Wide Web. Still not established as legitimate adjuncts of medical treatment, they are yet a part of the enormous self-care health market, worth billions of dollars in the USA alone. There is no greater understanding of how probiotics work, or indeed if they work, than there was 3 years ago when "*Probiotics: a critical review*" was published. Nevertheless, probiotics and prebiotics will continue to be produced, marketed, and bring satisfaction to a widespread clientele.

Probiotics and prebiotics have, however, generated new interest by the medical profession in the gut microflora. Gastroenterologists, respiratory tract physicians and paediatricians lead the way in this respect because of their need to understand the role of the gut microflora in the aetiopathogenesis of inflammatory bowel diseases and atopic diseases. Probiotic and prebiotic products of the future may be targeted for use in the prevention, or alleviation of symptoms, of specific diseases if "abnormal microfloras" can be recognised, and the safety of probiotics administered to the very young and to those who are immunologically dysfunctional (Wagner *et al.,* 1997) can be guaranteed. The development of such targeted products will doubtless be difficult and expensive, but worth the effort if the well-being of humans can truly be promoted by this means.

There is no doubt that we have the tools to carry out detailed and informative studies of the gut microflora. The utility of various nucleic acid-based methods of analysis has been demonstrated clearly during the past three years. Doubtless other techniques will be derived and validated. There is a danger, however, that the quest for new and better techniques becomes an end in itself. What we need now is the application of existing validated methods to answer fundamental questions about the gut microflora-host relationship.

We need to understand how the gut ecosystem functions
and how gut bacteria function in it.

References

Alander, M, Satokari, R., Korpela, R., *et al.,* 1999. Persistence of colonization of human colonic mucosa by a probiotic strain, *Lactobacillus rhamnosus* GG, after oral consumption. Appl. Environ. Microbiol. 65: 351-354.

Alexander, M. 1971. Microbial Ecology. John Wiley and Sons, Inc., New York.

Aranki, A., and Freter, R. 1972. Use of anaerobic glove boxes for the cultivation of strictly anaerobic bacteria. Am. J. Clin. Nutr. 25: 1329-1334.

Bateup, J.M., McConnell, M.A., Jenkinson, H.F., and Tannock, G.W. 1995. Comparison of *Lactobacillus* strains with respect to bile salt hydrolase activity, colonization of the gastrointestinal tract, and growth rate of the murine host. Appl. Environ. Microbiol. 61: 1147-1149.

Berg, D.J., Davidson, N., Kuhn, R., *et al.,* 1996. Enterocolitis and colon cancer in interleukin-10-deficient mice are associated with aberrant cytokine production and CD4(+) TH1-like responses. J. Clin. Invest. 98: 1010-1020.

Berg, R.D. 1983. Host immune repsonse to antigens of the indigenous intestinal flora. In: Human Intestinal Microflora in Health and Disease. Hentges, D.J., ed. Academic Press, New York. p. 101-126.

Berg, R.D., and Savage, D.C. 1972. Immunological responses to microorganisms indigenous to the gastrointestinal tract. Am. J. Clin. Nutr. 25: 1364-1371.

Bjorksten, B. 1997. The environment and sensitisation to allergens in early childhood. Pediatr Allergy Immunol. 8: 32-39.

Blem, C.R. 1976. Patterns of lipid storage and utilization in birds. Am. Zool. 16: 671-684.

Brouqu, P., and Raoult, D. 2001. Endocarditis due to rare and fastidious bacteria. Clinical Microbiol. Rev. 14: 177-207.

Bry, L., Falk, P.G., Midtvedt, T., and Gordon, J.I. 1996. A model of host-microbial interactions in an open mammalian ecosystem. Science 273: 1380-1383.

Burney, P.G., Chinn, S., and Rona, R.J. 1990. Has the prevalence of asthma increased in children? Evidence from the national study of health and growth 1973-1986. Brit. Med. J. 300: 1306-1310.

Burr, M.L., Butland, B.K., King, S., and Vaughan, W.E. 1989. Changes in asthma prevalence: two surveys 15 years apart. Arch. Dis. Childhood 61: 1452-1456.

Chadwick, V.S., and Chen, W. 1999. The intestinal microflora and inflammatroy bowel disease. In: Medical Importance of the Normal Microflora. Tannock, G.W., ed. Kluwer Academic Publishers, Dordrecht. p. 177-221.

Cohen, I.R., and Norins, L.C. 1966. Natural antibodies to gram-negative bacteria: immunoglobulins G, A, and M. Science 152: 1257-1259.

Crittenden, R.G. 1999. Prebiotics. In: Probiotics: A Critical Review. Tannock, G.W., ed. Horizon Scientific Press, Wymondham, UK. p. 141-156.

Cummings, J.H., and Macfarlane, G.T. 1991. The control and consequences of bacterial fermentation in the human colon. J. Appl. Bacteriol. 70: 443-459.

D'Andrea, A., Aste-Amezaga, M., Valiante, N.M., Ma, X., Kubin, M., and Trinchieri, G. 1993. Interleukin 10. IL-10. inhibits human lymphocyte interferon gamma-production by suppressing natural killer cell stimulatory factor/IL-12 synthesis in accessory cells. J. Exp. Med. 178: 1041-1048.

Drasar, B.S., and Barrow, P.A. 1985. Intestinal Microbiology. American Society for Microbiology., Washington, D. C.

Drasar, B.S., and Hill, M.J. 1974a. Human Intestinal Flora. Academic Press, London.

Drasar, B.S., and Hill, M.J. 1974b. Role of bacteria in the aetiology of cancer. In: Human Intestinal Flora. Academic Press, London. p. 193-225.

Drouault, S., Corthier, G., Ehrlich, S.D., and Renault, P. 1999. Survival, physiology, and lysis of *Lactococcus lactis* in the digestive tract. Appl. Environ. Microbiol. 65: 4881-4886.

Dubos, R., Savage, D., and Schaedler, R. 1966. Biological freudianism. Lasting effects of early influences. Pediatrics 38: 789-800.

Duchmann, R., Kaiser, I., Hermann, E., Mayet, W., Ewe, K., and Meyer zum Buschenfelde, K.-H. 1995. Tolerance exists towards resident intestinal flora but is broken in active inflammatory bowel disease. (IBD). Clin. Exp. Immunol. 102: 448-455.

Dunne, C., Murphy, L., Flynn, S., O'Mahory, L., O'Halloran, S.O., *et al.*, 1999. Probiotics: from myth to reality. Demonstration of functionality in animal models of disease and in human clinical trials. Antonie van Leewenhoek 76: 279-292.

Elson, C.O., Sartor, R.B., Tennyson, G.S., and Riddell, R.H. 1995. Experimental models of inflammatory bowel disease. Gastroenterol. 109: 1344-1367.

Erb, K.J., Holloway, J.W., Sobeck, A. Moll, H., and Le Gros, G. 1998. Infection of mice with *Mycobacterium bovis*-Bacillus Calmette-Guerin. BCG. suppresses allergen-induced airway eosinophilia. J. Exp. Med. 187: 561-569.

Evans, R.T., Spaeth, S., and Mergenhagen, S.E. 1966. Bactericidal antibody in mammalian serum to obligately anaerobic gram-negative bacteria. J. Immunol. 97: 112-119.

Farina, C., Arosio, M., Mangia, M., and Moioli, F. 2001. *Lactobacillus casei* subsp. *rhamnosus* sepsis in a patient with ulcerative colitis. J. Clin. Gastroenterol. 33: 251-252.

Farooqi, I.S., and Hopkin, J.M. 1998. Early childhood infection and atopic disorder. Thorax 53: 927-932.

Feighner, S.D., and Dashkevicz, M.P. 1987. Subtherapeutic levels of antibiotics in poultry feeds and their effects on weight gain, feed efficiency, and bacterial cholyltaurine hydrolase activity. Appl. Environ. Microbiol. 53: 331-336.

Finegold, S.M., and Sutter, V.L. 1978. Fecal flora in different populations, with special reference to diet. Am. J. Clin. Nutr. 31: S116-S122.

Finegold, S.M., Sutter, V.L., and Mathisen, G.E. 1983. Normal indigenous intestinal flora. In: Human Intestinal Microflora in Health and Disease. Hentges, D.J., ed. Academic Press, New York. p. 3-31.

Franks, A.H., Harmsen, H.J., Raangs, G.C., Jansen, G.J., Schut, F., and Welling, G.W. 1998. Variations of bacterial populations in human feces measured by fluorescent *in situ* hybridization with group-specific 16S rRNA-targeted oligonucleotide probes. Appl. Environ. Microbiol. 64: 3336-3345.

Fuller, R. 1989. Probiotics in man and animals. J. Appl. Bacteriol. 66: 365-378.

Fuller, R., Barrow, P.A., and Brooker, B.E. 1978. Bacteria associated with the gastric epithelium of neonatal pigs. Appl. Environ. Microbiol. 35: 582-591.

Fuller, R., and Brooker, B.E. 1974. Lactobacilli which attach to the crop epithelium of the fowl. Am. J. Clin. Nutr. 27: 1305-1312.

Galan, J.E. 1996. Molecular genetic bases of *Salmonella* entry into host cells. Mol. Microbiol. 20: 263-271.

Gibson, G.R. 1998. Dietary modulation of the human gut microflora using prebiotics. British J. Nutr. 80: S209-S212.

Gibson, G.R., Beatty, E.R., Wang, X., and Cummings, J.H. 1995. Selective stimulation of bifidobacteria in the human colon by oligofructose and inulin. Gastroenterol. 108: 975-982.

Gibson, G.R., and Roberfroid, M.B. 1995. Dietary modulation of the human colonic microbiota: introducing the concept of prebiotics. J. Nutr. 125: 1401-1412.

Gillespie, H.B., Steber, M.S., Scott, E.N., and Christ, Y.S. 1950. Serological relationships existing between bacterial parasites and their host. I. Antibodies in human blood serum for native intestinal bacteria. J. Immunol. 65: 105-113.

Goldin, B.R., and Gorbach, S.L. 1992. Probiotics for humans. In: Probiotics. The Scientific Basis. Fuller, R., ed. Chapman and Hall, London. p. 355-376.

Gorbach, S.L. 1967. Population control in the small bowel. Gut 8: 530-532.

Gordon, H.A., and Pesti, L. 1971. The gnotobiotic animal as a tool in the study of host microbial relationships. Bacteriol. Rev. 35: 390-429.

Groux, H., O'Garra, A., Bigler, M., *et al.*, 1997. A CD4+ T-cell subset inhibits antigen-specific T-cell responses and prevents colitis. Nature 389: 737-742.

Harmsen, H.J.M., Wildeboer-Veloo, A.C., Raangs, G.C., *et al.*, 2000. Analysis of intestinal flora development in breast-fed and formula-fed infants by using molecular identification and detection methods. J. Pediatr. Gastroenterol. Nutr. 30: 61-67.

Hoiby, N., and Hertz, J.B. 1979. Precipitating antibodies against *Escherichia coli, Bacteroides fragilis* ss. *thetaiotaomicron* and *Pseudomonas aeruginosa* in serum from normal persons and cystic fibrosis patients, determined by means of crossed electrophoresis. Acta Paediatrica. Scand. 68: 495-500.

Holdeman, L.V., and Moore, W.E. 1972. Roll-tube techniques for anaerobic bacteria. Am. J. Clin. Nutr. 25: 1314-1317.

Hooper, L.V., Falk, P.G., and Gordon, J.I. 2000. Analyzing the molecular foundations of commensalism in the mouse intestine. Curr. Opin. Microbiol. 3: 79-85.

Hooper, L.V., Wong, M.H., Thelin, A., Hansson, L., Falk, PG., and Gordon, J.I. 2001. Molecular analysis of commensal host-microbial relationships in the intestine. Science 291: 881-884.

Hooper, L.V., Xu, J., Falk, P.G., Midtvedt, T., Gordon, J.I. 1999. A molecular sensor that allows a gut commensal to control its nutrient foundation in a competitive ecosystem. Proc. Natl. Acad. Sci. 96: 9833-9838.

Hungate, R.E. 1966. The Rumen and its Microbes. Academic Press, New York.

Isolauri, E., Arvola, T., Sutas, Y., Moilanen, E., and Salminen, S. 2000. Probiotics in the management of atopic eczema. Clin. Exp. Allergy 30: 1605-1610.

Isolauri, E., Majamaa, H., Arvola, T., Rantala, I., Virtanen, E., and Arvilommi, H. 1993. *Lactobacillus casei* GG reverses increased intestinal permeability induced by cow milk in suckling rats. Gastroenterol. 105: 1643-1650.

Jepson, M.A., Clark, M.A., Simmons, N.L., and Hirst, B.H. 1993. Actin accumulation at sites of attachment of indigenous apathogenic segmented filamentous bacteria to mouse ileal epithelial cells. Infect. Immun. 61: 4001-4004.

Kandler, O., and Weiss, N. 1986. Genus *Lactobacillus* Beijerinck 1901, 212[AL]. In: Bergey's Manual of Systematic Bacteriology. Sneath, P.H.A., Mair, N.S., Sharpe, M.E., Holt, J.G., eds. Williams and Wilkins, Baltimore. p. 1209-1234.

Kawaguchi, M., Nanno, M., Umesaki, Y., *et al.*, 1993. Cytolytic activity of intestinal intraepithelial lymphocytes in germ-free mice is strain dependent and determined by T cells expressing γδ T-cell receptors. Proc. Natl. Acad. Sci. 90: 8591-8594.

Kimura, K., McCartney, A.L., McConnell, M.A., and Tannock, G.W. 1997. Analysis of fecal populations of bifidobacteria and lactobacilli and investigation of the immunological responses of their human hosts to the predominant strains. Appl. Environ. Microbiol. 63: 3394-3398.

Klaasen, H.L., Koopman, J.P., Poelma, F.G., and Beynen, A.C. 1992. Intestinal, segmented, filamentous bacteria. FEMS Microbiol. Rev. 8: 165-180.

Kopczynski, E.D., Bateson, M.M., and Ward, D.M. 1994. Recognition of chimeric small-subunit ribosomal DNAs composed of genes from uncultivated microorganisms. Appl. Environ. Microbiol. 60: 746-748.

Krinos, C.M., Coyne, M.J., Weinacht, K.G., Tzianabos, A.O., Kasper, D.L., and Comstock, L.E. 2001. Extensive surface diversity of a commensal microorganism by multiple DNA inversions. Nature 414: 555-558.

Langendijk, P.S., Schut, F., Jansen, G.J., *et al.*, 1995. Quantitative fluorescence *in situ* hybridization of *Bifidobacterium* spp. with genus-specific 16S rRNA-targeted probes and its application in fecal samples. Appl. Environ. Microbiol. 61: 3069-3075.

Macpherson, A.J., Gatto, D., Sainsbury, E., Harriman, G.R., Hengartner, H., and Zinkernagel, R.M. 2000. A primitive T cell-independent mechanism of intestinal mucosal IgA responses to commensal bacteria. Science 288: 2222-2226.

Madsen, K.L., Doyle, J.S., Tavernini, M.E., Jewel, L.D., Rennie, R.P., and Fedorak, R.N. 2000. Antibiotic therapy attenuates colitis in interleukin 10 gene-deficient mice. Gastroenterol. 118: 1094-1105.

Marteau, P., Pochart, P., Dore, J., Bera-Maillet, C., Bernalier, A., and Corthier, G. 2001. Comparative study of bacterial groups within the human cecal and fecal microbiota. Appl. Environ. Microbiol. 67: 4939-4942.

Martinez, F.D., and Holt, P.G. 1999. Role of microbial burden in aetiology of allergy and asthma. Lancet 354: 12-15.

McCartney, A.L., Wenzhi, W., and Tannock, G.W. 1996. Molecular analysis of the composition of the bifidobacterial and lactobacillus microflora of humans. Appl. Environ. Microbiol. 62: 4608-4613.

McConnell, M.A., and Tannock, G.W. 1991. Lactobacilli and azoreductase activity in the murine cecum. Appl. Environ. Microbiol. 57: 3664-3665.

McConnell, M.A, and Tannock, G.W. 1993a. Lactobacilli do not influence enzyme activities of duodenal enterocytes of mice. Microb. Ecol. Health Dis. 6: 315-318.

McConnell, M.A., and Tannock, G.W. 1993b. A note on lactobacilli and beta-glucuronidase activity in the intestinal contents of mice. J. Appl. Bacteriol. 74: 649-651.

McCracken, V.J., and Gaskins, H.R. 1999. Probiotics and the immune system. In: Probiotics: A Critical Review. Tannock, G.W. ed. Horizon Scientific Press, Wymondham. p. 85-111.

Mercenier, A. 1999. Lactic acid bacteria as live vaccines. In: Probiotics: A Critical Review. Tannock, G.W. ed. Horizon Scientific Press, Wymondham. p. 113-127.

Metchnikoff, E. 1907. The Prolongation of Life. Optimistic Studies. William. Heinemann, London.

Metchnikoff, E. 1908. The Nature of Man. Studies in Optimistic Philosophy. William. Heinemann, London.

Miller, C.E., Wong, K.H., Feely, J.C., and Forlines, M.E. 1972. Immunological conversion of *Vibrio cholerae* in gnotobiotic mice. Infect. Immun. 6: 739-742.

Mitsuoka, T. 1992. The human gastrointestinal tract. In: The Lactic Acid Bacteria. Wood B.J.B., ed. Elsevier Applied Science, London. p. 69-114.

Moore, W.E., Cato, E.P., and Holdeman, L.V. 1978. Some current concepts in intestinal bacteriology. Am. J. Clin. Nutr. 31: S33-42.

Moore, W.E., and Holdeman, L.V. 1974. Special problems associated with the isolation and identification of intestinal bacteria in fecal flora studies. Am. J. Clin. Nutr. 27: 1450-1455.

Morrissey, I., Charrier, K., Braddy, S., Liggitt, D., and Watson, J.D. 1993. CD4+ T cells that express high levels of CD45RB induce wasting disease when transferred into congenic severe combined immunodeficient mice. Disease development is prevented by cotransfer of purified CD4+ T cells. J. Exp. Med. 178: 237-244.

Moshfegh, A.J., Friday, J.E., Goldman, J.P., and Ahuja, J.K. 1999. Presence of inulin and oligofructose in the diets of Americans. J. Nutr. 129: 1407S-1411S.

Murosaki, S., Yamamoto, Y., Ito, K., Inokuchi, T., Kusaka, H., and Yoshikai, Y. 1998. Heat-killed *Lactobacillus plantarum* L-137 suppresses naturally fed antigen-specific IgE production by stimulation of IL-12 production in mice. J. Allergy Clin. Immunol. 102: 57-64.

Muyzer, G., and Smalla, K. 1998. Application of denaturing gradient gel electrophoresis (DGGE) and temperature gradient gel electrophoresis (TGGE) in microbial ecology. Antonie Van Leeuwenhoek 73: 127-141.

Neish, A.S., Gewirtz, A.T., Zeng, H., *et al.*, 2000. Prokaryotic regulation of epithelial responses by inhibition of IkappaB-alpha ubiquitination. Science 289: 1560-1563.

Nubel, U., Engelen, B., Felske, A., *et al.*, 1996. Sequence heterogeneities of genes encoding 16S rRNAs in *Paenibacillus polymyxa* detected by temperature gradient gel electrophoresis. J. Bacteriol. 178: 5636-5643..

O'Sullivan, D.J. 1999. Methods of analysis of the intestinal microflora. In: Probiotics: A Critical Review. Tannock, G.W. ed. Horizon Scientific Press., Wymondham, UK. p. 23-44.

Pauwels, R., Straeten, V.D.M., Platteau, B., and Bazin, H. 1983. The non-specific enhancement of allergy: *in vivo* effects of *Bordetella pertussis* vaccine on IgE synthesis. Allergy 38: 239-246.

Raskin, L., Capman, W.C., Sharp, R., Poulsen, L.K, and Stahl, D.A. 1997. Molecular ecology of gastrointestinal ecosystems. In: Gastrointestinal Microbiology. Vol. 2. Gastrointestinal Microbes and Host Interactions. Mackie, R.I., ed. Chapman and Hall, New York. p. 243-298.

Rath, H.C., Bender, D.E., Holt, L.C., *et al.*, 1995. Metronidazole attenuates colitis in HLA-B27/β2 microglobulin transgenic rats: a pathogenic role for anaerobic bacteria. Clin. Immunol. Immunopathol. 76.

Rath, H.C., Herfath, H.H., Ikeda, J.S., *et al.*, 1996. Normal luminal bacteria, especially *Bacteroides* species, mediate chronic colitis, gastritis, and arthritis in HLA-B27/human β2 microglobulin transgenic rats. J. Clin. Invest. 98: 945-953.

Rath, H.C., Schulz, M., Dieleman, L.A., *et al.*, 1998. Selective vs. broad spectrum antibiotics in the prevention and treatment of experimental colitis in two rodent models. Gastroenterol. 114: A1067.

Rath, H.C., Wilson, K.H., Mackie, R.I., and Sartor, R.B. 1999. Differential induction of colitis and gastritis in HLA-B27 transgenic rats selectively colonized with *Bacteroides vulgatus* or *Escherichia coli*. Infect. Immun. 67: 2969-2969.

Rautio, M., Jousimies-Somer, H., Kauma, H., Pietarinen, I. *et al.* 1999. Liver abscess due to a *Lactobacillus rhamnosus* strain indistinguishable from *L. rhamnosus* strain GG. Clin. Infect Dis. 28: 1159-1160.

Renner, R. 1965. Site of fat absorption in the chick. Poultry Sci. 44: 861-864.

Rettger, L.F., Levy, M.N., Weinstein, L., and Weiss, J.E. 1935. *Lactobacillus acidophilus* and its therapeutic application. Yale university press, New Haven.

Reysenbach, A.L., Giver, L.J., Wickham, G.S., and Pace, N.R. 1992. Differential amplification of rRNA genes by polymerase chain reaction. Appl. Environ. Microbiol. 58: 3417-3418.

Romagnani, S. 1997. Atopic allergy and other hypersensitivities. Interactions between genetic susceptibility, innocuous and/or microbial antigens and the immune system. Curr. Opin. Immunol. 9: 773-775.

Sack, R.B., and Miller, C.E. 1969. Progressive changes of vibrio serotypes in germfree mice infected with *Vibrio cholerae*. J. Bacteriol. 99: 688-695.

Sartor, R.B. 1997. The influence of normal microbial flora on the development of chronic mucosal inflammation. Res. Immunol. 148: 567-576.

Sartor, R.B., Rath, H.C., Lichtman, S.N., and van Tol, E.A. 1996. Animal models of intestinal and joint inflammation. Bailliere's Clin. Rheumatol. 10: 55-76.

Satokari, R.M., Vaughan, E.E., Akkermans, A.D.L., Saarela, M., and De Vos, W.M. 2001. Bifidobacterial diversity in human feces detected by genus-specific PCR and denaturing gradient gel electrophoresis. Appl. Environ. Microbiol. 67: 504-513.

Savage, D.C. 1977. Microbial ecology of the gastrointestinal tract. Annu. Rev. Microbiol. 31: 107-133.

Savage, D.C. 1983. Morphological diversity among members of the gastrointestinal microflora. Int. Rev. Cytol. 82: 305-334.

Scardovi, V. 1986. Genus *Bifidobacterium* Orla-Jensen 1924, 472[AL]. In: Bergey's Manual of Systematic Bacteriology. Sneath, P.H.A., Sharpe, M.E., and Holt, J.G. eds. Williams and Wilkins, Baltimore. p. 1418-1434.

Sellon, R.K., Tonkonogy, S., Schulz, M., *et al.*, 1998. Resident enteric bacteria are necessary for development of spontaneous colitis and immune system activation in interleukin-10-deficient mice. Infect. Immun. 66: 5224-5231.

Sghir, A., Gramet, G., Suau, A., Rochet, V., Pochart, P., and Dore, J. 2000. Quantification of bacterial groups within the human fecal flora by oligonucleotide probe hybridization. Appl. Environ. Microbiol. 66: 2263-2266.

Simpson, S.J., Hollander, G.A., Mizoguchi, E., *et al.*, 1997. Expression of pro-inflammatroy cytokines by TCRαβ+ and γδ+ T cells in an experimental model of colitis. Eur. J. Immunol. 27: 17-25.

Snel, J., Blok, H.J., Kengen, H.M.P., *et al.*, 1994. Phylogenetic characterization of the *Clostridium* related segmented filamentous bacteria in mice based on 16S ribosomal RNA analysis. System. Appl. Microbiol. 17: 172-179.

Spanhaak, S., Havenaar, R., and Schaafsma, G. 1998. The effect of consumption of milk fermented by *Lactobacillus casei* strain Shirota on the intestinal microflora and immune parameters in humans. European J. Clin. Nutr. 52: 899-907.

Stackebrandt, E., and Goebel, B.M. 1994. Taxonomic note: a place for DNA-DNA reassociation and 16S rRNA sequence analysis in the present species definition in bacteriology. International J. Syst. Bacteriol. 44: 846-849.

Steidler, L., Hans, W., Schotte, L., *et al.*, 2000. Treatment of murine colitis by *Lactococcus lactis* secreting interleukin-10. Science 25: 1311-1312.

Suau, A., Bonnet, R., Sutrem, M., *et al.*, 1999. Direct analysis of genes encoding 16S rRNA from complex communities reveals many novel molecular species within the human gut. Appl. Environ. Microbiol. 65: 4799-4807.

Summanen, P., Baron, E.J., Citron, D.M., Strong, C., Wexler, H.M., and Finegold, S.M. 1993. Wadsworth Anaerobic Bacteriology Manual, 5th edn. Star Publishing Company, Belmont.

Tannock, G.W., Dashkevicz, M.P., and Feighner, S.D. 1989. Lactobacilli and bile salt hydrolase in the murine intestinal tract. Appl. Environ. Microbiol. 55: 1848-1851.

Tannock, G.W., and McConnell, M.A. 1994. Lactobacilli inhabiting the digestive tract of mice do not influence serum cholesterol concentration. Microb. Ecol. Health Dis. 7: 331-334.

Tannock, G.W., Munro, K., Harmsen, H.J.M., Welling, G.W., Smart, J., Gopal, P.K. 2000. Analysis of the fecal microflora of human subjects consuming a probiotic containing *Lactobacillus rhamnosus* DR20. Appl. Environ. Microbiol. 66: 2578-2588.

Tannock, G.W., Tangerman, A., Van Schaik, A., and McConnell, M.A. 1994. Deconjugation of bile acids by lactobacilli in the mouse small bowel. Appl. Environ. Microbiol. 60: 3419-3420.

Tissier, H. 1905. Repartition des microbes dans l'intestin du nourrisson. Ann. Inst. Pasteur. Paris. 19: 109-123.

Tuohy, K.M., Kolida, S., Lustenberger, A.M., and Gibson, GR. 2001. The prebiotic effects of biscuits containing partially hydrolysed guar gum and fructo-oligosaccharides - a human volunteer study. Brit. J. Nutr. 86: 341-348.

Umesaki, Y., Okada, Y., Matsumoto, S., Imaoka, A., and Setoyama, H. 1995. Segmented filamentous bacteria are indigenous intestinal bacteria that activate intraepithelial lymphocytes and induce MHC class II molecules and fucosyl asialo GM1 glycolipids on the small intestinal epithelial cells in the ex-germ-free mouse. Microbiol. Immunol. 39: 555-562.

Umesaki, Y., Setomaya, S., Matsumoto, S., and Okada, Y. 1993. Expansion of the $\gamma\delta$ T cell receptor-bearing intestinal intraepithelial lymphocytes after microbial colonization in germfree mice, and its independence form the thymus. Immunol. 79: 32-37.

Wagner, R.D., Warner, T., Roberts, L., Farmer, J., and Balish, E.. 1997. Colonization of congenitally immunodeficient mice with probiotic bacteria. Infect. Immun. 65: 3345-3351.

Walter, J., Hertel, C., Tannock, G.W., Lis, C.M., Munro, K., and Hammes, W.P. 2001. Detection of *Lactobacillus*, *Pediococcus*, *Leuconostoc*, and *Weissella* species in human faeces by using group-specific PCR primers and denaturing gradient gel electrophoresis. Appl. Environ. Microbiol. 67: 2578-2585.

Welling, G.W., Elfferich, P., Raangs, G.C., Wildeboer-Veloo, A.C., Jansen, G.J., and Degener, J.E. 1997. 16S ribosomal RNA-targeted oligonucleotide probes for monitoring of intestinal tract bacteria. Scand. J. Gastroenterol. Suppl. 222: 17-19.

Wickens, K., Pearce, N., Crane, J., and Beasley, R.. 1999. Antibiotic use in early chidhood and the development of asthma. Clin. Exp. Allergy 29: 766-771.

Woese, C.R. 1987. Bacterial evolution. Microbiol. Rev. 51: 221-271.

Woo, P.C.Y., Fung, A.M.Y., Lau, S.K.P., and Yuen, K.-Y. 2002. Identification by 16S rRNA gene sequencing of *Lactobacillus salivarius* bacteremic cholecystitis. J. Clin. Microbiol. 40: 265-267.

Zoetendal, E.G. 2001. Molecular characterization of bacterial communities in the human gastrointestinal tract. In: PhD Thesis. Department of Microbiology. University of Wageningen, Wageningen, The Netherlands.

Zoetendal, E.G., Akkermans, A.D., and De Vos, W.M. 1998. Temperature gradient gel electrophoresis analysis of 16S rRNA from human fecal samples reveals stable and host-specific communities of active bacteria. Appl. Environ. Microbiol. 64: 3854-3859.

From: *Probiotics and Prebiotics: Where Are We Going?*
Edited by: Gerald W. Tannock

Chapter 2

Fluorescence *In Situ* Hybridization as a Tool in Intestinal Bacteriology

Hermie J.M. Harmsen and Gjalt W. Welling

Abstract

The human gut microflora has an important function in health and disease. During the last six years fluorescence *in situ* hybridization (FISH) has been increasingly used to analyze bacterial communities in the gastrointestinal tract. This method is easy and inexpensive, and it can be performed in every laboratory equipped with a fluorescence microscope. However, for large studies involving many samples the enumeration of the fluorescent bacteria can be laborious. Automation of this process made this method into a reliable tool to perform comparative studies of larger sets of samples in an objective way. In this chapter the developments that are going on in the field of (automated) FISH analysis of colonic and fecal samples are discussed (i.e. developments in automation, standardization and validation of the protocol) and the design of a useful set of probes covering the total microflora. Experiments with fecal samples of volunteers show that 90% of the total hybridizable bacteria are covered by the current probe set. Furthermore, it is shown that stability of the microflora is difficult to measure due to population dynamics and variations in the protocol that was used. However, the results do indicate that the microflora is characteristic for an individual even over a long period of time. FISH has become a valuable tool to study the dynamics of the gut microflora.

Introduction

The human colonic microflora comprises more than 400 bacterial species (Moore and Holdeman, 1974) some eucaryotic species and, as far as we know, one methanogenic archaeal species. The composition and functioning of this microflora plays an important role in the protection of the host against several pathogenic conditions such as colonic cancer (Goldin and Gorbach, 1977), gastroenteritis (Gorbach *et al.,* 1987), immunological disorders (Eckmann *et al.,* 1995) and development of atopy (Kalliomäki *et al.,* 2001). Because of the large influence of the gut microflora on the health of the host, accurate assessment of the composition of this bacterial ecosystem is receiving much attention in both clinical microbiology and food microbiology.

Historically, the assessment of the microflora composition is performed by means of selective culturing (Finegold *et al.,* 1983). The sample is diluted and plated on a specific medium. The bacterial count of the original sample is then determined by multiplying the number of colonies that develop by the degree of dilution. There are two problems using this technique. Firstly, the bacterial count depends on the culturability of a bacterial species. For many years it has been explicitly stated by several microbial ecologists (summarized in Ward *et al.,* 1992*)* that as long as not all bacteria can be cultured it will be impossible to define the microbial composition of the sample. Not all bacteria can be cultured and therefore this will lead to an underestimation of the quantitative contribution of certain groups of bacteria. Secondly, selective media are not truly "selective" and certain bacterial species may be counted more than once on different 'specific' media. This may lead to an overestimation of the quantitative contribution of certain genera. The net result is an inaccurate picture of the composition of the gut microflora and this method is therefore not suitable to study population dynamics in the intestinal tract.

Molecular Analysis of Colonic and Fecal Microflora

Advances in the field of molecular phylogeny using a ribosomal constituent, the 16S ribosomal RNA (16S rRNA), have made it possible to study bacterial populations by a culture-independent approach. Each cell contains 10,000-60,000 ribosomes. The 30S subunit of a ribosome contains 21 different structural proteins and a 16S rRNA molecule. This molecule has become an important tool in molecular phylogeny studies (Woese *et al.*, 1987). Comparison of sequences of different bacterial 16S rRNAs shows that the molecule contains segments with different degrees of variability. This made it possible to construct phylogenetic trees and it revealed evolutionary relationships between species (Olsen *et al.*, 1986, Amann *et al.*, 1995). The different degree of variability also led to another application of the sequence information. Currently, more than 20,000 16S rRNA sequences are available and this allows the design of oligonucleotide-probes and primer sets that hybridize with a particular sequence in the 16S rRNA molecule. Oligonucleotides were designed on the basis of rRNA sequences of pure cultures know to be present in the gut (Barcenilla *et al.*, 2000), on the basis of 16S rDNA clone libraries (Suau *et al.*, 1999) and sequences coming from DGGE/TGGE analysis (Zoetendal *et al.*, 1998). These were applied in molecular analysis of fecal samples by dot blot and PCR approaches (Sghir *et al.*, 1998; 2000, Doré *et al.*, 1998; Matsuki *et al.*, 1999)

Fluorescence *In Situ* Hybridization

Two major quantitative molecular techniques are available that are independent of culturability: quantitative PCR (Q-PCR) (Higgins *et al.*, 2001) and fluorescence *in situ* hybridization (FISH). Application of the latter method to quantitatively determine the bacterial composition of fecal samples is the subject of this chapter. An already extensive set of oligonucleotides probes for FISH analysis of colonic bacteria has been developed by several groups and these probes have been applied in fecal and colonic studies (Franks *et al.* 1998; Simmering *et al.*, 1999; Harmsen *et al.*, 2000a; 2000b; Schwiertz *et al.*, 2000; Hopkins *et al.*, 2001; Kalliomäki *et al.*, 2001; Kleessen *et al.*, 2001; Suau *et al.*, 2001), and this development is continuing (unpublished data).

Quantification of positively hybridized bacteria in a fecal preparation is, in most cases, performed by means of visual counting-procedures. Since this is a time-consuming process and heavily depends on the skills and experience of the technician, only moderate levels of accuracy are reached (Langendijk *et al.,* 1995). In order to overcome these problems, the counting procedure was automated and the procedure was published (Jansen *et al.*, 1999). This chapter will address developments, and difficulties in standardization of non-automated and automated FISH.

The FISH Protocol

A fresh fecal sample is homogenized by kneading in a stomacher for 10 min. From this sample, half a gram is suspended in 4.5 ml of filtered PBS, including 2-4 glass beads (diameter 4 mm), and homogenized on a vortex mixer for 3 min. The fecal suspension is then centrifuged for 1 min at $700 \times g$ to remove large particles from the suspension. From the supernatant 1 ml is collected and fixed in 3 ml freshly prepared 4% paraformaldehyde solution. After fixation overnight at 4°C, the cells are divided in aliquots of 0.5 ml in eppendorf tubes and stored at - 80°C until use. After thawing, dilutions suitable for counting are made in PBS, *e.g.* 1600x for counting with the Bact338 probe, 160x for the bifidobacterial probe. Dilutions are applied to custom-made slides with six square-shaped wells (1x1 cm) (Nutacon, Leimuiden, NL). In order to enhance adhesion of fecal bacteria to the slide, slides are pre-treated by soaking in a gelatin-suspension (0.1% gelatin, 0.01% $KCr(SO4)_2.12H_2O$, Sigma) for 30 min and allowed to dry at room temperature. Subsequently, 10 µl of the proper dilution is carefully spread over the total surface of each separate well. After drying at room temperature, the slides are fixed for 10 min using 96% ethanol (v/v). Dilutions of the probe and DAPI are made in milli-Q to a concentration of 50 ng/µl and 500 ng/µl, respectively, and then stored at -20°C. Prior to use, the diluted probe solutions are further diluted in 50.0°C preheated hybridization buffer (20 mM Tris-HCl, 0.9M NaCl, 0.1% SDS, pH 7.2) to a concentration of 10 ng/µl. A coverslip is applied to the slides with hybridization mixture and the slides are incubated in a dark moist chamber at 50.0°C for 3 h in the case of Gram negative bacteria and for 16 h in the case of Gram positive bacteria. Subsequently, the slides are rinsed in preheated washing buffer (20 mM Tris-HCl, 0.9M NaCl, pH 7.2) for 30 min at 50.0°C. After briefly rinsing in milli-Q, the slides are dried and mounted with 6 µl of Vectashield (Vector laboratories, Burlingame, USA) on each well and a coverslip. Prior to staining with DAPI, the DAPI is diluted in PBS to a concentration of 1.25 ng/µl. Slides are incubated with this DAPI solution for 5 min at room temperature. Slides are then washed at room temperature in PBS for 10 min. After rinsing in milli-Q, slides are dried and mounted with Vectashield.

Automated FISH Analysis

Jansen *et al.,* (1999) developed an automated microscopic method for quantification of fluorescently labeled bacteria. This method is routinely used in our laboratory. Recent improvements include (i) upgrading of the hardware configuration and optimization of the software and (ii) miniaturization. With respect to (i) the improved method now makes use of the Leica Q550 image analysis system, a Kodak MegaPlus camera model 1.4 and a servo-controlled Leica DM/RXA ultra-violet microscope, referred to as Quantimet system. The performance of the method was validated by measuring 20 fecal samples from healthy human volunteers using a set of 4 fluorescent oligonucleotide probes: a universal probe for the detection of all bacterial species (Bact338), one probe specific for *Bifidobacterium* spp. (Bif164), a di-genus-probe specific for *Bacteroides* spp. and *Prevotella* spp. (Bac303) and a tri-genus-probe specific for *Ruminococcus* spp., *Clostridium* spp. and *Eubacterium* spp. (Erec482). A nucleic acid stain, 4',6-diamidino-2-phenylindole (DAPI), was also included in the validation. A series of performance-parameters was quantified. The results of the validation are listed in Table 1. Future developments will include faster autofocusing using another digital camera (Coolsnap FX) and miniaturization. With respect to the miniaturization, a new lay-out of Teflon-coated microscopic slides was designed and manufactured. These new slides contain 18 wells in a 3x6 configuration. Each separate well has a surface area of 25 mm^2. The measure-and-control software will be updated so that it can handle this new slide format.

Comparative Assessment of Selective Culturing and FISH

New methods used to quantify the microbial structure of the human gut microflora require thorough validation before they can be applied rationally. Quantitative FISH was compared with selective, anaerobic culturing. For FISH we used a set of 4 fluorescent oligonucleotide probes: a universal probe for the detection of all bacterial species, one probe specific for *Bifidobacterium* spp., a di-genus-probe specific for *Bacteroides* spp. and *Prevotella* spp. and a probe specific for *Escherichia coli.* For culturing, the following selective media, which are presumed to have about the same selectivity as the probes used in FISH, have been used: Wilkens-Chalgren agar for the total number of anaerobes, BMS-agar for the number of *Bacteroides*, BF-agar for the number of bifidobacteria and MacConkey for the number of *Escherichia coli.* The selectivity of the media was first tested

Table 1. Performance of the automated counting procedure

Parameter	value
mean focusing time	20 sec
mean image analysis and storage time	5 sec
mean time needed for change of wells	1 sec
mean time needed for change of slides	1 sec
fraction of fatal out-of-focus incidents	< 0.01
number of images per well	25

by colony-hybridization of colonies using the probes (minimally 40 colonies per medium). The results of the selectivity testing are listed in Table 2. This table shows that the specificity of the media is acceptable except for the BF-medium. About 2 out of 3 colonies belong to the *Bifidobacterium* genus. All other colonies consisted of bacteria belonging to the *Lactobacillus* genus. To quantify the assay error, one fecal sample was measured twenty times using each separate probe. The error in each of these data sets (*i.e.* the assay error) was expressed as the coefficient of variation (*i.e.* standard deviation/mean; see paragraph on accuracy). The composite coefficient of variation measured this way was then corrected by arithmetical subtraction of the assay error. The remaining value or the interindividual error is also expressed as a coefficient of variation. The results of this study are listed in Table 3. From this table it can be concluded that the assay errors of the FISH method are all smaller than the assay error of culturing. The relatively low value of the assay-error of the FISH method indicates a large discriminating power to distinguish real differences between volunteers. This seems especially true in the case of the total numbers of bacteria. When comparing the actual numbers obtained with the two methods, it appears that only the numbers of *E. coli* and bifidobacteria are significantly the same when using culture or FISH. The total number of bacteria is not significantly identical. This is

Table 2. Specificity of culturing

Probe	Medium	Specificity (%)
BACT338	Wilkins-Chalgren	100
BAC303	BMS	96.3
BIF164	BF	66.4
EC1531*	MacConkey	99.4

* Probe directed against the 23S rRNA

Table 3. Comparison of FISH and culturing

Target	Method	CV-assay[a]	CV-individual[b]
Bacteria	FISH	0.12	0.53
	Culture	0.47	0.01
Bacteroides	FISH	0.07	0.66
	Culture	0.30	0.65
Bifidobacteria	FISH	0.14	0.64
	Culture	0.22	0.56
E. coli	FISH	n.d.	2.84[c]
	Culture	n.d.	1.58[c]

[a] Coefficient of variation in slide preparation
[b] Coefficients of variation between volunteers corrected for the CV-assay
[c] Coefficients of variation not corrected for CV-assay
n.d. = not determined

probably due to the fact that only a fraction of the total number of bacteria present in the microflora can grow under the culturing conditions used. The number for *Bacteroides* obtained with culture and FISH is not identical. This may very well be due to the differences in specificity between the probe (which also detects *Prevotella*) and the medium.

Measurement of Variation in Intestinal Microflora

When studying microflora dynamics with FISH it is very important to measure as accurately as possible, otherwise significant changes might be lost due to the variation caused by the assay. Furthermore, we usually measure on a non-logarithmic scale. This facilitates detection of small variations, but also puts a lot of pressure on the reproducibility of the experiments; small experimental errors can influence the results significantly. Diet, age of the individual, sampling time, volume and consistency of the sample and perhaps other factors cause biological variations in the samples. Some of these variations are those we want to measure, others are unwanted. For instance, having large chunks of undigested material in a sample gives variation in the microbial content of the sample. Taking larger sample volumes can circumvent this problem. However, there are also variations introduced by the assay itself. These variations can be caused by several factors.

Table 4. Determination of the coefficient of variation due to the assay error (CV_{assay}) by repeated sample processing and measurement (25 frames) using the protocol described in this chapter and image analysis procedures. The top ten values (1-10) were obtained after a low spin of 2000 rpm, the bottom two (11, 12) with a low spin of 1000 rpm.

	DAPI staining	CVC^a	cell/g w/w x 10^{10}	Probe Erec482	CVC	cell/g w/w x 10^9
2000 rpm 700 x *g*	Mean counts/frame			Mean Counts/frame		
1	45	5	7.1	29	8	11.4
2	51	6	8.2	23	6	9.3
3	41	7	6.5	24	6	9.5
4	49	5	7.8	24	6	9.7
5	45	7	7.2	23	7	9.2
6	41	8	6.6	23	9	9.3
7	28	6	4.5	17	4	6.7
8	45	5	7.1	31	7	12.3
9	39	8	6.2	26	7	10.4
10	43	7	6.9	22	7	8.6
Mean			6.8			9.6
Standard deviation			1.0			1.5
CV_{assay}			0.15			0.16
1000 rpm	Mean Counts/frame	CVC		Mean Counts/frame	CVC	
11	46	6	7.3	28	8	11.3
12	44	8	7.1	30	5	12.1
Mean			7.2			11.7

^{a} Cumulative variation coefficient

(i) Processing of the samples. This includes sampling and sample storage before processing, sample weighing, low spin centrifugation to remove debris, fixation of the sample with 4% paraformaldehyde, storage of the fixed sample. (ii) Preparation of the slides including slide cleaning and coating, preparation of dilutions and application to the slide. The last factor especially is influenced by the skills of the investigator. (iii) Hybridization procedures and quality of the probes. In our lab we try to standardize the hybridization protocols as much as possible and store the samples in frozen aliquots, so that they are thawed only once. (iv) Evaluation and counting of the sample, which includes Quantimet settings and determination of fluorescence detection limit. In most cases fluorescence intensity of the cells in a fecal sample may vary from very bright to very weak. Based on intensity and color of the fluorescent signal it is normally easy for an operator to discriminate positive from negative signals. The operators have to set a threshold value for the Quantimet system to let the machine judge which fluorescence should be considered negative or positive. This is of course a subjective parameter and relies very much on the experience of the operator. However, once this parameter is set, it can be used to enumerate fluorescent cells objectively in 25 fields per sample of 48 samples per run.

Coefficient of Variation Due to the Assay

To determine the accuracy of our measurement we regularly determine the coefficient of variation due to the assay (CV_{assay}). Table 4 shows a typical example of such a measurement. Portions of feces, in this case 2.5 grams, were processed 12 times, 10 times using a low spin centrifugation at 2000 rpm (700 x g) and to test the effect of the spin, twice using a low spin centrifugation at 1000 rpm. The complete process of weighing the portions, dilution, centrifugation, fixation, slide preparation, hybridization and automated counting is repeated.

Cumulating Variation Coefficient

Table 4 lists, besides the cells /g wet weight, the mean counts per image frame analyzed by the software, and the cumulating variation coefficient (CVC) (Bloem *et al.,* 1992). These parameters are an indication of the quality of the measurement. For good reliable enumeration of the cells/g (w/w) 25 image frames (1 frame is about $^1/_4$ of a microscopic field) are measured. Ideally, the mean number per image frame should be more than 10 and less than 100. If this mean number is too low, accurate values cannot be obtained. If it is too high, the cells overlap in the images and the numbers are underestimated. The CVC is calculated as the standard deviation of the mean/ (mean x square root of the number of measurements) x 100%. This parameter is very useful as an indication of the quality of the mean number. It can be calculated after each measurement of a frame. After a certain number of frames the CVC drops below 10% variation and the measurements are considered statistically valid. In some cases this number is reached after 10 frames, usually after 15 to 25 frames. Sometimes the CVC does not drop below 10 after 25 frames. Usually the mean counts are then also below 10, and the measurement should be repeated at a different dilution. Otherwise, the counts are regarded as being under the detection limit of the Quantimet-system and should be evaluated visually, because this allows counting of the whole field of view. A high CVC with a mean count above 10 indicates that the cells are not spread homogeneously on the slide. This is often caused by clumps in the sample. Solutions to this problem may be counting more frames (e. g. up to 49) or preparation of new slides after homogenization of the sample.

Quantification of the Total Microflora by FISH

With the group-specific probes we aim at detecting all bacteria in the human gut. In the paper of Franks *et al.* (1998) a probe set was described that was able to count 65% of the total microflora as counted by DAPI staining and by Bact338 hybridization. At that time we expected a future coverage of 90% of the total cells would be feasible. Since then our methods have improved and we made new additional probes. Table 5 lists a set of probes that is currently used in our laboratory. The hybridization procedure is now performed on slides, while in the paper by Franks *et al.* (1998) it was performed on filters. In general, this slightly altered the counts of our hybridization results. We do not count more DAPI stained cells, but less total hybridizable bacteria (Table 5). This is probably due to reduced background coming from the filter, and improved settings of the detection limits for the automated counting on the Quantimet image analysis system. The probe set used in our last experiment (Table 5, see footnote a) can detect 90% of the total bacterial microflora enumerated by Bact338 hybridization, which seems a good coverage. However, 62 % of the total cells, as determined by DAPI-staining, are detected by the Bact338 probe and 56% by the sum of all the specific probes currently used. This would indicate that 44% of cells still remain undetected by the specific probes. Although clone library studies indicate that the probe set used does not cover all gut bacteria, we do believe it is indeed more than 90% of the total. We have given this problem much thought, and came up with a model to explain this (Figure 1). Assuming all cells are detected by DAPI staining, only 62% are detected by Bact338. The remaining cells are not detected due to the fact that they are *Archaea* or *Eucarya,* dead cells, not permeable or not metabolically active. The specific probes detect bacteria that are also detected by Bact338. In addition there are bacteria that need lysozyme permeabilization for hybridization, such as ruminococci and streptococci. These bacteria are detected by the specific

Table 5. Probes, target groups, references and average percentages of total microflora determined by DAPI staining, as they are enumerated in groups of volunteers between 20 and 55 years of age.

Probe	Target genera or species	Reference	Percentage of total cells (DAPI) in feces
S-D-Bact-0338-a-A-18	Domain Bacteria	Amann *et al.*, 1990	62 [a]
S-D-Bact-0338-a-S-18	Negative control of above	Amann *et al.*, 1990	0
S-*-Ato-0291-a-A-17	*Atopobium* group	Harmsen *et al.*, 2000	7 [a]
S-S-Bdis-0656-a-A-18	*Bacteroides distasonis*	Franks *et al.*, 1998 ⎫ combined 20
S-*-Bfra-0602-a-A-19	*Bacteroides fragilis* group	Franks *et al.*, 1998 ⎭
S-G-Bac303-a-A-16	*Bacteroides/Prevotella*	Manz *et al.*, 1996	18 [a]
S-G-Bif-0164-a-A-18	*Bifidobacterium* genus	Langendijk *et al.*, 1995	3 [a]
S-*-Chis-0150-a-A-23	*Clostridium histolyticum* group	Franks *et al.*, 1998	< 0.1
S-*-Clit-0135-a-A-19	*Clostridium lituseburense* group	Franks *et al.*, 1998	< 0.1
S-*-Cor-0653-a-A-18	*Coriobacterium* subgroup	Harmsen *et al.*, 2000	1
L-S-Eco-1531-a-A-21	*Escherichia coli*	Poulsen *et al.*, 1995	0.2 [a]
S-*-Ecyl-0386-a-A-18	*Eubacterium cylindroides* group	Harmsen *et al.*, 2002	0.9 [a]
S-*-Erec-0482-a-A-19	*Eubacterium rectale-Clostridium coccoides* group	Franks *et al.*, 1998	14 [a]
S-*-Fprau-0645-a-A-23	*Fusobacterium prausnitzii* group	Suau *et al.*, 2001	16
S-*-LowGC2P-a-A-18	*Fusobacterium prausnitzii*-like spp.	Wilson & Blitchington, 1996	7 [a]
S-G-Lab-0158-a-A20	*Lactobacillus/Enterococcus* group	Harmsen *et al.*, 1999	0.01 [a]
S-*-Strc-0493-a-A-19	*Lactococcus-Streptococcus* group	Franks *et al.*, 1998	< 0.1
S-*-Phasco-0741-a-19	*Phascolarctobacterium* group	Harmsen *et al.*, 2002	0.5 [a]
S-*-Rbro-0730-a-A-18	*Ruminococcus bromii* subcluster	Harmsen *et al.*, 2002 ⎫ combined 6 [a]
S-*-Rfla-0729-a-A-18	*Ruminococcus flavefaciens* subcluster	Harmsen *et al.*, 2002 ⎭
S-G-Veil-0223-a-A-18	*Veillonella*	Harmsen *et al.*, 2002	0.01 [a]

[a] Data of eleven dutch volunteers according to the protocol for automated measurements as presented in this chapter (Harmsen *et al.*, 2002), other data are taken from the reference stated in the preceding column.

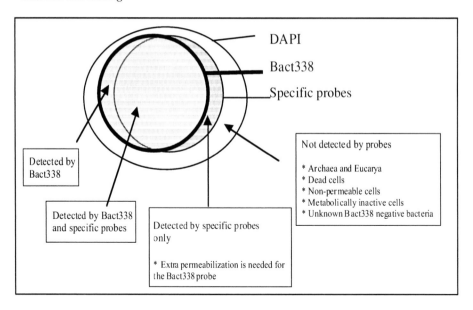

Figure 1. Model explaining the different parts of the microflora detected by the various methods.

probes, since we use lysozyme in the protocol, but not by the Bact338 probe where no lysozyme treatment is used. If lysozyme treatment is used in the Bact338 hybridization protocol other bacteria, presumably gram negative, are lysed and will become undetected. In most cases lysozyme treatment in the Bact338 protocol leads to a decrease in the number of fluorescent cells. Exceptions to this rule are found in some fecal samples, for instance of pre-term newborn babies, where enterococci and streptococci can become dominant, and which are better detected when lysozyme is used in the protocol.

There is also a principal reason for the differences in DAPI staining and Bact338 hybridization originating from differences in the molecules that are stained. DAPI intercalates with the minor groove of double stranded nucleic acids, therefore mainly with chromosomal DNA. The hybridization probes are aimed at ribosomal RNA. DAPI-stained DNA will mainly fluoresce from the central part of the bacterial cell where the chromosomal DNA is located. The probes will stain that part of the cell where the ribosomes are present, mostly the complete cytoplasm (and occasionally the part where DNA is located will remain unstained) and will fluoresce more globally from the cell. This has clear consequences for our image analysis procedure.

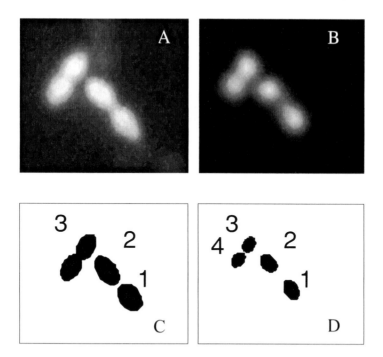

Figure 2. Schematic model illustrating how the different fluorescent staining methods can influence the enumeration by the image analysis processing. A: epifluorescent image of hybridization with a fluorescent probe directed against ribosomal RNA. B: epifluorescent image after DAPI staining the chromosomal DNA of the cells. C: a binary image after image analysis of image A, the software can separate cells #1 and #2 but not the dividing cell #3, in total three cells are counted by the software. D: a binary image of B, the image analysis software is capable of counting each individual cell and therefore four cells are counted. All images representing the same four cells were created artificially.

Some dividing cells are counted as one in the case of hybridization and are counted as individual cells in case of DAPI staining, especially with long chains of bacteria, this can strongly influence the counts. Fortunately, most cells in feces seem to be present as single or pairs of cells, which limits the bias. This is illustrated in Figure 2.

Dynamics of the Human Microflora

For study of the dynamics of the microflora over time, we compared the results of three volunteers over a 3.5-year period. Table 6 lists the percentage of the total DAPI-stained cells of the different bacterial groups. We compared data from 1997 obtained using hybridization with the filter method as described in Franks *et al.* (1998) with data from 1998 measured by the

Table 6. Comparison of the determined percentages of the indicated bacterial groups enumerated using the probes indicated in brackets, relative to the total DAPI stained cells on three different sampling dates over a 3.5-year period.

Volunteer	Sample date	Total bacteria (Bact338)	Bacteroides (Bac303)	Clostridia (Erec482)	Bifidobacteria (Bif164)	Eubacteria low G+C (Elgc01)
1	*11/06/'97*[a]	*86.8*	*18.1*	*21.5*	*11.2*	*9.2*
	10/01/'98	74.4	16.7	14.9	1.9	7.0
	03/28/'01	60.2	16.5	15.8	3.5	4.7
2	*05/28/'97*	*60.4*	*25.7*	*19.9*	*1.6*	*21.8*
	10/01/'98	55.1	17.2	23.7	4.1	12.9
	03/28/'01	58.0	18.8	25.2	4.1	19.6
3	*08/09/'97*	*104.7*	*7.7*	*39.9*	*1.4*	*10.1*
	10/01/'98	86.0	33.1	26.8	1.3	8.0
	03/29/'01	66.9	21.1	17.2	1.5	8.6

[a] Percentage in bold italics are determined by hybridization in cups and counting on filters, others are determined by hybridization and counting on slides.

improved method as described by Jansen *et al.* (1999). The last (2001) samples were measured on slides with basically the same method as described by Jansen *et al.* (1999), with the exception that the samples were diluted in phosphate buffered saline (PBS), not in 5% Tween80. This did not influence the counts but made the spreading of the dilutions on the slides easier. Table 6 shows that some percentages of 1998 and 2001 are quite different and some are remarkably similar and that these similarities seem to be specific for an individual. Most of the 1997 counts seem to be different from the 1998 and 2001 counts, which is probably caused by the different method used. However, these data show that it remains difficult to compare data that have not been measured with the same method.

As a concluding remark we can say that the FISH method has developed into a valuable tool for the analysis of gut microflora. Applying quantitative FISH with a defined collection of probes targeted at bacteria in feces or within the intestine will help to refine our knowledge about the spatial distribution and dynamics of gut bacteria and their interaction with the host in health and disease.

References

Amann, R.I., Binder, B.J., Olson, R.J., Chrisholm, S.W., Devereux, R., and Stahl, D.A. 1990. Combination of 16S rRNA-targeted oligonucleotide probes with flow cytometry for analyzing mixed microbial populations. Appl. Environ. Microbiol. 56: 1919-1925.

Amann, R.I., Ludwig, W., and Schleifer, K.-H. 1995. Phylogenetic identification and *in situ* detection of individual microbial cells without cultivation. Microbiol. Rev. 59: 143-169.

Barcenilla, A., Pryde, S.E., Martin, J.C., Duncan, S.H., Stewart, C.S., Henderson, C., and Flint, H.J. 2000. Phylogenetic relationships of butyrate-producing bacteria from the human gut. Appl. Environ. Microbiol. 66: 1654-1661.

Bloem, J., van Mullem, D.K., and Bolhuis, P.R. 1992. Microscopic counting and calculation of species abundances and statistics in real time with an MS-Dos personal computer, applied to bacteria in soil smears. J. Microbiol. Methods. 16: 203-213.

Doré, J., Sghir, A., Hannequart-Gramet, G., Corthier, G., and Pochart, P. 1998. Design and evaluation of a 16S rRNA-targeted oligonucleotide probe for specific detection and quantitation of human faecal *Bacteroides* populations. Syst. Appl. Microbiol. 21: 65-71.

Eckman, L., Kagnoff, M.F., and Fierer, J. 1995. Intestinal epithelial cells as watchdogs for the natural immune system. Trends Microbiol. 3: 118-120.

Finegold, S.M., Sutter, V.L., and Mathisen, G.E. 1983. Normal indigenous intestinal flora. In: Human intestinal microflora in health and disease. D.J. Hentges ed. Academic Press, New York. p. 3-31.

Franks, A.H., Harmsen, H.J.M., Raangs, G.C., Jansen, G.J., Schut, F., and Welling, G.W. 1998. Variations of bacterial populations in human feces quantified by fluorescent *in situ* hybridization with group-specific 16S rRNA-targeted oligonucleotide probes. Appl. Environ. Microbiol. 64: 3336-3345

Goldin, B.R, and Gorbach, S.L. 1977 Alterations in fecal microflora enzymes related to diet, age, lactobacillus supplements, and dimethylhydrazine. Cancer. 40: 2421-2426.

Gorbach, S.L., Chang, T.W., and Goldin, B. 1987. Successful treatment of relapsing *Clostridium difficile* colitis with *Lactobacillus* GG. Lancet. ii: 1519.

Harmsen, H.J.M., Elfferich, P., Schut F., and Welling, G.W. 1999. A 16S rRNA-targeted probe for detection of lactobacilli and enterococci in fecal samples by fluorescent *in situ* hybridization. Microbiol. Ecol. Health Dis. 11: 3-12.

Harmsen, H.J.M., Raangs, G.C., He, T., Degener, J.E., Welling, G.W. 2002. Extensive set of 16S rRNA-based probes for detection of bacteria in human feces. Appl. Environm. Microbiol. 68: In press.

Harmsen, H.J.M., Wildeboer-Veloo, A.C.M., Grijpstra, J., Knol, J., Degener, J.E., and Welling, G.W. 2000a. Development of 16S rRNA-based probes for the *Coriobacterium* group and the *Atopobium* cluster and their application for enumeration of *Coriobacteriaceae* in human feces from volunteers of different age groups. Appl. Environ. Microbiol. 66: 4523-4527.

Harmsen, H.J.M., Wildeboer-Veloo, A.C.M., Raangs, G.C., Wagendorp, A.A., Klijn, N., Bindels, J.G., and Welling, G.W. 2000b. Analysis of intestinal flora development in breast-fed and formula-fed infants by using molecular identification and detection methods. J. Pediatr. Gastroenterol. Nutr. 30: 61-67.

Higgins J.A., Fayer R., Trout J.M., Xiao L., Lal A.A., Kerby S., and Jenkins M.C. 2001. Real-time PCR for the detection of *Cryptosporidium parvum*. J. Microbiol. Methods. 47: 323-337.

Hopkins, M.J., Sharp, R., and Macfarlane, G.T. 2001. Age and disease related changes in intestinal bacterial populations assessed by cell culture, 16S rRNA abundance, and community cellular fatty acid profiles. Gut. 48: 198-205.

Jansen, G.J., Wildeboer-Veloo, A.C.M., Tonk, R.H.J., Franks, A.H., and Welling G.W. 1999. Development and validation of an automated, microscopy-based method for enumeration of groups of intestinal bacteria. J. Microbiol. Meth. 37: 215-221.

Kalliomäki, M., Kirjavainen, P., Eerola, E., Kero, P., Salminen, S., and Isolauri, E. 2001. Distinct patterns of neonatal gut microflora in infants in whom atopy was and was not developing. J. Allergy Clin. Immunol. 107: 129-134.

Kleessen, B, Hartmann, L, and Blaut, M. 2001. Oligofructose and long-chain inulin: influence on the gut microbial ecology of rats associated with a human faecal flora. Br. J. Nutr. 86: 291–300.

Langendijk, P.S., Schut, F., Jansen, G.J., Raangs, G.C., Kamphuis, G.R., Wilkinson, M.H.F., and Welling, G.W. 1995. Quantitative fluorescence *in situ* hybridization of *Bifidobacterium* spp with genus-specific 16S rRNA-targeted probes and its application in faecal samples. Appl. Environ. Microbiol. 61: 3069-3075.

Manz, W., Amann, R., Ludwig, W., Vancanneyt, M., and Schleifer, K.-H. 1996. Application of a suite of 16S rRNA-specific oligonucleotide probes designed to investigate bacteria of the phylum cytophaga-flavobacter-bacteroides in the natural environment. Microbiology 142: 1097-1106.

Matsuki, T., Watanabe, K., Tanaka, R., Fukuda, M., and Oyaizu, H. 1999. Distribution of bifidobacterial species in human intestinal microflora examined with 16S rRNA-gene-targeted species-specific primers. Appl. Environ. Microbiol. 65: 4506-4512.

Moore, W.E., and Holdeman, L.V. 1974. Human fecal flora: the normal flora of 20 Japanese-Hawaiians. Appl. Microbiol. 27: 961-979.

Olsen, G.J., Lane, D.J., Giovannoni, S.J., Pace, N., and Stahl, D.A. 1986. Microbial ecology and evolution: a ribosomal RNA approach. Annu. Rev. Microbiol. 40: 337-365.

Poulsen, L.K., Licht, T.R., Rang, C., Krogfelt, K.A., and Molin, S. 1995. Physiological state of *E. coli* BJ4 growing in the large intestines of streptomycin-treated mice. J. Bact. 177: 5840-5845.

Schwiertz, A., Le Blay, G., and Blaut, M. 2000. Quantification of different *Eubacterium* spp. in human fecal samples with species-specific 16S rRNA-targeted oligonucleotide probes. Appl. Environ. Microbiol. 66: 375-382.

Sghir, A., Gramet, G., Suau, A., Rochet, V., Pochart, P., and Doré, J. 2000. Quantification of bacterial groups within human fecal flora by oligonucleotide probe hybridization. Appl. Environ. Microbiol. 66: 2263-2266.

Simmering, R., Kleessen, B., and Blaut, M. 1999. Quantification of the flavonoid-degrading bacterium *Eubacterium ramulus* in human fecal samples with a species-specific oligonucleotide hybridization probe. Appl. Environ. Microbiol. 65: 3705-3709.

Suau, A., Bonnet, R., Sutren, M., Godon, J.J., Gibson, G.R., Collins, M.D., and Doré, J. 1999. Direct analysis of genes encoding 16S rRNA from complex communities reveals many novel molecular species within the human gut. Appl. Environ. Microbiol. 65: 4799-4807.

Suau, A., Rochet, V., Sghir, A., Gramet, G., Brewaeys, S., Sutren, M., Rigottier-Gois, L., and Doré, J. 2001. *Fusobacterium prausnitzii* and related species represent a dominant group within the human fecal flora. Syst. Appl. Microbiol. 24: 139-145.

Ward, D.M., Bateson, M.M., Weller, R., and Ruff-Roberts, A.L. 1992. Ribosomal RNA analysis of microorganisms as they occur in nature. Adv. Microb. Ecol. 12: 219-286.

Welling, G.W., Elfferich, P., Raangs, G.C., Wildeboer-Veloo, A.C., Jansen, G.J., and Degener, J.E. 1997. 16S ribosomal RNA-targeted oligonucleotide probes for monitoring of intestinal tract bacteria. Scand. J. Gastroenterol. Suppl. 222: 17-19

Wilson, K.H., and Blitchington, R.B. 1996. Human colonic biota studied by ribosomal DNA sequence analysis. Appl. Environ. Microbiol. 62: 2273-2278.

Woese, C.R. 1987. Bacterial evolution. Microbiol. Rev. 51: 221-271.

Zoetendal, E.G., Akkermans, A.D.L., and De Vos, W.M. 1998. Temperature gradient gel electrophoresis analysis of 16S rRNA from human fecal samples reveals stable and host-specific communities of active bacteria. Appl. Environ. Microbiol. 64: 3854-3859.

From: *Probiotics and Prebiotics: Where Are We Going?*
Edited by: Gerald W. Tannock

Chapter 3

From Composition to Functionality of the Intestinal Microbial Communities

Sergey R. Konstantinov,
Nora Fitzsimons, Elaine E. Vaughan,
and Antoon D.L. Akkermans

Abstract

The mammalian gastrointestinal (GI) tract harbours a large bacterial community that has an essential role in creating optimum health conditions for the host. This chapter focuses on the use of molecular fingerprinting tools to describe the taxonomic and functional diversity of the microbial community in the GI tract. Special attention is given to the composition analysis of microbial communities based on 16S rDNA sequence diversity. Basic principles and new developments of several PCR-based methods, such as denaturing gradient gel electrophoresis (DGGE) and related fingerprint methods as well as methods to analyse these fingerprints are described. Advantages and drawbacks of DGGE are described and compared with the terminal restriction fragment length polymorphism (T-RFLP) method. In addition to methods investigating the taxonomic diversity of microbial communities in the GI tract, we also address the recent progress to describe the functional diversity of bacterial communities in the GI tract. Although relatively little information is available yet, we anticipate that our insight in the occurrence and activity of functional bacterial genes in the GI tract will rapidly expand in the next decade due to the enormous increase in sequence information and developments in microarrays technology.

Introduction

The mammalian gastrointestinal (GI) tract represents a dynamic ecosystem containing a complex community of microaerophilic and anaerobic microbes that are involved in the fermentative conversion of ingested food and the components secreted by the host into the intestinal tract. Insight into the structure and function of the GI tract microbial communities and into the activity of a specific microbial species within this ecosystem is necessary for the development of functional foods such as probiotics and prebiotics. It is now well accepted that many microbes from natural environments including the GI tract have yet to be isolated and characterised (Amann *et al.*, 1995; Vaughan *et al.*, 1999; 2000). Comparison between microscopic and plate counts, group-specific dot-blot hybridisation and fluorescent *in situ* hybridisation (FISH) has indicated that a significant fraction of the microflora can escape cultivation (Finegold *et al.*, 1983; Langendijk *et al.*, 1995; Suau *et al.*, 1999; Sghir *et al.*, 2000). Similar results have been reported for bacterial inhabitants in a variety of animal models including pigs (Pryde *et al.*, 1999; Leser *et al.*, 2002). This is understandable considering the challenges for the microbiologist studying the GI tract. Firstly, the anaerobic conditions in the human colonic ecosystems require extra facilities and care in handling samples. Secondly, the substantial diversity of GI microbes necessitates the use of suitable selective media and incubation times are longer for anaerobes. Realising the limitation of the traditional methods, such as microscopy and cultivation to gain insight into the structure and activity of the bacterial communities, new molecular microbial techniques have been developed. For the most part these novel methods are based on particular molecular markers, such as the 16S rRNA or its encoding gene, and are being increasingly used to explore the microbial diversity of bacterial communities. Denaturing Gradient Gel Electrophoresis (DGGE) is a fingerprinting technique used to bypass cultivation and visualise complex bacterial communities without prior knowledge of the composition.

In this review DGGE will be described with special focus on the application to the intestinal tract bacterial communities. Other novel fingerprinting techniques including Terminal-Restriction Fragment Length Polymorphism (T-RFLP) and DNA arrays for intestinal diversity will also be briefly described and compared to DGGE. The potential of further molecular methods for analysing diversity and functionality will also be addressed.

Estimation of Diversity by Fingerprinting Techniques

Gradient Gel Electrophoresis

Several types of gradient gel electrophoresis have been proposed to describe microbial diversity. DGGE was first applied in microbial ecology to study the bacterial diversity in a marine ecosystem (Muyzer *et al.*, 1993). Since the first publication a cascade of studies in microbial ecology have used DGGE and Temperature Gradient Gel Electrophoresis (TGGE), and only occasionally Temporal Temperature Gradient Gel Electrophoresis (TTGE) (Zhu *et al.*, 2002). These techniques are particularly suitable for rapidly comparing bacterial communities from different environments and/or monitoring the activity of a specific community over time. A general strategy for use of these techniques is presented in Figure 1. Initially nucleic acids (DNA or RNA) are extracted followed by Polymerase Chain Reaction (PCR) amplification of genes encoding the 16S rRNA and subsequent separation of the PCR products by DGGE, TGGE or TTGE. The band identification in TGGE and DGGE may be done either by screening of 16S rDNA clone libraries, or by directly excising the bands from the gel following by reamplification, and DNA sequencing. These techniques are presently used for a variety of applications: describing the bacterial community complexity and stability over time in numerous ecosystems, monitoring the enrichment and isolation of bacteria, comparing different DNA/RNA extracting protocols, screening of clone libraries, cloning biases and heterogeneity in rDNA genes (Muyzer and Smalla, 1998; Muyzer *et al.*, 1998; 1999).

Principal of Denaturing Electrophoresis Technique

DGGE is a technique that allows the separation of DNA molecules differing by single base changes (Myers *et al.*, 1985; 1987). The separation of DNA fragments of the same length is based on the sequence-specific melting behaviour in a polyacrylamide gel containing either a linear gradient of

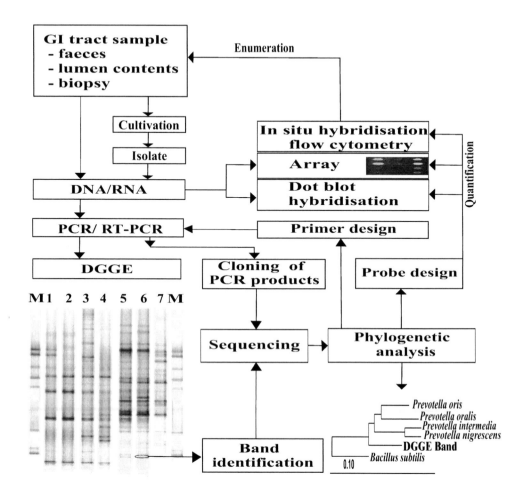

Figure 1. Scheme for analysis of GI tract samples including fingerprinting techniques (see text for details).

chemical denaturant during DGGE (Fischer and Lerman, 1979), or a linear temperature gradient as in TGGE (Rosenbaum and Riesner, 1987; Riesner *et al.*, 1992). Alternatively, the separation is based on a temporal temperature gradient that increases gradually in a linear fashion over the length of the electrophoresis time during TTGE (Yoshino *et al.*, 1991). During electrophoresis, each of the 16S rDNA PCR products (or amplicons) starts to melt in so-called melting domains. Sequence variations within such domains causes the melting temperatures of the amplicons to differ. The incorporation of a GC-clamp, i.e. a 30 to 50 bp GC-rich domain, into the amplicons during PCR by adding it at the 5'-end of one of the primers, prevents complete melting of the amplicons. In principle, all single base differences at each position can be separated for PCR products of up to 500 bp (DGGE) (Myers *et al.*, 1985; Sheffield *et al.*, 1989). Therefore, a mix of fragments with different sequences will essentially stop migrating at different positions in the denaturing gradient. To obtain the best possible separation of different DNA fragments, it is necessary to experimentally optimise either chemical or temperature gradient, and the duration of the electrophoresis (Muyzer *et al.*, 1993; 1998).

TTGE differs from TGGE in that TGGE has a fixed temperature gradient from the top to the bottom of the gel. In TTGE the temperature at any location in the gel is the same at any particular point in time, but changes with the progression of time (temporal temperature). TTGE, a modified parallel form of DGGE, does not require the preparation of a chemical denaturant gradient gel and can be performed without a GC clamp (Chen *et al.*, 1999).

Analysis and Interpretation of Complex Fingerprints

Electrophoretic patterns obtained by DGGE, TGGE and TTGE are often highly complex. Therefore, numerical analysis of banding profiles is essential to obtain an objective interpretation, which by visual evaluation is very difficult to achieve. Interpretation is greatly facilitated by the use of computer assisted pattern analysis using specialised computer software packages. Figure 2 shows a general strategy of multivariate statistical analysis of DGGE/TGGE.

DGGE/TGGE patterns can be analysed based on distinct bands or on densimetric curves. Using a band-based method, a collection of fingerprints, irrespective of their complexity, can be transformed into a matrix of binary variables; bands present in a profile at a particular position are designated as

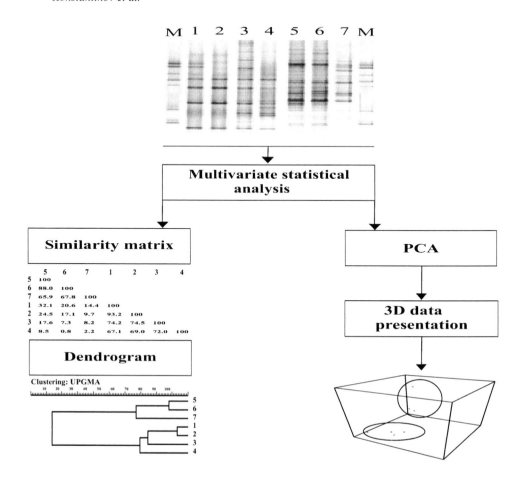

Figure 2. Cluster analysis of the DGGE/TGGE fingerprints can be achieved by different ways: for example by a dendrogram using the similarity matrix based on Pearson product-moment correlation coefficient between DGGE fingerprints, or by 3 dimensional (3D) data presentation as in Principle Component Analysis (PCA).

1, and those absent designated 0 (Rademaker *et al.*, 1999). This binary system is representative only of the number of the bands and their position in the fingerprint; consequently, it is not sufficient to express the more complex attributes of profiles such as band intensities, the band area under the peaks or different ratios in the peak heights.

The analysis of fingerprint patterns generally requires a simplification of the original data via the generation of a proximity matrix based on dissimilarity or similarity criteria. Such proximity matrices can be generated using a wide range of coefficients (Rademaker *et al.*, 1999). The Sorenson's coefficient sometimes refereed to as the Dice coefficient (Dice, 1945) is a pairwise similarity coefficient used to compare species compositions of different ecosystems (Magurran, 1988; Murray *et al.*, 1996; Gillan *et al.*, 1998; Simpson *et al.*, 1999; 2000; Leser *et al.*, 2000). The Sorenson's coefficient is defined as: $Cs=[2nAB/(nA+nB)]x\,100$, where A is the number of bands in lane 1, B represents the number of bands in lane 2, and nAB is the number of common DGGE bands (Murray *et al.*, 1996; Gillan *et al.*, 1998; Simpson *et al.*, 1999).

DGGE or TGGE fingerprints can also be compared using the Pearson product-moment correlation coefficient (Pearson, 1926; Zoetendal *et al.*, 2001a), that is directly applied to the array of densimetric values forming the fingerprint. The product moment correlation coefficient is insensitive to the relative concentrations of bands between fingerprints, to the background and it is also insensitive to differences in overall intensity of profiles (Rademaker *et al.*, 1999). In general, the Pearson product-moment coefficient is better suited for identification of DNA fingerprints profiles than band matching algorithms (Häne *et al.*, 1991).

Cluster analysis of the DGGE/TGGE fingerprints can be achieved by different ways, usually by a dendrogram (Jardine and Sibson, 1971) or 3D data presentation as in Principle Component Analysis (PCA) (Cooley and Lohnes, 1971; Yang *et al.*, 2001; Röling *et al.*, 2001). Several algorithms are available for clustering analysis leading to generation of dendrograms. The unweighted pair group method using arithmetic averages (UPGMA) (Sneath and Socal, 1973) and the method of Ward (Ward, 1963; Sneath and Socal, 1973) are the most widely used clustering algorithms to process DGGE/TGGE profiles. The selection of clustering methods depends on the nature of the original data and the purpose of the analysis. Moreover, the clustering methods of the fingerprints are useful to describe and explain the data, but they do not constitute statistical tests to prove or reject a hypothesis (Rademaker *et al.*, 1999).

The complexity of the DGGE/TGGE profiles can also be expressed by diversity indices (Shannon's, Simpson's and Hill's) calculated using peak number and area or relative intensities of bands in an individual lane (Shannon and Waver, 1963; Magurran, 1988; Eicher *et al.*, 1999). Thus, a distinct diversity value for each sample is obtained and changes in community diversity over time may be observed.

Insight Into GI Tract Microflora Using PCR and DGGE/ TGGE

D/TGGE analysis of amplified 16S rDNA fragments of the mammalian GI tract is now recognised as a fast and reliable method to access profiles from the intestinal microbial communities. The ability to analyse a large set of samples in a short time allows the establishment, persistence and composition of the bacteria over time in the GI tract to be monitored (Tannock, 1999; Vaughan *et al.*, 1999; 2000; Akkermans *et al.*, 2000). This has already resulted in substantial knowledge concerning factors that affect the community structure such as environmental disturbances, physiological conditions and the genetic background of the host as described below. Comparison of the TGGE fingerprints from faecal samples of adult humans has demonstrated that the composition of the predominant microflora is host specific and stable over time (Zoetendal *et al.*, 1998; Zoetendal 2001; Zoetendal *et al.*, 2001a; 2001b). By detailed analysis of PCR amplicons generated from the V6 to V8 regions of the 16S rRNA gene separated by TGGE, it was shown that each individual harbours a unique faecal bacterial population suggesting a strong host influence. A similar observation has been made for the DGGE banding profiles of piglets (Simpson *et al.*, 2000). DGGE was also used to investigate the effect of weaning and potential prebiotics on the microflora of piglets (S.R. Konstantinov, unpublished). The hypothesis that the genetic background strongly influences the composition of the GI tract bacterial community has been recently tested in a comparative study of DGGE profiles from adults with a different genetic relatedness varying from unrelated persons to monozygotic twins (Zoetendal *et al.*, 2001a). Statistical analysis based on the Pearson product-moment coefficient of pairwise comparison showed that the similarity between DGGE profiles of monozygotic twins was significantly higher than that for unrelated individuals, indicating that the genetic background indeed affects the composition of the predominant microflora. These findings could partly explain why in controlled trials a probiotic strain has little or no effect on the dominant microbial community in adults relative to the control subjects (de Vos *et al.*, 1998; Vaughan *et al.*, 1999; 2000).

Moreover PCR-DGGE has been used to investigate the contribution of non-cultivable bacteria to gut disorders such as necrotizing enterocolitis (NEC) for the specific case of pre-term infants (Millar *et al.*, 1996). PCR-DGGE study of human GI tract combined with 16S rDNA sequence analysis have been especially valuable to monitor the establishment of the bacterial community in the new-born intestinal ecosystem (Favier *et al.*, 2002). The successive changes in the predominant bacteria after birth could be visualised using DGGE and were further identified by comparative 16S rDNA sequence analysis. Screening of 16S rDNA products from human baby and adult faeces showed that the majority of the dominant bands present in the profiles were derived from sequences with less than 98% identity with known sequences in the databases (Zoetendal *et al.*, 1998; Favier *et al.*, 2002).

Another important application of the DGGE to the GI tract microbial ecology is the analysis of the response of bacterial community structure to environmental stress such as changes in the diet, weaning, antibiotic treatment or the introduction of an exogenous strain to the bacterial community (Simpson *et al.*, 2000; Satokari *et al.*, 2000; Tannock *et al.*, 2000, Zoetendal *et al.*, 2001b). DGGE in combination with FISH and flow cytometry was also used to describe the microflora attached to human biopsies (Zoetendal, 2001).

Lactobacillus and *Bifidobacterium* Group Specific PCR and DGGE

It has been reported that DGGE or TGGE are sensitive enough to visualise as a band a population of bacteria that constitutes up to 1% of the total bacterial community (Muyzer *et al.*, 1993; Zoetendal *et al.*, 1998). A combination of flow cytometry and PCR-DGGE of dilution series of pure cultures estimated the detection limit of 10^5 cells ml^{-1} bacteria, which was influenced by the DNA isolation protocols (Zoetendal *et al.*, 2001c). The sensitivity of these techniques may be vastly improved by combining specific primers for groups of interest. An overview of frequently used 16S rDNA-targeted primers in gradient gel electrophoresis studies of the GI tract microflora are presented in Table 1. Bifidobacteria which constitute approximately 3% of the total microflora of adults, and up to 90 % in breast-fed babies, usually appear as a single band towards the end of a denaturing gel, presumably due to the high GC content of their DNA (Favier *et al.*, 2002). Bifidobacterial specific primers were used to PCR the 16S rDNA sequences from human bifidobacterial populations in feces, and following optimisation of the denaturing gradient, their diversity could be visualised by a DGGE profile (Satokari *et al.*, 2001a). This method has recently been used to monitor fecal bifidobacterial populations in a prebiotic and probiotic feeding trial (Satokari *et al.*, 2001a; 2001b).

However, other important species are present in even lower numbers in the GI tract, such as the *Lactobacillus* species which have been shown to

Table 1. 16S rDNA-targeted primers used for DGGE, TGGE and TTGE of domain Bacteria, *Lactobacillus* group and *Bifidobacterium* species found in the human and animal intestine

Primer name	Specificity	Primer sequence (5'-3')	Reference
F 341-GC	Eubacterial 16S rDNA	(GC clamp)-CCT ACG GGA GGC AGC AG	Muyzer *et al.*, 1993
R 534	Universal	ATT ACC GCG GCT GCT GG	Muyzer *et al.*, 1993
R 907	Universal	CCG TCA ATT CCT TTR (A/G)GT TT	Muyzer *et al.*, 1998
F 968-GC	Eubacterial 16S rDNA	(GC clamp)-AAC GCG AAG AAC CTT AC	Nübel *et al.*, 1996
R 1401	Eubacterial 16S rDNA	CGG TGT GTA CAA GAC CC	Nübel *et al.*, 1996
HDA1-GC	Eubacterial 16S rDNA	(GC clamp)- AC TCC TAC GGG AGG CAG CAG T	Tannock *et al.*, 2000
HDA2	Eubacterial 16S rDNA	GTA TTA CCG CGG CTG CTG GCA C	Tannock *et al.*, 2000
aS-G-Lab-0677-a-A-17	*Lactobacillus* group*	CAC CGC TAC ACA TGG AG	Heilig *et al.*, 2002
aS-G-Lab-0159-a-S-20	*Lactobacillus* group*	GGA AAC AG(A/G) TGC TAA TAC CG	Heilig *et al.*, 2002
F Lac1	*Lactobacillus* group*	AGC AGT AGG GAA TCT TCC A	Walter *et al.*, 2001

R Lac 2-GC	*Lactobacillus* group*	(GC clamp)-ATT (C/T)CA CCG CTA CAC ATG	Walter et al., 2001
F Lacto #1-GC	*Lactobacillus reuteri* group	CGC CCG CGC GCG GCG GGC GGG GCG GGG GCA CGG GGG GGT CGA (A/G)CG (A/C)AC TGG CCC	Simpsom et al., 2000
R Lacto #2	*Lactobacillus reuteri* group	GCT GCC TCC CG(A/G) AGG AGT	Simpsom et al., 2000
F Bif164-GC	*Bifidobacterium*	CGC CCG GGG CGC GCC CCG GGC GGG GCG GGG GCA CGG GGG G-GGG TGG TAA TGC CGG ATG	Satokari et al., 2001
R Bif662-GC	*Bifidobacterium*	CGC CCG CGC GCG GCG GGC GGG GCG GGG GCA CGG GGG G-CCA CCG TTA CAC CGG GAA	Satokari et al., 2001
F Bif164	*Bifidobacterium*	GGG TGG TAA TGC CGG ATG	Langendijk et al., 1995; Satokari et al., 2001
R Bif662	*Bifidobacterium*	CCA CCG TTA CAC CGG GAA	Langendijk et al., 1995; Satokari et al., 2001

[a]Nomenclature according to Alm et al. (1996)

*Includes genera *Lactobacillus, Leuconostoc, Pediococcus* and *Weissella*.

(GC clamp)- 5'-CGC CCG CCG CGC CCG GCG GGC GGG GCG GGG GCA CGG GGG G-3' (39)

F- forward primer

R- reverse primer

constitute less than 1% of the total bacterial community (Sghir *et al.*, 2000). Therefore, using universal bacterial primers it is nearly impossible to monitor this group. Recently, two sets of *Lactobacillus* group-specific PCR primers in combination with DGGE, have been used to monitor the molecular diversity of *Lactobacillus* spp. and related lactic acid bacteria in the human intestine (Walter *et al.*, 2001; Heilig *et al.*, 2002). The approach has also been tested to study the persistence in faecal samples of an emerging *Lactobacillus paracasei* strain F19 administered to children during a clinical trial. In addition to tracking of the F19 strain, the DGGE profiles supported the natural presence of this strain within the intestinal community of a proportion of individuals (Crittenden *et al.*, 2002; Heilig *et al.,* 2002). These authors concluded that DGGE in combination with group-specific PCR analysis of 16S rDNA products allows the characterisation of bacteria present in low numbers in the human GI tract. Simpson *et al.* (2000) used an exogenous *Lactobacillus* strain to assess its affect on the faecal bacterial population of piglets. Analysis of DGGE band intensity after a specific PCR revealed an antagonistic relationship between the exogenous strain and another indigenous *Lactobacillus*. DGGE/TGGE are considered to be semiquantitative methods, but the absolute quantification of band intensity by competitive PCR as already applied in environmental studies (Felske *et al.*, 1998), still remains to be done for the complex bacterial communities in the GI tract.

Developments and Improvements in Denaturing Gradient Techniques

PCR-DGGE or TGGE are well established methods to study the bacterial diversity and the community structure in the GI tract. They are fast, reliable and sensitive tools to demonstrate differences between bacterial communities from different individuals and to monitor changes in time. By using universal bacterial primers, the spatial and temporal distribution of as yet uncultured species could be analysed. However, bands with an identical position within a DGGE or TGGE profile do not always belong to the same 16S rRNA and further analysis is required. The potential drawbacks can be overcome and the sensitivity of the techniques can be improved by a combination of group- or species-specific PCR or by subsequent hybridisation analysis (Muyzer *et al.*, 1999). The possibility of identification of existing bands by sequencing or by hybridisation with a specific probe is a strong advantage compared to other fingerprinting methods (see below). Moreover, following cloning and sequencing, probes may be designed and used in hybridisation analysis precluding the need for phylogenetic information (Heuer *et al.*, 1999). In addition to the general limitations of all PCR based methods (Von Wintzinger *et al.*, 1997; Schmalenberger *et al.*, 2001), DGGE and TGGE have their own specific restrictions (Muyzer and Smalla, 1998; Muyzer, 1999). DGGE analyses may be affected by heteroduplex molecules introduced during the PCR, DNA fragments produced from different rDNA operons from one organism (Nübel *et al.*, 1996), the comigration of closely related sequences, and only relatively small DNA fragments can be separated (~500 bp). Difficulties in standardising the casting of DGGE and TGGE gels, and maintenance of constant electrophoresis conditions over the long running period (10-16 h) makes comparison of different gels difficult, and requires reasonable experience on the part of the researcher. Nevertheless, the use of a DNA standard ladder on each gel has allowed effective comparison between samples run on different gels (Simpson *et al.*, 2000). Recently, a new development of the DGGE technique called constant-denaturant capillary electrophoresis (CDCE) in combination with a novel quantitative PCR (QPCR) approach, accurately enumerated microbial cells at abundance ranging from approximately 10 to 10^4 cells ml^{-1} (Lim *et al.*, 2001).

Terminal-Restriction Fragment Length Polymorphism

Terminal-Restriction Fragment Length Polymorphism (T-RFLP) is a molecular method for rapid analysis of microbial community diversity. The technique is a development of RFLP where one of the 16S-specific PCR primers is labelled with a fluorescent dye to allow detection of the amplified product. The PCR products are then digested with restriction enzyme and the DNA sequencer is employed to analyse the size of the restriction fragments (Avaniss-Aghajani *et al.*, 1994) (Figure 3). Different fluorescently labeled terminal restriction fragments (TRF) in theory correspond to different bacterial species since the nucleotide sequence of the 16S rRNA gene is unique for each bacterial species. The raw data are automatically converted to a digitised form that can be analysed with a variety of multivariate statistical tools. The identification of specific TRF is possible by comparison to entries in an on-line database provided by Michigan State University, Center for Microbial Ecology (http: //www.cme.msu.edu/RDP/html) or by comparison to a clone library. The method is used for characterisation of complex microbial communities and assessing community dynamics in a variety of ecosystems (Liu *et al.*, 1997; Marsh, 1999; Kitts, 2001). TRF patterns are recognised as having better resolution than other DNA based techniques and have already been applied to investigate complex microbial communities in intestines (Leser *et al.*, 2000; Kapplan *et al.*, 2001).

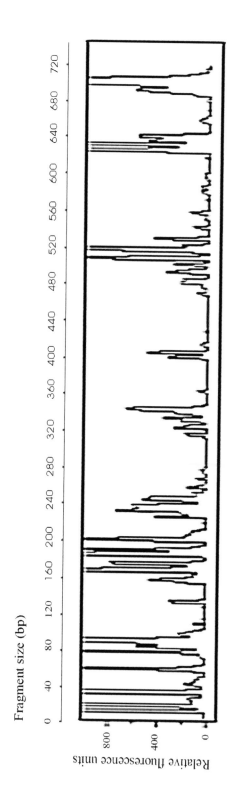

Figure 3. T-RFLP analysis of a pig colon lumen sample. The different fluorescently labeled terminal restriction fragments (TRF) in theory correspond to different bacterial phylotypes (bacteria defined only by their 16S rDNA sequences). The identification of a specific TRF may be achieved by comparison to a clone library or by comparison to an on-line database.

Table 2. Advantages and drawbacks of various techniques for analysis of intestinal microflora

Method	Advantages	Drawbacks
DGGE	Rapid and reproducible diversity	Semi-quantitative assessment
TGGE	Easy comparative analysis	PCR biases, ~500 bp fragment
TTGE	Analysis of uncultured bacteria	Cloning and sequencing is required for bands identification
	Allow band extraction from the gel	Comparison between different gels is difficult
	Southern blot hybridisation on TGGE/DGGE	Comigration of DGGE fragments
	Group-specific PCR for minor bacterial population	Insensitive for minor bacterial groups by using universal bacterial primers
T- RFLP	Fast and sensitive fingerprinting method	Semi-quantitative
	Standardised comparison between gels	PCR biases
	Size fragment analysis	Cloning and sequencing is required for bands identification
	Analysis of uncultured bacteria	Southern blot hybridisation is not possible
		Densimetric curve is sensitive to the background
		Differences between predicted and actual fragment size

Perspectives

Compositional Analysis

Developing PCR primer sets for combination with DGGE/TGGE target the main bacterial groups of the intestinal microflora could provide further insight into this complex ecosystem. Direct analysis of 16S rDNA genes from human fecal samples and porcine intestine revealed that the majority of sequences have low homology with sequences in the databases (Suau *et al.*, 1999; Leser *et al.*, 2002), and strongly suggests that novel and more universal bacteria primers would benefit fingerprinting techniques. Furthermore, functional genes as molecular markers can be used to separate or differentiate species in a genus, or ecologically different populations, where the 16S rDNA sequences are highly conserved. This concert of methods will enable us to address fundamental questions on the intestinal microbial community structure at very different levels of complexity, from single cells to major phylogenetic groups. Despite the power of the fingerprinting techniques described above none of the methods are perfect. Table 2 summarises some of the advantages and disadvantages of the DGGE/TGGE/TTGE as well as T-RFLP. While T-RFLP allows fast and standardised analysis, DGGE methods are easier to combine with sequencing analysis.

Diversity and Functional DNA Microarrays

More recently microbial diversity can also be determined by designing DNA arrays (reviewed by Rick *et al.*, 2001). Microarray technology allows a fast analysis of RNA abundance and DNA homology of genes in a single experiment. Two major factors are facilitating the developing of the DNA array technology – the large scale genome sequencing and the ability to immobilise thousands of DNA fragments on a surface, such as coated glass slides or membrane. An entire microbial genome can be easily represented in a single array, allowing genome-wide analysis. By using genome fragments and DNA microarray technology the similarity or differences in genetic contents between species can be also estimated. Currently, a number of bacterial species ranging from pathogens, laboratory strains and environmental isolates are being studied by DNA chips technology.

DNA microarrays are basically a form of dot blot, but in a high-throughput format. According to the size of the fragments spotted on the array there are two major types of DNA microarrays. One is the oligonucleotide based array and the other is the PCR product-based array. A DNA microarray experiment includes array fabrication, probe preparation, hybridization and data analysis (Rick *et al.*, 2001). DNA microarrays are becoming a common tool in many areas, including microbial physiology, pathogenesis, phylogeny and ecology. Although, these techniques are still not applied for monitoring of the diversity of the complex bacterial community in the intestine, they posses a great potential for composition analysis and phylogenetic studies of the microflora in the GI tract. Recently, glass slide microarrays containing DNA fragments from different functional genes for monitoring various environmental processes were developed and referred to as a functional gene array (Wu *et al.*, 2001). In conclusion, the future design of a functional chip-containing unique 16S rRNA, 23S rRNA sequences for most of the representative GI bacteria and also genotype markers of the major biochemical pathways known from the intestine will reveal the relationships between the members of the intestinal microflora and the host microbial interactions in the gut.

Functional Analysis

Although the aforementioned methods have been developed to determine the composition and diversity of the GI tract microflora (Zoetendal *et al.*, 1998; Suau *et al.*, 1999; Leser *et al.*, 2002), the functionality of its constituent populations also needs to be investigated. The functional efficacy of a particular organism in the complex microbial environment of the GIT is dependent on its numerical abundance, survival, competitiveness and metabolic activity. Assessment of the functionality of a particular microorganism in the gastrointestinal milieu necessitates the use of a molecular approach to specifically detect expression of a particular gene or genes. Although the development of techniques to detect *in vivo* bacterial gene expression is underway, such as the use of a green fluorescent protein (*gfp*) gene from *Aequoria victoria* as a marker for a *Lactobacillus* strain to detect its *in vivo* expression and location in the mouse GI tract (Geoffroy *et al.*, 2000), there is a need to develop additional methods to detect prokaryotic gene expression in complex ecosystems.

Most of the currently available methodology for transcriptional analysis is applicable only to poly (A)-tailed messenger RNA (mRNA), which is rarely present in prokaryotes. Since only 4-5% of total bacterial RNA is comprised of mRNA, detecting transcription of genes, particularly those that are poorly expressed or that have unstable messengers, remains technically challenging. Specific prokaryotic gene expression has mainly been monitored in pure cultures using techniques such as Reverse Transcriptase-PCR (RT-PCR) (Hernandez *et al.*, 2000), *in situ* RT-PCR (ISRT-PCR) coupled with *in situ* hybridisation (ISH) and flow cytometry (Chen *et al.*, 2000), ISRT-PCR (Holmstrom *et al.*, 1999; Lange *et al.*, 2000) or ISH using labelled mRNA probes (Wagner *et al.*, 1998) coupled with epifluoresence or phase contrast microscopy, or customised amplification libraries (Alland *et al.*, 1998). Assessment of bacterial gene expression has been achieved within complex ecosystems such as the expression pattern of *Helicobacter pylori* during its infection of the gastric mucosae using real time RT-PCR (Rokbi *et al.*, 2001), lignin-degrading communities by ISRT with microscopy (Chen *et al.*, 1997), in low biomass sediments with RT-PCR (Ogram *et al.*, 1995), and soils by Northern hybridisation (Tsai *et al.*, 1991). However, in terms of microbial density, such environments are less complex relative to feces by several orders of magnitude (Suau *et al.*, 1999), and its varying composition and presence of inhibitors present a challenge to the extraction of PCR-quality mRNA. A molecular protocol is being developed in our laboratory to assess bacterial activity by detection of mRNA by reverse transcription-PCR (RT-PCR) in GI tract samples (N. Fitzsimons, unpublished). This methodology has potential applicability to the assessment of the *in vivo* gene expression/ activity of a specific microbe in a complex ecosystem, such as a probiotic culture in the GI tract, where its efficacy in conferring health-promoting/ immunopotentiating effects to the host is a function of it is competitiveness and metabolic activity in its microbial environment.

Monitoring differential gene expression in micro- and higher organisms has been greatly facilitated by the development of DNA microarrays. This technology was exploited to monitor global intestinal transcriptional responses to colonisation of germ-free mice with *Bacteroides thetaiotamicron*, a prominent member of the normal murine and human intestinal microflora (Hooper *et al.*, 2001). The cellular location of selected responses was determined by laser-capture microdissection. Their results showed that this commensal was able to modulate expression of host genes participating in diverse and fundamental physiological functions, including nutrient absorption, mucosal barrier fortification, xenobiotic metabolism, angiogenesis and postnatal intestinal maturation. To determine if similar host responses were elicited by other members of the GI tract microflora, the mice were colonised with *Bifidobacterium infantis* and *E. coli*. The resultant species selectivity for some of the colonisation-associated changes in gene expression highlights how changes in host physiology can be affected by changes in the composition of the indigenous microflora.

The development of ultra-fast genome sequencing techniques which could allow a human genome to be sequenced in days (Adam, 2001), and advances in data mining technology mean that detection of differences in global gene expression is within our grasp and will be the ultimate means of determining the efficacy and functionality of our intestinal microbes and probiotics *in vivo*. In particular this will aid in the unravelling of the mechanisms underlying the postulated health-promoting effects of probiotics and uncover novel functionalities of our microflora.

Acknowledgements

This work has been carried out with financial support of the European Community specific RTD programme "Quality of Life and Management of Living Resources" research projects: HEALTHYPIGUT (QLK5-LT2000-00522) and Microbe Diagnostics (QLK1-2000-00108). Thomas D. Leser is gratefully acknowledged for his hospitality to S. R. Konstantinov in Danish Veterinary Institute, Department of Microbiology, Copenhagen V, Denmark and for excellent discussions on T-RFLP analysis.

References

Adam, D. 2001. Individual genomes targeted in sequencing revolution. Nature 411: 402.

Akkermans, A.D.L., Zoetendal, E.G., Favier, C.F., Heilig, G.H.J., Akkermans-van Vliet, W.M., and de Vos, W.M. 2000. Temperature and denaturing gradient gel electrophoresis analysis of 16S rRNA from human faecal samples. Bioscience Microflora 19: 93-98.

Alland, D., Kramnik, I., Weisbrod, T.R., Otsubo, L., Cerny, R., Miller, L.P., Jacobs Jr., W.R., and Bloom, B.R. 1998. Identification of differentially expressed mRNA in prokaryotic organisms by customized amplification libraries (DECAL): the effect of isoniazid on gene expression in *Mycobacterium tuberculosis*. Proc. Natl. Acad. Sci. USA. 95: 13227-13232.

Alm, E.W., Oerther, D.B., Larsen, N., Stahl, D.A., and Raskin, L. 1996. The oligonucleotide probe database. Appl. Environ. Microbiol. 62: 3557-3559.

Amann, R.I., Ludwig, W., and Schleifer, K.-H. 1995. Phylogenetic identification and *in situ* detection of individual microbial cells without cultivation. Microbiol. Rev. 59: 143-169.

Avaniss-Aghajani, E., Jones, K., Chapman, D., and Brunk, C. 1994. A molecular technique for identification of bacteria using small subunit ribosomal RNA sequences. BioTechniques 17: 144-149

Chen, F., Binder, B., Hodson, R.E. 2000. Flow cytometric detection of specific gene expression in prokaryotic cells using *in situ* RT-PCR. FEMS Microbiol. Letts. 184: 291-296.

Chen, F., Gonzalez, J.M., Dustman, W.A., Moran, M.A., and Hodson, R.E. 1997. *In situ* reverse transcription, an approach to characterize genetic diversity and activities of prokaryotes. Appl. Environ. Microbiol. 63: 4907-4913.

Chen, T-J., Boles, R.G., and Wong, L-J,C. 1999. Detection of mitochondrial DNA mutations by temporal temperature gradient gel electrophoresis. Clin. Chem. 45: 1162-1167.

Cooley, W.W., Lohnes, P.R. 1971. Multivariance data analysis. John Wiley and Sons, New York, NY.

Crittenden, R., Saarela, M., Mättö, J., Ouwehand, A.C., Salminen, S., Pelto, L., Vaughan, E.E., de Vos, W.M., von Wright, A., Fondén, R., and Mattila-Sandholm, T. 2002. *Lactobacillus paracasei* F19: survival, ecology and safety in the human intestinal tract. Microbial Ecol. Health Dis. S3: 22-26.

Dice, L.R. 1945. Measures of the amount of ecological association between the species. J. Ecology. 26: 297-302.

Eichner, C.A., Erb, R.W., Timmis, K.N., and Wagner-Döbler, I. 1999. Thermal gradient gel electrophoresis analysis of bioprotection from pollutant shocks in the activated sludge microbial community. Appl. Environ. Microbiol. 65: 102-109

Favier, C.F., Vaughan, E.E., de Vos, W.M. and Akkermans, A.D.L. 2002. Molecular monitoring of succession of bacterial communities in human neonates. Appl. Environ. Microbiol. 68: 219-226.

Felske, A., Akkermans, A.D.L., and de Vos, W.M. 1998. Quantification of 16S rRNAs in complex bacterial communities by multiple competitive reverse transcription-PCR in temperature gradient gel electrophoresis fingerprints. Appl. Environ. Microbiol. 64: 4581-4587.

Finegold, S. M., Sutter, V.L., and Mathisen, G.E. 1983. Normal indigenous intestinal flora. In: Human Intestinal Microflora in Health and Disease. D. J. Hentges,ed. Academic Press, New York. p. 3-31.

Fischer, S.G., and Lerman, L.S. 1979. Length-independent separation of DNA restriction fragments in two-dimensional gel electrophoresis. Cell 16: 191-200.

Geoffroy, M.-C., Guyard, C., Quatannens, B., Pavan, S., Lange, M. and Mercenier, A. 2000. Use of green fluorescent protein to tag lactic acid bacterium strains under development as live vaccine vectors. Appl. Environ. Microbiol. 66: 383-391.

Gillan, D., Speksnijder, A., Zwart, G., and de Ridder, C. 1998. Genetic diversity in the biofilm covering *Montacuta ferruginosa* (Mollusca, Bivalvia) as evaluated by denaturing gradient gel electrophoresis analysis and cloning of PCR-amplified gene fragments coding for 16s rRNA. Appl. Environ. Microbiol. 64: 3464-3472.

Häne, B.G., Jäger, K., and Drexler, H. 1993. The Pearson product-moment coefficient is better suited for identification of DNA fingerprinting profiles than band matching algorithms. Electrophoresis 14: 967-972.

Heilig, H.G.H.J., Zoetendal, E.G., Vaughan, E.E., Marteau, P., Akkermans, A.D.L. and de Vos, W.M. 2002. Molecular diversity of *Lactobacillus* spp., and other lactic acid bacteria in the human intestine as determined by specific amplification of 16S ribosomal DNA. Appl. Environ. Microbiol. 68: 114-123.

Hernandez, A., Figueroa, A., Rivas, L. A., Parro, V., and Mellado, R. P. 2000. RT-PCR as a tool for systematic transcriptional analysis of large regions of the *Bacillus subtilis* genome. Microbiol. 146: 823-828.

Heuer, H., Hartung, K., Wieland, G., Kramer, I., and Smalla, K. 1999. Polynucleotide probes that target a hypervariable region of 16S rRNA genes to identify bacterial isolates corresponding to bands of community fingerprints. Appl. Environ. Microbiol. 65: 1045–1049.

Holmstrom, K., Tolker-Nielsen, T., and Molin, S. 1999. Physiological states of individual *Salmonella typhimurium* cells monitored by *in situ* reverse transcription-PCR. J. Bacteriol. 181: 1733-1738.

Hooper, L. V., Wong, M.H., Thelin, A., Hansson, L., Falk, P. G. and Gordon, J. I. 2001. Molecular analysis of commensal host-microbial relationships in the intestine. Science. 291: 881-884.

Jardine, N., and Sibson, R. 1971. Mathematical Taxonomy, John Wiley & Sons, New York. NY.

Kaplan, C.W., Astaire, J.C., Sanders, M.E., Reddy, B.S., and Kitts, C.L. 2001. 16S ribosomal DNA terminal restriction fragment pattern analysis of bacterial communities in faeces of rats fed *Lactobacillus acidophilus* NCFM. Appl. Environ. Microbiol. 67: 1935-1939.

Kitts, L.C. 2001. Terminal restriction fragment patterns: a tool for comparing microbial communities and assessing community dynamics. Curr. Issues Intest. Microbiol. 2: 17-25.

Lange, M., Tolker-Nielsen, T., Molin, S., and Ahring, B.K. 2000. *In situ* reverse transcription-PCR for monitoring gene expression in individual *Methanosarcina mazei* S-6 cells. Appl. Environ. Microbiol. 66: 1796-1800.

Langendijk, P.S., Schut, F., Jansen, G.J., Raangs, G.C., Kamphuis, G., Wilkinson, M.H.F., and Welling, G.W. 1995. Quantitative fluorescence *in situ* hybridisation of *Bifidobacterium* spp. with genus-specific 16S rRNA targeted probe and its application in faecal samples. Appl. Environ. Microbiol. 61: 3069-3075.

Leser, T.D., Vindecrona, R.H., Jensen, T. K., Jensen, B.B., Moller, K. 2000. Changes in the colon of pigs fed different experimental diet and after infection with *Brachyspira hyodysenteriae*. Appl. Environ. Microbiol. 66: 3290-3296.

Leser T.D., Amenuvor, J.Z., Jensen, T.K., Lindecrona, R.H., Boye, M., and Møller, K. 2002. Culture-independent analysis of gut bacteria: the pig gastrointestinal tract microbiota revisited. Appl. Environ. Microbiol. 68: 673-690.

Lim, E.L., Tomita, A.V., Thilly, W.G., and Polz, M.F. 2001. Combination of competitive quantitative PCR and constant-denaturant capillary electrophoresis for high-resolution detection and enumeration of microbial cells. Appl. Environ. Microbiol. 67: 3897-3903.

Liu, W-T, Marsh, L., Cheng, H., and Forney, L.J. 1997. Characterization of microbial diversity by determining terminal restriction fragment length polymorphism of genes encoding 16S rRNA. Appl. Environ. Microbiol. 63: 4516–4522.

Magurran, A. 1988. Diversity indices and species abundance models. In: Ecological Diversity and Its Measurement. Princeton University Press, Princeton, N.J. p.8-45.

Marsh, T.L. 1999. Terminal restriction fragment length polymorphism (T-RFLP): an emerging method for characterizing diversity among homologous populations of amplification products. Curr. Opin. Microbiol. 2: 323-327.

Millar, M.R., Linton, C.J., Cade, A., Glancy, D., Hall, M., and Jalal, H.1996. Application of 16S rRNA gene PCR to study bowel flora of preterm infants with and without necrotizing enterocolitis. J. Clin. Microbiol. 34: 2506-2510.

Murray, A.E., Hollibaugh, J.T., and Orrego, G. 1996. Phylogenetic composition of bacterio plankton from two California estuaries compared by denaturing gradient gel electrophoresis of 16S rDNA fragments. Appl. Environ. Microbiol. 62: 2676-2680.

Muyzer, G., de Waal, E.C., and Uitterlinden, G.A. 1993. Profiling of complex populations by denaturing gradient gel electrophoresis of polymerase chain reaction-amplified genes coding for 16S rRNA. Appl. Environ. Microbiol. 59: 695-700.

Muyzer G., and Smalla, K. 1998. Application of denaturing gradient gel electrophoresis (DGGE) and temperature gradient gel electrophoresis (TGGE) in microbial ecology. Antonie van Leeuwenhoek 73: 127-141.

Muyzer, G., Brinkhoff, T., Nübel, U., Santegoeds, C., Schäfer, H., and Wawer, C. 1998. Denaturing gradient gel electrophoresis (DGGE) in microbial ecology. In: Molecular Microbial Ecology Manual. A. D. L. Akkermans, J. D. van Elsas, and F. J. de Bruijn, eds. Kluwer Academic Publishers, The Netherlands. vol. 3.4.4. p. 1-27.

Muyzer, G. 1999. DGGE/TGGE a method for identifying genes from natural ecosystem. Curr. Opin. Microbiol. 2: 317-322.

Myers, R. M., Fischer, S. G., Lerman, L. S., and Maniatis, T. 1985. Nearly all single base substitutions in DNA fragments jointed to a GC- clamp can be detected by denaturing gradient gel electrophoresis. Nucleic Acid Res. 13: 3131-3145.

Myers, R.M., Maniatis T., and Lerman L.S. 1987. Detection and localization of single base changes by denaturing gradient gel electrophoresis. Methods Enzymol. 155: 501-527.

Nübel U., Engelen, B., Felske, A., Snaidr, J., Wieshuber, A., Amann, R.I., Ludwig, W. and Backhaus, H. 1996. Sequence heterogeneities of genes encoding 16S rRNA in *Paenibacillus polymixa* detected by temperature gradient gel electrophoresis. J. Bacteriol. 178: 5636-5643.

Ogram, A, Sun, W., Brockman, F. J., and Fredrickson, J. K. 1995. Isolation and characterization of RNA from low-biomass deep-subsurface sediments. Appl. Environ. Microbiol. 61: 763-768.

Pearson, K. 1926. On the coefficient of racial likeness. Biometrika 18: 105-117.

Pryde, S.E., Richardson, A.J., Stewart, C.S., and Flint, H.J. 1999. Molecular analysis of the microbial diversity present in the colonic wall, colonic lumen, and cecal lumen of a pig. Appl. Environ. Microbiol. 65: 5372-5377.

Rademaker, J.L.W., Louws, F., Rossbach, U., Vinuesa, P. and de Bruijn, F. J. 1999. Computer- assisted pattern analysis of molecular fingerprints and database construction. In: Molecular Microbial Ecology Manual. A.D.L. Akkermans, J.D. van Elsas, and F.J. de Brijn, eds. Kluwer Academic Publishers. The Netherlands. vol. 7.1.3/1. p. 1-33

Rick, W.Y., Wang, T., Bedzyk, L., and Croker, K. 2001. Application of DNA microarrays in microbial systems. J. Microbiol. Meth. 47: 257-272.

Riesner, D., Steger, G., Wiese, U., Wulfert, M., Heiby, M., and Henco, K. 1992. Temperature –gradient gel electrophoresis for the detection of polymorphic DNA and for quantitative polymerase chain reaction. Electrophoresis 13: 632-636.

Rokbi, B., Seguin, D., Guy, B., Mazarin, V., Vidor, E., Mion, F., Cadoz, M., and Quentin-Millet, M. J. 2001. Assessment of *Helicobacter pylori* gene expression within mouse and human gastric mucosae by real-time reverse transcriptase PCR. Infect. Immun. 69: 4759-4766.

Rosenbaum, V., and Riesner, D. 1987. Temperature-gradient gel electrophoresis thermodynamic analysis of nucleic acids and proteins in purified form and in cellular extracts. Biophys. Chem. 26: 235-246.

Röling, W.F.M, van Breukelen, B.M., Braster, M., Lin, B., and van Verseveld, H. W. 2001. Relationships between microbial community structure and hydrochemistry in a landfill leachate- polluted aquifer. Appl. Environ. Microbiol. 67: 4619-4629.

Satokari, R. M., Vaughan, E.E., Akkermans, A.D.L., Saarela, M., and de Vos, W. M. 2001a. Bifidobacterial diversity in human faeces detected by genus-specific PCR and denaturing gradient gel electrophoresis. Appl. Environ. Microbiol. 67: 504-513.

Satokari, R. M., Vaughan, E.E., Akkermans, A.D.L., Saarela, M., and de Vos, W.M. 2001b. Polymerase chain reaction and denaturing gel electrophoresis monitoring of faecal *Bifidobacterium* populations in prebiotics and probiotics feeding trial. System. Appl. Microbiol. 24: 227-231.

Schmalenberger, A., Schwieger, F., and Tebbe, C.C. 2001. Effect of primers hybridizing to different evolutionarily conserved regions of the small-subunit rRNA gene in PCR-based microbial community analyses and genetic profiling. Appl. Environ. Microbiol. 67: 3557-3563.

Sghir, A., Gramet, G., Suau, A., Rochet, V., Pochart, P., and Doré, J. 2000. Quantification of bacterial groups within the human faecal flora by oligonucleotide probe hybridisation. Appl. Environ. Microbiol. 66: 2263-2266.

Shannon, C.E., and Weaver, W. 1963. The mathematical theory of communication. University of Illinois Press, Urbana.

Sheffield, V.C., Cox, D.R., and Myers, R.M. 1989. Attachment of 40-bp G + C rich sequences (GC-clump) to genomic DNA fragments by polymerase chain reaction results in improved detection of single-base changes. Proc. Natl. Acad. Sci. USA 86: 232 – 236.

Simpson, J.M., McCracken, V.J., White, B.A., Gaskins, H.R., and Mackie, R.I. 1999. Application of denaturant gradient gel electrophoresis for the analysis of the porcine gastrointestinal microbiota. J. Microbiol. Meth. 36: 167-179.

Simpson, J.M., McCracken, J.V., Gaskins, H.R., and Mackie, R.I. 2000. Denaturing gradient gel electrophoresis analysis of 16S ribosomal DNA amplicons to monitor changes in faecal bacterial populations of weaning pigs after introduction of *Lactobacillus reuteri* strain MM53. Appl. Environ. Microbiol. 66: 4705-4714.

Sneath, P.H.A., and Sokal, R.R. 1973. Numerical taxonomy: The principles and practice of numerical classification. W. H. Freeman and Company, San Francisco, CA.

Suau, A., Bonnet, R., Sutren, M., Godon, J.J., Gibson, G.R., Collins, M.D., and Doré, J. 1999. Direct analysis of genes encoding 16S rRNA from complex communities reveals many novel molecular species within the human gut. Appl. Environ. Microbiol. 65: 4799-4807.

Tannock, G.W. 1999. Analysis of the intestinal microflora: a renaissance. Antonie van Leeuwenhoek 76: 265-278.

Tannock, G.W., Munro, K., Harmsen, H.J.M., Welling, G.W., Smart, J., and Gopal, P.K. 2000. Analysis of the faecal microflora of human subjects consuming a probiotic product containing *Lactobacillus rhamnosus* DR20. Appl. Environ. Microbiol. 66: 2578-2588.

Tsai, Y.-L., Park, M.J., and Olson, B.H. 1991. Rapid method for direct extraction of mRNA from seeded soils. Appl. Environ. Microbiol. 57: 765-768.

Vaughan, E.E., Heilig, H.G.H.J., Zoetendal, E.G., Satokari, R., Collins, J.K., Akkermans, A.D.L., and de Vos, W.M. 1999. Molecular approaches to study probiotic bacteria. Trends Food Sci. Tech. 10: 400-404.

Vaughan, E.E., Schut, F., Heilig, H.G.H.J., Zoetendal, E.G., de Vos, W.M., and Akkermans, A.D.L. 2000. A molecular view of the intestinal ecosystem. Curr. Issues Intest. Microbiol. 1: 1-12.

von Wintzingerode F, Göebal, U. B., and Stackebrandt E. 1997. Determination of microbial diversity in environmental samples: pitfalls of PCR-based rRNA analysis. FEMS Microbiol. Rev. 21: 213–229.

de Vos, W.M., Zoetendal, E.G., Polewijk, E., Heilig, H. and Akkermans, A.D.L. 1998. Molecular tools for analysing the functionality of probiotic properties of microorganisms. In: Proceedings of the 25th International Dairy Congress. A. Ravn (Ed.), IDF, Aarhus DK, 323-328.

Wagner, M., Schmid, M., Juretschko, S., Trebesius, K.-H., Bubert, A., Goebel, W., and Schleifer, K.-H. 1998. In situ detection of a virulence factor mRNA and 16S rRNA in *Listeria monocytogenes*. FEMS Microbiol. Letts. 160: 159-168.

Walter, J., Hertel, C., Tannock, G.W., Lis, C.M., Munro, K., and Hammes, W.P. 2001. Detection of *Lactobacillus, Pediococcus, Leuconostoc*, and *Weissella* species in human faeces by using group-specific PCR primers and denaturing gradient gel electrophoresis. Appl. Environ. Microbiol. 67: 2578-2585.

Ward, J. H. 1963. Hierarchical grouping to optimise an objective function. J. Am. Statist. Assoc. 58: 236-244.

Wu, L., Thomson, D.K., Li, G., Hurt, R.A., Tiedje, J.M., and Zhou, J. 2001. Development and evaluation of functional gene analysis for detection of selected genes in the environment. Appl. Environ. Microbiol. 67: 5780-5790.

Yang, C.-H., Crowley, D.E., Merge, J.A.. 2001. 16S rDNA fingerprinting of rhizospfere bacteria communities associated with healthy and Phytophora infected avocado roots. FEMS Microbiol. Ecol. 35: 129-136.

Yoshino K., Nishigaki, K., Husimi, Y. 1991. Temperature sweep gel electrophoresis: a simple method to detect point mutations. Nucleic. Acids. Res. 19: 3153.

Zhu X. Y., T. Zhong, Y. Pandya, and R. D. Joerger. 2002. 16S rRNA –based analysis of microbiota from the cecum of broiler chickens. Appl. Environ. Microbiol. 68: 124-137.

Zoetendal, E.G., 2001. Molecular characteriazation of bacterial communities in the human gastrointestinal tract. PhD thesis. Wageningen University, The Netherlands. ISBN: 90-5808-520-1.

Zoetendal, E. G., Akkermans, A.D.L. and. de Vos, W.M. 1998. Temperature gradient gel electrophoresis analysis of 16S rRNA from human faecal samples reveals stable and host specific communities of active bacteria. Appl. Environ. Microbiol. 64: 3854-3859.

Zoetendal, E.G., Akkermans, A.D.L., Akkermans-van Vliet, W.M., de Visser, J.A.G.M., and de Vos, W. M. 2001a. The host genotype affects the bacterial community in the human gastrointestinal tract. Microbial Ecol. Health Dis. 13: 129-134.

Zoetendal, E.G., Akkermans, A.D.L., and. de Vos, W.M. 2001b. Molecular characterisation of microbial communities based on 16S rRNA sequence diversity. In: New Approaches for the Generation and Analysis of Microbial Fingerprints. Chapter 12. L. Dijkshoorn, K. Towner, and M. Struelens, eds. Elsevier. p. 267-298.

Zoetendal, E.G., Kaouther, K.B-A., Akkermans, A.D.L., Abee, T., and de Vos, W.M. 2001c. DNA isolation protocols affect the detection limit of PCR approaches of bacteria in samples from the human gastrointestinal tract. System. Appl. Microbiol. 24: 405-410.

From: *Probiotics and Prebiotics: Where Are We Going?*
Edited by: Gerald W. Tannock

Chapter 4

Genus- and Species-specific PCR Primers for the Detection and Identification of Bifidobacteria

Takahiro Matsuki, Koichi Watanabe,
and Ryuichiro Tanaka

Abstract

16SrDNA-targeted genus- and species-specific PCR primers have been developed and used for the identification and detection of bifidobacteria. These primers cover all of the described species that inhabit the human gut, or occur in dairy products. Identification of cultured bifidobacteria using PCR primer pairs is rapid and accurate, being based on nucleic acid sequences. Detection of bifidobacteria can be achieved using DNA extracted from human faeces as template in PCR reactions. We have found that, in adult faeces, the *Bifidobacterium catenulatum* group was the most commonly detected species, followed by *Bifidobacterium longum*, *Bifidobacterium adolescentis*, and *Bifidobacterium bifidum*. In breast-fed infants, *Bifidobacterium breve* was the most frequently detected species, followed by *Bifidobacterium infantis*, *B. longum* and *B. bifidum*. It was notable that the *B. catenulatum* group was detected with the highest frequency in adults, although it has often been reported that *B. adolescentis* is the most common species. Real time, quantitative PCR using primers targeting 16S rDNA shows promise in the enumeration of bifidobacteria in faecal samples. The approach to detect the target bacteria with quantitative PCR described in this chapter will contribute to future studies of the composition and dynamics of the intestinal microflora.

Introduction

In the study of intestinal microflora, it is important to clarify the relationship between the host and the microbial community. The structure of the human intestinal microflora has been investigated in great detail by anaerobic culture techniques (Moore and Holdeman, 1974; Finegold *et al.*, 1974). These intensive investigations provided significant information about the microflora. However traditional culture methods raise the following problems: the methods are labor-intensive and time-consuming; only easily cultivable organisms are counted; and classification and identification based on phenotypic traits does not always provide clear-cut results, and is sometimes unreliable. Therefore there is a need for practical techniques that enable rapid and accurate analysis of the gut microflora. Recent research has led to rapid advances in the application of molecular techniques based on 16S and 23S rRNA gene sequences to study the microbial diversity in ecosystems (Amann *et al.*, 1995). In the analysis of intestinal microflora, 16S rRNA targeted oligonucleotide probes have been applied in fluorescent *in situ* hybridization (FISH) as a culture independent method (Langendijk *et al.*, 1995; Franks *et al.*, 1998). The 16S rDNA-cloning-library method (Wilson and Blitchington, 1996; Suau *et al.*, 1999) and DGGE/TGGE (denaturing/ temperature gradient gel electrophoresis) method (Millar *et al.*, 1996; Zoetendal *et al.*, 1998) have also provided an efficient strategy for exploring the biodiversity of the microflora. Although these studies provide significant information, the approaches are time-consuming and detect 90-99% of the members of the microflora.

On the other hand, PCR with specific 16S rDNA-based oligonucleotide primers is a powerful method for the detection of target bacteria within complex ecosystems. As the method is rapid, accurate and sensitive, it has frequently been used in the detection and identification of pathogenic bacteria (Satake *et al.*, 1997; Gumerlock *et al.*, 1991). Wang *et al.* (1996) first applied the PCR method to the analysis of intestinal microflora. They prepared species-specific primers based on 16S rRNA gene sequences for 12 bacteria that are predominant in the human intestinal tract, and showed that the PCR procedure is a powerful tool in the detection of these species. Specific oligonucleotide primers have been designed for many bacterial species that are known to be present in the intestinal tract. So far, species-specific primers for *Bifidobacterium* (Matsuki *et al.*, 1999), *Ruminococcus* (Wang *et al.*, 1997), *Eubacterium* (Kageyama and Benno, 2001), and *Lactobacillus* (Walter *et al.*, 2000, Song *et al.*, 2000) have been developed and applied successfully. The specific PCR techniques do not provide information on bacteria that are not the target of the primers. Therefore it is necessary to derive specific primers for the major genera or groups present in the gut as well as for bacterial species. Moreover, the development of reliable quantitative PCR methods is necessary. Research on quantitative detection is already proceeding, and the establishment of optimal procedures and application of Real-Time quantitative PCR is expected.

This chapter details the identification of cultured strains using species-specific PCR primers and the specific detection of bacteria from fecal DNA, with particular emphasis on *Bifidobacterium* species.

Phylogenetic Relationship of the Genus *Bifidobacterium*

The genus *Bifidobacterium* includes gram-positive pleomorphic strict anaerobes and some species are commonly detected in the human gut. Bifidobacteria are phylogenetically grouped in the actinomycete branch of gram-positive bacteria with high G+C content and presently are assigned to 32 species (Dong *et al.*, 2000b; Euzeby, 1997). The use of rRNA sequences as indicators of phylogenetic analysis is now widespread (Woese, 1987). In the genus *Bifidobacterium*, 16S rDNA (Frothinfham *et al.*, 1992; Miyake *et al.*, 1998; Leblond-Bourget *et al.*, 1996), 16S to 23S internal transcribed spacer (ITS) sequences (Leblond-Bourget *et al.*, 1996), and the HSP60 gene sequence (Jian *et al.*, 2001) were used to analyze the phylogenetic relationship. Although some differences were reported, the phylogenetic trees constructed from ITS and HSP60 sequences were basically similar to that of 16S rDNA. Therefore, this section describes the phylogenetic relationship of the genus *Bifidobacterium* on the basis of 16S rDNA sequences.

A phylogenetic tree was constructed from 16S rDNA sequences of 32 bifidobacterial species and 10 related bacteria with accordance as described

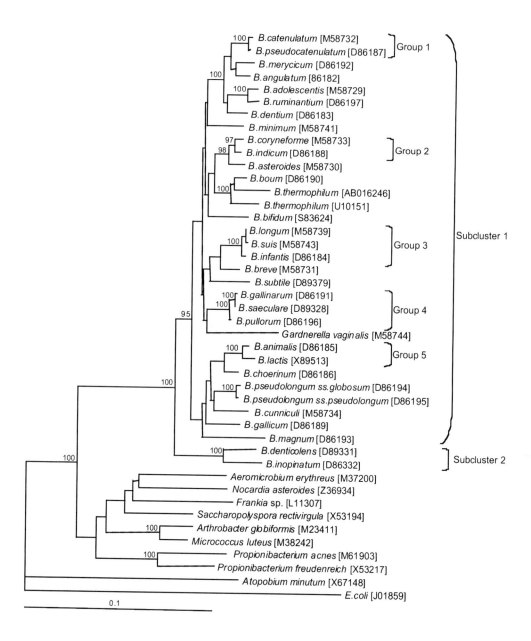

Figure. 1. Phylogenetic tree based on 16S rDNA sequences, showing the relationships of *Bifidobacterium* species (Miyake *et al.*, 1998, with some modification). The tree was rooted with *E. coli* and constructed using neighbor-joining method with bootstrap values calculated from 100 trials. Only values above 95% are indicated. Bar = 10% sequence divergence.

by Miyake *et al.*, (1998)(Figure 1). *Bifidobacterium* is a relatively coherent genus, and forms two subclusters. The two species isolated from human dental caries, *Bifidobacterium denticolens* and *Bifidobacterium inopinatum*, formed subcluster 2; all of the other members, including *Gardnerella vaginalis,* formed subcluster 1.

Subcluster 1 was composed of 30 bifidobacterial species and *Gardnerella vaginalis.* There are five groups that were characterized by high bootstrap values (>97%) and 16S rDNA sequence similarity (>98.9%). The closely related groups of species are (1) *Bifidobacterium catenulatum* and *Bifidobacterium pseudocatenulatum* (16S rDNA similarity 99.5%), (2) *Bifidobacterium indicum* and *Bifidobacterium coryneforme* (similarity 99.1%), (3) *Bifidobacterium longum, Bifidobacterium infantis,* and *Bifidobacterium suis* (similarities 99.1-99.2 %), (4) *Bifidobacterium gallinarum, Bifidobacterium pullorum*, and *Bifidobacterium saeculare* (similarities 99.3-99.9%), and (5) *Bifidobacterium animalis* and *Bifidobacterium lactis* (similaritiy 98.9%)(Miyake *et al.,* 1998). These groups also show moderate (40 to 60%) to high (>70%) relatedness in DNA-DNA hybridization tests (Scardovi, 1984; Wayne *et al.,* 1987). Cai *et al.* (2000) reported that the levels of DNA-DNA hybridization between the type strains of *B. lactis* and *B. animalis* ranged from 85.5 to 92.3%, indicating that *B. lactis* should be included in *B. animalis.*

When 16S rDNA is used as an index, it is difficult to differentiate *Gardnerella* and *Bifidobacterium* in subcluster 1. However, it is reported that the G+C% of *G. vaginalis* (42%) is significantly lower than that of subcluster 1 (55-67%)(Scardovi, 1984). In the phylogenetic analysis based on HSP60 sequences, *G. vaginalis* and the *Bifidobacterium* subcluster 1 formed another cluster (Jian *et al.,* 2001). Thus, the phylogenetic position of *G. vaginalis* is still under discussion, and further studies are expected.

B. denticolens and *B. inopinatum* were characterized not only by the 16S rDNA-based phylogenetic analysis, but also by their low G+C content (Miyake *et al.,* 1998). It was also suggested from the analysis of HSP60 that these two species should be removed from the genus *Bifidobacterium* and established as new genera in the family *Bifidobacteriaceae* (Jian *et al.,* 2001). Further studies, such as the analysis of sequences encoding other conserved molecules, are needed.

Table 1. Currently used genus- and species-specific primers for *Bifidobacterium*

Target bifidobacteria	Name of primers	Sequence	Product size (bp)	References
Bifidobacterium	Bif164-PCR Bif662-PCR	GGGTGGTAATGCCGGATG CCACCGTTACACCGGGAA	523	Kok *et al.*, 1996
Bifidobacterium	lm26 Lm3r	GATTCTGGCTCAGGATGAACG CGGGTGCTICCCACTTTCATG	1.35k	Kaufmann *et al.*, 1997
B. adolescentis	BiADO-1 BiADO-2	CTCCAGTTGGATGCATGTC CGAAGGCTTGCTCCCAGT	279	Matsuki *et al.*, 1998
B. angulatum	BiANG-1 BiANG-2	CAGTCCATCGCATGGTGGT GAAGGCTTGCTCCCCAAC	275	Matsuki *et al.*, 1998
B. bifidum	BiBIF-1 BiBIF-2	CCACATGATCGCATGTGATTG CCGAAGGCTTGCTCCCAAA	278	Matsuki *et al.*, 1998
B. breve	BiBRE-1 BiBRE-2	CCGGATGCTCCATCACAC ACAAAGTGCCTTGCTCCCT	288	Matsuki *et al.*, 1998
B. catenulatum group	BiCATg-1 BiCATg-2	CGGATGCTCCGACTCCT CGAAGGCTTGCTCCCGAT	285	Matsuki *et al.*, 1998
B. longum group	BiLON-1 BiLON-2	TTCCAGTTGATCGCATGGTC TCSCGCTTGCTCCCCGAT	277	Matsuki *et al.*, 1998
B. longum	BiLON-1 BiLON-2	TTCCAGTTGATCGCATGGTC GGGAAGCCGTATCTCTACGA	831	Matsuki *et al.*, 1999
B. infantis	BiINF-1 BiINF-2	TTCCAGTTGATCGCATGGTC GGAAACCCCATCTCTGGGAT	828	Matsuki *et al.*, 1999
B. dentium	BiDEN-1 BiDEN-2	ATCCCGGGGGTTCGCCT GAAGGGCTTGCTCCCGA	387	Matsuki *et al.*, 1999

Species	Primer	Sequence	Size (bp)	Reference
B. gallicum	BiGAL-1 BiGAL-2	TAATACCGGATGTTCCGCTC ACATCCCGAAAGGACGC	303	Matsuki et al., 1999
B. lactis	Bflac2 Bflac5[a]	GTGGAGACACGGTTTCCC CACACCACACAATCCAATAC	680	Ventura et al., 2001
B. breve	BreU3 BreL4	CTCCAGCTCGACTGTCGC GCACTTTGTGTTGAGTGTACCTTTCG	811	Roy et al., 1996
B. infantis	InfU5 InfL6	CCATCTCTGGGATCGTCGG TATCGGGGAGCAAGCGTGA	565	Roy et al., 1996
B. longum	LonU7 LonL8	GCCGTATCTCTACGACCGTCG TATCGGGGAGCAAGCGAGAG	567	Roy et al., 1996
B. adolescentis	BIA-1 BIA-2	GGAAAGATTCTATCGGTATGG CTCCCAGTCAAAAGCGGTT	244	Wang et al., 1996
B. longum	BIL-1 BIL-2	GTTCCCGACGGTCGTAGAG GTGAGTTCCCGGCATAATCC	153	Wang et al., 1996
B. breve	L Bre-BV.R [a]	GCTGGATCACCTCCTTTCT GCAAGAACGAGGAATCAAAGG	340	Brigidi et al., 2000
B. bifidum	PBI245f[b]	GCTTGTTGGTGAGGTAACGGCT	1180	Dong et al., 2000a
B. breve	PBR442f[b]	AGGGAGCAAGGCACTTTGTGT	991	Dong et al., 2000a
B. infantis	PIN710f[b]	CTGTTACTGACGCTGAGGAGCT	723	Dong et al., 2000a
B. adolescentis	PAD805f[b]	GTGGGGACCATTCCACGGTC	628	Dong et al., 2000a
B. longum	PLO965f[b]	TCCCGACGGTCGTAGAGATAC	467	Dong et al., 2000a
	Lm3r	CGGGGTGCTGCCCACTTTCATG	—	Kaufmann et al., modified

[a] Specific primers based on 16S to 23S internal transcribed spacer sequences
[b] These specific primers were used in the multiplex PCR system with genus-specific primer Lm3r.

Currently Used Genus-and Species-specific Primers

It has been reported that *Bifidobacterium adolescentis, B. catenulatum, B. pseudocatenulatum, B. longum,* and *B. bifidum* are major bifidobacterial species of the adult intestinal microflora (Biavati *et al.,* 1986; Mutai and Tanaka, 1987), and that *B. infantis* and *Bifidobacterium breve* are predominant species in the intestinal tract of human infants (Benno *et al.,* 1984; Biavati *et al.,* 1984). In addition, *Bifidobacterium angulatum, Bifidobacterium dentium,* and *Bifidobacterium gallicum* have been also reported as human intestinal bifidobacteria (Scardovi and Crociani., 1974; Lauer, 1990). *B. longum, B. breve, Bifidobacterium bifidum, B. infantis*, and *B. lactis* are frequently used in probiotic products (Lee *et al.,* 1999).

For the identification and detection on these species, genus-and species-specific primers have been developed. Table 1 shows the currently used genus- and species-specific primers for *Bifidobacterium.* Most of them are designed on the basis of 16S rDNA sequences. Depending on the target sequence chosen, 16S rDNA-targeted specific primers can be used to detect bacteria on different phylogenetic levels. Since *B. catenulatum* and *B. pseudocatenulatum* are closely related in their 16S rDNA similarity (Miyake *et al.,* 1998) and DNA-DNA relatedness (Lauer and Kandler, 1983), these two species are treated as the members of the *B. catenulatum* group (Matsuki *et al.,* 1998). In addition to the 16S rDNA-targeted specific primers, species- and strain-specific primers have been derived on the basis of 16S to 23S internal transcribed spacer (ITS) sequences (Brigidi *et al.,* 2000; Ventura *et al.,* 2001).

Identification

Until recently, isolates of *Bifidobacterium* have been identified to the genus level by using Gram-staining, morphological observations, end-product analysis of glucose metabolism, and fructose-6-phosphate conversion ability as indices. Species identification has also been made on the basis of phenotypic traits (Scardovi, 1984). However, these identification methods require a large amount of time and labour, and do not always provide accurate results. On the other hand, the identification methods using the 16S rRNA sequence are less subjective, being nucleic acid sequence-based. For the 16S rDNA-based identification, three methods have been suggested; (1) sequence analysis, (2) species-specific hybridization method (Yamamoto *et al.,* 1992), and (3) the specific PCR techniques. When these approaches are compared, PCR provides identification results with the least effort and in the shortest time. For the precise identification of bacterial species, the DNA-DNA homology should be determined to confirm the existence of more than 70% homology with the type strain, but the species-specific primers are capable of accurate identification of bifidobacteria (Matsuki *et al.,* 1998). This section describes the isolation of bifidobacteria and their identification using species-specific primers.

Isolation of *Bifidobacterium* Strains

Feces or dairy products are sampled, and serial 10-fold dilutions are prepared anaerobically (Holdman *et al.,* 1977). Suitable aliquots of dilutions are smeared onto bifidobacteria-selective media (Tanaka and Mutai, 1980, Nebra and Blanch, 1999), and are anaerobically incubated in a glove box or in anaerobic jars. Colonies formed on the media are picked with a sterile toothpick, and suspended in 50 μl of TE (10 mM Tris-HCl, 1mM EDTA; pH 8.0). For verification, 1 μl of the suspension is smeared onto a slide glass and Gram-stained. The bacterial suspension is heated at 95°C for 10 minutes, and 1 μl of the supernatant is subjected to PCR.

Identification of the Isolates with Species-specific Primers

For the identification of the isolates, species-specific primers designed by Matsuki *et al.* (1999) should basically be used for the following reasons: the primers cover all of the bifidobacterial species that have been isolated from, and detected in, the human intestinal tract; specificity has been confirmed for all of the *Bifidobacterium* species; the PCR reactions can be conducted using the same thermal cycling conditions. PCR is carried out in a total volume of 25 μl of reaction mixture containing 10 mM of Tris-HCl (pH 8.3), 50 mM of KCl, 2.5 mM of $MgCl_2$, 200 μM of dNTP mixture, 25 μM of each primer, 0.45 U Taq DNA polymerase, and 1 μl of template DNA. The PCR amplification program consisted of one cycle of 94°C for 5 minutes, then 35 cycles of 94°C for 20 seconds, 55°C for 20 seconds, and 72°C for 30 seconds, and finally one cycle of 72°C for 5 minutes. The products are separated by electrophoresis in 1% agarose gels, and ethidium bromide staining is performed to observe the presence of bands under UV transillumination. The use of Real-Time PCR equipment, which is discussed later, makes it possible to identify bacterial species from the melting curve analysis without conducting electrophoresis (Ririe *et al.,* 1997), thus making identification simpler in our laboratory.

It is useful to use genus-specific primers together with the species-specific primers because the bifidobacteria media are not completely selective. For the identification of bifidobacteria isolated from human adult feces, it is recommended to use primers for *B. adolescentis, B. longum*, the *B. catenulatum* group, and *B. bifidum* at first. In our study, more than 90% of the isolates were identified as one of these four species. Unidentified isolates were then examined with the primers targeting the remaining five species. For isolates from infant feces, it is efficient to apply the primers for *B. breve, B. infantis, B. longum*, and *B. bifidum* at first, and then test with the remaining primers. More than 80% of the bifidobacterial isolates from infants are identified as one of these four species. Specific primers for *B. breve, B. bifidum, B. longum*, and *B. infantis* are also capable of identifying isolates from dairy products. But it is necessary to use them in combination with the *B. lactis*-specific primers prepared by Ventura *et al.* (2001).

In addition to the primers described here, the multiplex PCR method reported by Dong *et al.* (2000a) is effective in the identification of cultured bifidobacteria. The reaction mixture has a total volume of 25 µl and contains 50-100 ng of template DNA, 100 pmol of dNTP, 2.5 mM of $MgCl_2$, 10 pmol each of PBR442f and PAD805f, 20 pmol each of PBI245f and PLO965f, 30 pmol of PIN, 50 pmol of lm3r, and 7.5U Taq of polymerase. The condition of the thermal cycler is as follows: one cycle consisting of 95°C for 5 minutes, followed by 30 cycles consisting of 94°C for 30 seconds, 63°C for 30 seconds, and 72°C for 60 seconds. Amplification products are detected by electrophoresis on 1% agarose, followed by ethidium bromide staining. This method can simultaneously identify the five species - *B. adolescentis, B. bifidum, B. breve, B. longum*, and *B. infantis* - based on the differences in PCR product sizes. The reported protocol needed a larger amount of Taq polymerase than is used in conventional PCR.

The specific primers for *B. infantis, B. breve*, and *B. longum*, which were prepared by Roy *et al.* (1996), and the *B. breve*-specific primers prepared by Bridgidi *et al.* (2000) are also available. These primers should be used in combination with the primers described by Matsuki *et al.* (1999). The primers prepared by Wang *et al.* (1997) have not been sufficiently tested for specificity, and their conditions were optimized by the capillary PCR method. Experiments using the primers under general thermal-cycling conditions require some caution. Brigidi *et al.* (2000) reported that the *B. longum*-specific primers BIL-1 and BIL-2 cross-react with some strains of *B. breve*.

Specific PCR Detection of *Bifidobacterium*

The species-specific detection of bifidobacteria without cultivation is possible using the DNA extracted from feces as a PCR template. This section describes the DNA preparation method, specific PCR detection using fecal DNA, the distribution of bifidobacterial species in the human intestinal microflora.

DNA Preparation Method

Fecal samples are difficult specimens with which to perform PCR analysis because (1) some bacterial cells are difficult to lyse (Zoetendal *et al.,* 1998) and, (2) the presence of multiple substances that inhibit the polymerase enzyme (Wilson, 1997; Satake *et al.,* 1997). Although various procedures have been employed in recent molecular studies to overcome these problems (Wang *et al.,* 1996; Kageyama *et al.,* 2000), it is necessary to standardize the DNA extraction procedure for future analyses.

Mechanical procedures, which shake the sample vigorously in the presence of glass beads, have been shown to be effective in disrupting bacterial cells from a variety of ecosystems (Zoetendal *et al.,* 1998). This method has been employed in an increasing number of studies (Zoetendal *et al.,* 1998; Ventura *et al.,* 2001). However, in our experience, PCR inhibitors are not completely removed by this procedure. The DNA solution can be diluted for use as the template, but the detection sensitivity is influenced by the degree of dilution. In order to remove PCR inhibitors, fecal samples should be washed in buffers before breaking the bacterial cells. The combination of the washing step and the mechanical disruption can provide DNA that is suitable for use in PCR analysis. The protocol of our DNA extraction method is described below.

A portion of a fecal sample is taken and uniformly suspended in nine parts by volume of PBS buffer. Then, 200 µl of the one-tenth feces dilution is placed in a 2-ml tube and stirred together with 1 ml of PBS and several glass beads (2-3 mm diameter). After centrifugation at 15,000 rpm for five minutes, the supernatant is removed. The procedure is repeated two additional times in the same manner. These steps will produce a pellet that is the equivalent of 20 mg of feces. At this point the pellet can be cryopreserved, if necessary. The pellet is suspended in 500 µl of lysate buffer (100 mM of Tris-HCl, 40 mM of EDTA, and 1% of SDS, pH 9.0), 0.3 g of small glass beads (0.1 mm diameter), and 500 µl of buffer-saturated phenol. The mixture is vigorously shaken for 30 seconds with FastPrepFP120 at a power level of 5.0 to physically break the bacterial cells. The suspension is then centrifuged, and 400 µl of the supernatant is transferred to another tube. Then, 400 µl of phenol-chloroform-isoamyl alcohol (25:24:1) is added and the mixture is shaken again, and centrifuged. Next, 250 µl of the supernatant is transferred to another tube, and subjected to ethanol precipitation. If these steps are followed, 50% of the DNA fraction, or DNA equivalent to 10 mg of feces, will be collected. After being air-dried, the DNA is dissolved in 1 ml of TE, and 1 µl of the solution is subjected to PCR. Instead of fecal homogenate, bacterial cultures or dairy products can also be used as the sample.

Specific PCR Detection From Fecal DNA

The use of species-specific primers prepared by Matsuki *et al.* (1999) is recommended for the reasons given in the section on bacterial-species identification. In our protocol, PCR is conducted by repeating 35 cycles of three steps each, as described above. It is also recommended that the cycle number be increased to 40, as this will improve the detection limit by yet another order of magnitude. Under these conditions, the target bifidobacterial species were detected at a concentration of 10 cells per PCR assay (equivalent to 10^6 cells per g of feces) (Matsuki *et al.*, 1999).

It has been reported by Dong *et al.* (2000) that, with the multiplex PCR method, it is difficult to detect multiple species simultaneously. It would therefore be difficult to detect bifidobacteria in fecal samples by the multiplex PCR method. The specific primers prepared by Roy *et al.* (1996) and Bridgidi *et al.* (2000) can be used in combination with our primers.

The advantage of this specific PCR technique is its high detection sensitivity. The conventional method, which identifies bacterial strains isolated from bifidobacteria-selective media, can only analyze the more numerous bacteria in the sample. As the predominant *Bifidobacterium* species are usually present at the level of 10^9 to 10^{10} cells per gram in human feces (Mitsuoka *et al.*, 1974; Moore and Moore, 1995; Mutai and Tanaka., 1987), the detection limit of the culture method for minor bifidobacterial species is about 10^8 cells per g. In FISH analysis, the lower detection limit for reliable detection is 10^8 cells per gram (Franks *et al.*, 1998). Therefore, the specific PCR method has a sensitivity approximately 100 times greater than that of the culture and FISH methods. Other advantages include easy sample handling and simple operations. With the culture method, it is necessary to use fresh samples and perform operations after sampling, such as preparation of dilutions and culture, under strict anaerobic conditions. On the other hand, the PCR method does not require anaerobic conditions, and DNA can be preserved in the freezer, and it is possible to transport the DNA internationally. The scope of studies using PCR will increase in the future because of these factors. Establishment of the procedure for quantitative detection of bacteria is a future task and current progress will be described later in the chapter.

Satokari *et al.* (2001) combined the genus-specific primers with DGGE (denaturing gradient gel electrophoresis) to achieve species-specific detection of bifidobacteria in human feces. With the DGGE method, the PCR products were electrophoresed in polyacrylamide gels with a concentration gradient of DNA denaturant (urea and formamide) gradient to separate PCR products based on differences in their base sequences. When a fecal sample is analyzed with the bifidobacteria genus-specific primers, the 16S rRNA gene of each species can be detected as a separate band. This method has the advantage of being effective in the analysis of bifidobacterial species that are not targeted by the species-specific primers and it is considered effective in monitoring individual microfloras over time. The method has several problems, however, such as the fact that the formation of heteroduplexes may exhibit false bands, DNA fragments with different sequences may migrate to the same position, the copy number of 16S rDNA varies, and numerically subdominant species cannot be detected.

Distribution of Bifidobacterial Species in the Human Gut Microflora

Examination of the bifidobacterial species distribution in the human intestinal tract was accomplished with the species-specific PCR method (Table 1)(Matsuki *et al.*, 1999). In adult intestinal tracts, the *B. catenulatum* group was the most common taxon, followed by *B. longum, B. adolescentis*, and *B. bifidum* (Table 2). In breast-fed infants, *B. breve* was the most frequently found species, followed by *B. infantis, B. longum* and *B. bifidum*. It was a notable finding that the *B. catenulatum* group inhabited human adults with the highest frequency, although it has often been reported that *B. adolescentis* is the most common species (Mutai and Tanaka., 1987; Finegold *et al.*, 1974; Mitsuoka *et al.*, 1974). This contradiction may be explained by the difference in the identification methods used. It has been reported that *B. adolescentis, B. catenulatum*, and *B. pseudocatenulatum* are difficult to differentiate based on the usual carbohydrate fermentation pattern (Matsuki

Table 2. Distribution of the species of *Bifidobacterium* in human adults and infants

	No. of positive samples (% of total)	
	Adults (n=48)	**Infants (n=27)**
B. adolescentis	29 (60%)	2 (7.4%)
B. angulatum	2 (4.2%)	1 (3.7%)
B. bifidum	18 (38%)	6 (22%)
B. breve	6 (13%)	19 (70%)
B. catenulatum group	44 (92%)	5 (19%)
B. longum	31 (65%)	10 (37%)
B. infantis	0 (0%)	11 (41%)
B. dentium	3 (6.3%)	3 (11%)
B. gallicum	0 (0%)	0 (0%)

The adults were 38.8 ± 8.9 years old (mean ± standard deviation), and the infants were 31.2 ± 4.5 days old (Matsuki *et al.*, 1999).

et al., 1998). Therefore, the *B. catenulatum* group may have been confused with *B. adolescentis* in some studies. It is also interesting to note that *B. breve* was detected in adult fecal samples, even though it has been recognized as a typical infantile bifidobacterial species. This may be explained by the difference between the detection limits of the culture method and the specific PCR method. It will be interesting to determine the source of the bifidobacteria detected in infants. The highly sensitive specific PCR detection method stands a good chance of elucidating the route of transmission of bifidobacteria to infants.

Quantitative PCR Detection

In the analysis of a microbial community structure, it is important to detect the target bacteria quantitatively. Although the PCR analyses have primarily been conducted qualitatively, the quantitative PCR analysis of some bacterial communities is already underway. There are two strategies for quantitative PCR detection: (1) the competitive PCR method using internal standards, and (2) Real-Time PCR detection using external standards. This section briefly explains these principles, and describes the quantitative PCR method using external standards that we have investigated.

Competitive PCR Method

Until recently, quantitative PCR methods have been conducted primarily by the competitive PCR method using internal standards. With the competitive PCR method, competitive DNA fragments against the same primers are added at different concentrations to PCR solutions. Following PCR, the quantities of the PCR products from the target DNA and the competitive DNA fragments are compared to determine the quantity of the target DNA. In the case of the gastrointestinal microflora, the competitive PCR method has been used to count the number of *Clostridium proteoclasticum* and its related species present in rumen contents (Reilly and Attwood, 1998). This report suggested that the method was applicable to human fecal microflora studies. However, the method requires the preliminary preparation of competitive DNA, and the amplification efficiency of the competitive DNA and target DNA must be the same. There are many other measurement limits and thus establishment of the method requires considerable effort.

Quantitative PCR With External Standards

In recent years, Real-Time PCR equipment has become commercially available. This equipment provides a quantification method using external standards; the quantity of PCR products is measured in every cycle using fluorescent substances such as SYBR Green I and TaqMan probe, and is quantified based on an amplification curve. There have already been some reports on quantitative PCR detection using bacterial species-specific primers (Morrison *et al.*, 1999; Pahl *et al.*, 1999; Tajima *et al.*, 2001; Kageyama *et al.*, 2000). We studied a quantitative PCR detection method for the genus *Bifidobacterium* using a Light Cycler, which is described below.

Quantitative PCR Detection of the Genus *Bifidobacterium*

This method analyzes DNA extracted from feces, using DNA extracted from pure cultures and enumerated bifidobacteria as an external standard. In our study, SYBR Green I was used as the fluorescent substance, and a Light Cycler were used as the detection equipment for PCR. The reaction solution had a total volume of 10 μl and contained 10 mM of Tris-HCl, pH 8.3; 50 mM of KCl; 1.5 mM of MgCl$_2$; 500 ng/μ l of bovine serum albumin; 200 μM of each dNTP; template DNA; 1:30,000 dilution of SYBR Green I (Molecular Probes, Eugene, OR); 11 ng/μl of TaqStart™ antibody (ClonTech, Palo Alto, CA); 0.05 U/μl of Taq DNA polymerase (Takara, Tokyo, Japan); and 0.25 μM of the *Bifidobacterium* group-specific primers (g-Bifid-F [5'-CTC CTG GAA ACG GGT GG-3'] and g-Bifid-R [5'-GGT GTT CTT CCC GAT ATC TAC A-3']). The PCR was conducted as follows: 40 cycles of three steps each, comprised of heating at 20°C/sec to 95°C with a 0-second hold, cooling at 20°C/sec to 55°C with a 10-second hold, and then heating at 20°C /sec to 72°C with a 20-second hold. Fluorescent product was detected in the last step of each cycle. After amplification, a melting curve was obtained by heating at 20°C/sec to 96°C, cooling at 20°C/sec to 60°C, and slowly heating at 0.2°C/sec to 96°C, with fluorescence collection at intervals of 0.2°C. Melting curves were used to determine the specificity of the PCR. Comparison of bifidobacterial count obtained by culture and real time PCR showed that there was good correlation (Requena *et al.*, 2002).

The approach to detect the target bacteria with quantitative PCR described here will contribute to future studies of the composition and dynamics of the intestinal microflora. In addition, automation of DNA extraction and distribution of PCR solutions will be necessary to reduce the labor required.

Conclusion

This chapter describes the advantages of PCR methods in the analysis of the gut ecosystem with emphasis on the species-specific primers for bifidobacteria. The PCR methods allow the highly sensitive detection of specific bacteria and will have a significant effect on the analysis of gut community structure. The next significant challenges for researchers are the selection of bacterial genera and species for the preparation of further primers, and the selection of appropriate subjects for analysis.

References

Amann, R.I., Ludwig, W., and Schleifer, K.H. 1995. Phylogenetic identification and *in situ* detection of individual microbial cells without cultivation. Microbiol. Rev. 59: 143-69.

Benno, Y., Sawada, K., and Mitsuoka, T. 1984. The intestinal microflora of infants: composition of fecal microflora in breast-fed and bottle-fed infants. Microbiol. Immunol. 28: 975-986.

Biavati, B., Castagnoli, P., Crociani, F., and Trovatelli, L.D. 1984. Species of the *Bifidobacterium* in the feces of infants. Microbiologica 7: 341-345.

Biavati, B., Castagnoli, P., and Trovatelli, L.D. 1986. Species of the genus *Bifidobacterium* in the feces of human adults. Microbiologica 9: 39-45.

Brigidi, P., Vitali, B., Swennen, E., Altomare, L., Rossi, M., and Matteuzzi, D. 2000. Specific detection of *Bifidobacterium* strains in a pharmaceutical probiotic product and in human feces by polymerase chain reaction. Syst. Appl. Microbiol. 23: 391-399.

Dong, X., Cheng, G., and Jian, W. 2000a. Simultaneous identification of five *Bifidobacterium* species isolated from human beings using multiple PCR primers. Syst. Appl. Microbiol. 23: 386-390.

Dong, X., Xin, Y., Jian, W., Liu, X., and Ling, D. 2000b. *Bifidobacterium thermacidophilum* sp. nov., isolated from an anaerobic digester. Int. J. Syst. Evol. Microbiol. 50: 119-125.

Euzeby, J.P. 1997. List of bacterial names with standing in nomenclature: A folder available on the Internet. Int. J. Syst. Bacteriol. 47: 590-592.

Finegold, S.M., Attebery, H.R., and Sutter, V.L. 1974. Effect of diet on human fecal microflora: comparison of Japanese and American diets. Am. J. Clin. Nutr. 27: 1456-1469.

Franks, A.H., Harmsen, H.J., Raangs, G.C., Jansen, G.J, Schut, F., and Welling, G.W. 1998. Variations of bacterial populations in human feces measured by fluorescent *in situ* hybridization with group-specific 16S rRNA-targeted oligonucleotide probes. Appl. Environ. Microbiol. 64: 3336-3345.

Frothinfham, R., Duncan, A.J., and Wilson, K.H. 1992. Ribosomal DNA sequences of bifidobacteria: implications for sequence-based identification of the human colonic microflora. Microbial. Ecol. Health Dis. 6: 23-27.

Gumerlock, P.H., Tang, Y.J., Meyers, F.J., and Silva, Jr., J. 1991. Use of the polymerase chain reaction for the specific and direct detection of *Clostridium difficile* in human feces. Rev. Infect. Dis. 13: 1053-1060.

Holdman, L.V., Cato, E.P., and Moore, W.E.C. 1977. Anaerobe laboratory manual, 4th edition. Southern Printing Co., Blacksburg, Virginia.

Jian, W., Zhu, L., and Dong, X. 2001. New approach to phylogenetic analysis of the genus *Bifidobacterium* based on partial HSP60 gene sequences. Int. J. Syst. Evol. Microbiol. 51: 1633-1638.

Kageyama, A., and Benno, Y. 2001. Rapid detection of human fecal *Eubacterium* species and related genera by nested PCR method. Microbiol. Immunol 45: 315-318.

Kageyama, A., Sakamoto, M., and Benno, Y. 2000. Rapid identification and quantification of *Collinsella aerofaciens* using PCR. FEMS Microbiol. Lett 183: 43-47.

Langendijk, P.S., Schut, F., Jansen, G.J., Raangs, G.C., Kamphuis, G.R., Wilkinson, M.H., and Welling, G.W. 1995. Quantitative fluorescence *in situ* hybridization of *Bifidobacterium* spp. with genus-specific 16S rRNA-targeted probes and its application in fecal samples. Appl. Environ. Microbiol. 61: 3069-7305.

Lauer, E. 1990. *Bifidobacterium gallicum* sp. nov. isolated from human feces. Int. J. Syst. Bacteriol. 40: 100-102.

Lauer, E., and Kandler, O. 1983. DNA-DNA homology, murein types and enzyme patterns in the type strains of the genus *Bifidobacterium*. Syst. Appl. Microbiol. 4: 42-64.

Leblond-Bourget, N., Philippe, H., Mangin, I., and Decaris, B. 1996. 16S rRNA and 16S to 23S internal transcribed spacer sequence analyses reveal inter- and intraspecific *Bifidobacterium* phylogeny. Int. J. Syst. Bacteriol. 46: 102-111.

Lee, Y.-K., Nomoto, K., Salminen, S., and Gorbach, S.L. 1999. Probiotic microorganisms. In: Handbook of Probiotics. A Wiley-Interscience Publication, John Wiley and Sons, Inc, New York. p. 4-17.

Matsuki, T., Watanabe, K., Tanaka, R., and Oyaizu, H. 1998. Rapid identification of human intestinal bifidobacteria by 16S rRNA-targeted species- and group-specific primers. FEMS Microbiol. Letts. 167: 113-121.

Matsuki, T., Watanabe, K., Tanaka, R., Fukuda, M., and Oyaizu, H. 1999. Distribution of bifidobacterial species in human intestinal microflora examined with 16S rRNA-gene-targeted species-specific primers. Appl. Environ. Microbiol. 65: 4506-4512.

Millar, M.R., Linton, C.J., Cade, A., Glancy, D., Hall, M., and Jalal, H. 1996. Application of 16S rRNA gene PCR to study bowel microflora of preterm infants with and without necrotizing enterocolitis. J. Clin. Microbiol. 34: 2506-2510.

Mitsuoka, T., Hayakawa, K., and Kimura, N. 1974. Die Faekalmicroflora bei Menschen. II. Mitteilung: Die Zusammensetzung der Bifidobakerien microflora der verschiedenen Altersgruppen. Zentralbl. Bakteriol. Hyg. I. Orig. A226: 469-478.

Miyake, T., Watanabe, K., Watanabe, T., and Oyaizu, H. 1998. Phylogenetic analysis of the genus *Bifidobacterium* and related genera based on 16S rDNA sequences. Microbiol. Immunol. 42: 661-667.

Moore, W.E., and Holdeman, L.V.. 1974. Human fecal microflora: the normal microflora of 20 Japanese-Hawaiians. Appl. Microbiol. 27: 961-979.

Moore, W.E., and Moore, L.H.. 1995. Intestinal microfloras of populations that have a high risk of colon cancer. Appl. Environ. Microbiol. 61: 3202-3207.

Morrison, T.B., Ma, Y., Weis, J.H., and Weis, J.J. 1999. Rapid and sensitive quantification of *Borrelia burgdorferi*-Infect.ed mouse tissues by continuous fluorescent monitoring of PCR. J. Clin. Microbiol. 37: 987-992.

Mutai, M., and Tanaka, R. 1987. Ecolgy of *Bifidobacterium* in the human intestinal microflora. Bifidobacteria Microflora 6: 33-41.

Nebra, Y., and Blanch, A.R. 1999. A new selective medium for *Bifidobacterium* spp. Appl. Environ. Microbiol. 65: 5173-5176.

Pahl, A., Kuhlbrandt, U., Brune, K., Rollinghoff, M., and Gessner, A. 1999. Quantitative detection of *Borrelia burgdorferi* by real-time PCR. J. Clin. Microbiol. 37: 1958-63.

Reilly, K., and Attwood, G.T. 1998. Detection of *Clostridium proteoclasticum* and closely related strains in the rumen by competitive PCR. Appl. Environ. Microbiol. 64: 907-913.

Requena, T., Burton, J., Matsuki, T., Munro, K., Simon, M.A., Tanaka, R., Watanabe, K., and Tannock, G.W. 2002. Identification, detection and enumeration of *Bifidobacterium* species by PCR targeting the transaldolase gene. Appl. Environ. Microbiol. 68: 2420-2427.

Ririe, K.M., Rasmussen, R.P., and Wittwer, C.T. 1997. Product differentiation by analysis of DNA melting curves during the polymerase chain reaction. Anal. Biochem. 245: 154-610.

Roy, D., Ward, P., and Champagne, G. 1996. Differentiation of bifidobacteria by use of pulsed-field gel electrophoresis and polymerase chain reaction. Int. J. Food Microbiol. 29: 11-29.

Satake, S., Clark, N., Rimland, D., Nolte, F.S., and Tenover, F.C. 1997. Detection of vancomycin-resistant enterococci in fecal samples by PCR. J. Clin. Microbiol. 35: 2325-2330.

Satokari, R.M., Vaughan, E.E., Akkermans, A.D., Saarela, M., and de Vos, W.W. 2001. Bifidobacterial diversity in human feces detected by genus-specific PCR and denaturing gradient gel electrophoresis. Appl. Environ. Microbiol. 67: 504-513.

Scardovi, V. 1984. Genus *Bifidobacterium* Orla-Jensen, 1924, 472, In: Bergey's Manual of Systematic Bacteriology, Vol. I. N. R. Krieg and J. G. Holt, eds. The Williamd & Wilkins Co., Baltimore. p. 1418-1434.

Scardovi, V., and Crociani, F. 1974. *Bifidobacterium catenulatum, Bifidobacterium dentium,* and *Bifidobacterium angulatum*: three new species and their deoxyribonucleic acid homology relationships. Int. J. Syst. Bacteriol. 24: 6-20.

Song, Y., Kato, N., Liu, C., Matsumiya, Y., Kato, H., and Watanabe, K. 2000. Rapid identification of 11 human intestinal *Lactobacillus* species by multiplex PCR assays using group- and species-specific primers derived from the 16S-23S rRNA intergenic spacer region and its flanking 23S rRNA. FEMS Microbiol. Lett. 187: 167-173.

Suau, A., R. Bonnet, M. Sutren, J. J. Godon, G. R. Gibson, M. D. Collins, and J. Dore. 1999. Direct analysis of genes encoding 16S rRNA from complex communities reveals many novel molecular species within the human gut. Appl. Environ. Microbiol. 65: 4799-4807.

Tajima, K., Aminov, R.I., Nagamine, T., Matsui, H., Nakamura, M., and Benno, Y. 2001. Diet-dependent shifts in the bacterial population of the rumen revealed with real-time PCR. Appl. Environ. Microbiol. 67: 2766-2774.

Tanaka, R., and Mutai, M. 1980. Improved medium for selective isolation and enumeration of *Bifidobacterium*. Appl. Environ. Microbiol. 40: 866-869.

Ventura, M., Reniero, R., and Zink, R. 2001. Specific identification and targeted characterization of *Bifidobacterium lactis* from different environmental isolates by a combined multiplex-PCR approach. Appl. Environ. Microbiol. 67: 2760-2765.

Walter, J., Tannock, G.W., Tilsala-Timisjarvi, A., Rodtong, S., Loach, D.M., Munro, K., and Alatossava, T. 2000. Detection and identification of gastrointestinal *Lactobacillus* species by using denaturing gradient gel electrophoresis and species-specific PCR primers. Appl. Environ. Microbiol. 66: 297-303.

Wang, R.F., Cao, W.W., and Cerniglia, C.E. 1996. PCR detection and quantitation of predominant anaerobic bacteria in human and animal fecal samples. Appl. Environ. Microbiol. 62: 1242-1247.

Wang, R.F., Cao, W.W., and Cerniglia, C.E. 1997. PCR detection of *Ruminococcus* spp. in human and animal faecal samples. Mol. Cell Probes 11: 259-265.

Wayne, L.G., Brenner, D.J., Colwell, R.R., Grimont, P.A.D., Kandler, O., Krichevsky, M.I., Moore, L.H., Moore, W.E.C., Murray, R.G.E., Stackebrandt, E., Starr, M.P., and Truper, H.G. 1987. Report of the ad hoc committee on reconciliation of approaches to bacterial systematics. Int. J. Syst. Bacteriol. 37: 463-464.

Wilson, I. G. 1997. Inhibition and facilitation of nucleic acid amplification. Appl. Environ. Microbiol. 63: 3741-3751.

Wilson, K. H., and Blitchington, R.B. 1996. Human colonic biota studied by ribosomal DNA sequence analysis. Appl. Environ. Microbiol. 62: 2273-2278.

Woese, C. R. 1987. Bacterial evolution. Microbiol. Rev. 51: 221-271.

Yamamoto, T., Morotomi, M., and Tanaka, R. 1992. Species-specific oligonucleotide probes for five *Bifidobacterium* species detected in human intestinal microflora. Appl. Environ. Microbiol. 58: 4076-4079.

Zoetendal, E. G., Akkermans, A.D., and De Vos, W.W. 1998. Temperature gradient gel electrophoresis analysis of 16S rRNA from human fecal samples reveals stable and host-specific communities of active bacteria. Appl. Environ. Microbiol. 64: 3854-3859.

From: *Probiotics and Prebiotics: Where Are We Going?*
Edited by: Gerald W. Tannock

Chapter 5

Prebiotic Oligosaccharides: Evaluation of Biological Activities and Potential Future Developments

Robert A. Rastall and Glenn R. Gibson

Abstract

Prebiotics are recognised for their ability to increase levels of 'health promoting' bacteria in the intestinal tract of humans or animals. This normally involves targeting the activities of bifidobacteria and/or lactobacilli. Non digestible oligosaccharides such as fructo-oligosaccharides, lactulose and *trans*-galacto-oligosaccharides seem to be efficacious prebiotics in that they confer the degree of selective fermentation required. Other oligomers are used as prebiotics in Japan e.g. xylo-oligosaccharides, soybean-oligosaccharides, isomalto-oligosaccharides.

To determine prebiotic functionality, various *in vitro* systems may be used. These range from simple batch culture fermenters to complex models of the gastrointestinal tract. The definitive test however is an *in vivo* study. The advent of molecular based procedures in gut microbiology has alleviated many concerns over the reliability of microbial characterisation, in response to prebiotic intake. Techniques such as DNA probing and molecular fingerprinting are now being applied to both laboratory and human studies. These will help to further identify prebiotics that can be added to the diet and thereby fortify 'beneficial' bacteria. Such robust technologies can also be used in structure-function assays to identify the mechanisms behind prebiotic effects.

Considerable research effort is currently being expended in developing so called 'second generation' prebiotics. These are forms that have multiple biological activity that attempts health enhancement properties beyond the genus level stimulation of bifidobacteria or lactobacilli within the gut microbflora. Examples include higher molecular weight oligomers than is conventional for prebiotics, such that targeted activities in the distal colon are feasible (the left side of the human large gut being the frequent area for colonic disorder). Glycobiology is also developing anti-adhesive prebiotics that incorporate receptor sites for common gut pathogens and/or their activities. Through the use of reverse enzyme technology, as applied to β-galactosidase activity in probiotics, oligosaccharides that enhance a lactic microflora at the species, rather than genus, level are possible.

This review gives an account of how second generation prebiotics may be manufactured, through a variety of biotechnological techniques, and tested for their biological activity. The health attributes of such molecules as well as existing prebiotics is also discussed, with reference to specific target populations.

The prebiotic concept is a much more recent development in dietary intervention for enhanced gut function than is probiotics. Not surprisingly therefore, research developments are proceeding quickly. Because oligosaccharides can be added to a wide variety of foodstuffs, new functional food developments are continuing. It is important that these are tested using reliable methodologies and that any health effects are underpinned by realistic mechanisms of effect.

Table 1. Emerging prebiotics on the global market

Emerging Prebiotics
Isomalto-oligosaccharides IMO.
Soybean oligosaccharides SOS.
Lactosucrose LS.
Gentio-oligosaccharides GEOS.
Xylo-oligosaccharides XOS.

Background

Prebiotics

At the present time the only molecules known to act as prebiotics are carbohydrates (Gibson and Roberfroid, 1995; Roberfroid *et al.*, 1998). In particular, they are non-digestible oligosaccharides: the first requirement of a prebiotic being that it is not metabolised in the human small intestine and passes to the colon quantitatively – the main site of microbial action in the body. When the prebiotic arrives in the colon certain members of the indigenous microflora must ferment it selectively. The usual targets for prebiotic action are the probiotic bacterial genera, *Bifidobacterium* and *Lactobacillus* (Gibson and Roberfroid, 1995).

Currently, there is a range of prebiotics in use around the world mainly in Japan (Crittenden, 1999; Crittenden and Playne, 1996; Playne and Crittenden, 1996). However, in Europe and the USA there are only three types used in food manufacturing. These are the fructans, inulin and fructo-oligosaccharides (FOS) and galacto-oligosaccharides (GOS). The lactose derivative lactulose is well known as a laxative but it also displays prebiotic behaviour in the colon at sub-laxative doses (Salminen and Salminen, 1997). It is not currently used in foods, however. The range of oligosaccharides used as prebiotics in the Japanese market is much more extensive and it is only a matter of time before these ingredients find application on a global scale. These can, therefore, be considered as "emerging prebiotics" (Table 1).

Such ingredients have not yet been tested as rigorously as the current generation of prebiotics and we have an incomplete picture of their fermentation and technological properties. It is possible that the molecules night have better nutritional properties and more desirable properties to food processors than the established prebiotics.

Comparing Prebiotics

If we are to maximise the potential of the prebiotic approach towards modulation of the colonic microflora, we need to understand much more about the properties of the various prebiotic oligosaccharides. A detailed comparison will allow optimal use of the ingredients in food products.

The best means of comparing prebiotics is to use *in vitro* methods prior to human volunteer testing. Human trials are expensive, and the *in vitro* studies will provide information to design optimised trials. There are various *in vitro* approaches for studying prebiotic metabolism (Molly *et al.*, 1993; McBain and Macfarlane, 1997).

Pure Culture

Pure cultures are the simplest *in vitro* model. Substrates are added to a basal growth medium and growth of a selection of gut micro-organisms determined during time course incubations (Gibson and Wang, 1994a). When working with gut micro-organisms, however, the approach involves a significant challenge in that anaerobic growth conditions must be maintained together with standard microbiological aseptic techniques. Growth of the micro-organisms is usually monitored by measuring optical density of the culture or viable counts on agar plates.

The pure culture approach gives a reasonable comparative assessment of metabolism in mono-culture, and can be used to perform mechanistic studies on prebiotic metabolism. It does not, however, induce any element of competition. The gut microflora has at least 500 described species (Cummings and Macfarlane, 1991). For this reason the approach cannot identify true selectivity of fermentation of a particular substrate.

A more biologically meaningful approach is to use mixed culture experiments with selected gut microbial species. This simulates some element of competition between species and can be used to examine specific interactions, for instance between probiotics and pathogens. It does not, however, adequately model complex interactions that occur in the human gut.

Mixed Culture Bacterial Fermenters

A common means of modelling the microbial activities of the gut is to use batch culture fermenters inoculated with faeces (Wang and Gibson, 1993). Anaerobic conditions are maintained by the infusion of oxygen free nitrogen, pH is controlled and the fermenters are well mixed. They are, however, closed systems with limiting substrate concentrations and they are best suited for short time course experiments.

Substrate concentration decreases with time ultimately resulting in a stationary, then a declining culture. The colon, however, is constantly fed with microbial nutrients resulting in very different growth kinetics (Cummings *et al.*, 1989).

For this reason, a more physiologically relevant approach is continuous culture, whereby nutrients are continuously supplied and various physiological parameters such as dilution rate can be controlled (Gibson and Wang, 1994a). The most frequently used method is the chemostat. One substrate is generally provided in growth limiting concentrations thereby resulting in an enrichment of bacteria that can metabolise the limiting nutrient, and these will eventually predominate in the system. This is a very effective method of evaluating the metabolism of candidate prebiotic oligosaccharides.

In Vitro Gut Models

The human large intestine is a complex environment supporting a heterogeneous microbial ecosystem (Tannock, 1997). The predominant nutrient source is diet with the small intestine semi-continuously feeding residual foodstuffs into the colon. The large gut also comprises various anatomically distinct regions. These include the caecum, ascending colon (the proximal area), transverse colon, descending colon and sigmoid rectum (the latter two forming the distal area). The normal adult colon contains approximately 200g of contents, fairly evenly spread throughout the different areas. Around 60% (dry weight basis) of the colonic contents are micro-organisms.

The proximal colon has a ready supply of substrates resulting in rapid bacterial growth and an acidic pH as a result of the formation of short chain fatty acids as metabolic end products. The nutrient supply to the distal colon is diminished resulting in slower bacterial growth. Consequently, the environment is more pH neutral. A meaningful *in vitro* model of the human gut must take into account these characteristics.

A particularly useful model is the three-phase continuous culture system which simulates different anatomical areas of the large gut, such as the right, transverse and left sides (Macfarlane *et al.*, 1992; McBain and Macfarlane, 1997; Macfarlane *et al.*, 1998). This has been validated against samples of colonic contents taken at autopsy (Macfarlane *et al.*, 1998) and is a very close model of the lumenal microbiology of the colon. The system consists of three vessels, of increasing size, aligned in series such that a sequential feeding with growth medium occurs. The pH in the vessels is regulated to reflect pH values found *in vivo*. Vessel 1 has high substrate availability, rapid bacterial growth and an acidic pH, similar to the proximal colon. Vessel 3 resembles the neutral pH, slow bacterial growth and low substrate availability characteristic of the distal colon. The model is inoculated with faeces and an equilibration period allowed to achieve steady state growth conditions in each vessel. Once steady state is achieved, then feeding of the candidate prebiotic begins. Once a new steady state is achieved, the vessels are sampled and microbial groups enumerated.

Similar systems based on five-stage continuous fermenters have been developed to simulate the intestinal tract from the jejunum to the descending colon (Molly *et al.*, 1993).

Such gut models have given very important data on the fermentation properties of prebiotics. Developments of the gut model are envisaged in the future. In particular, the use of intestinal cell tissues in such models will allow modelling of the adherent biofilm communities present in the gut.

Animal Methods

Rats and mice have frequently been used to investigate the prebiotic properties of substrates (Rowland and Tanaka, 1993). Conventional animals, gnotobiotic (germ-free) rats or rats inoculated with one or a limited number of micro-organisms have been used, although the conditions in these animals do not resemble the normal human gut. Rats may be associated with a human faecal microflora, known as human microflora-associated (HFA) rats, and these give a closer representation of the human colonic situation; the intestinal physiology is not, however, the same. A major drawback with laboratory animal experiments is the differing gut anatomy as well as the rodent practice of coprophagy.

Table 2. Selected growth media and target bacterial groups for fermentation studies on the faecal microflora

Selective Growth Media[a]	Dilutions Plated	Target Group[b]
Nutrient agar	10^{-1} - 10^{-5}	Total aerobes
MacConkey agar No.3	10^{-1} - 10^{-4}	Coliforms
Wilkens-Chalgren agar	10^{-4} - 10^{-7}	Total anaerobes
Bacteroides agar[c]	10^{-4} - 10^{-7}	*Bacteroides* spp.
Beeren's agar[d]	10^{-4} - 10^{-7}	*Bifidobacterium* spp.
Azide/Crystal violet agar[e]	10^{-1} - 10^{-4}	Facultatively anaerobic Gram-positive cocci *Enterococcus* spp.
Rogosa agar	10^{-0} - 10^{-7}	*Lactobacillus* spp.
Fusobacterium agar[f]	10^{-2}- 10^{-5}	*Fusobacterium* spp.
Raffinose-Bifidobacterium agar[g]	10^{-4}- 10^{-7}	*Bifidobacterium* spp.
Clostridia agar[h]	10^{-3} - 10^{-6}	*Clostridium* spp.

[a] All agar media purchased in dry powder form from Oxoid.
[b] Target bacterial group of selective growth media, colonies being described through growth requirements and morphotype.
[c] Brucella agar with kanamycin 75 mg/l., vancomycin 7.5 mg/l., hemin 5 mg/l., vitamin K_1 10 mg/l. and laked horse blood 50 ml/l.
[d] Columbia agar with agar 5 g/l., glucose 5 g/l., cysteine HCl 0.5 g/l. and propionic acid 5 ml/l. to pH 5.
[e] Azide blood agar base, laked horse blood 50 ml/l. , crystal violet 0.0002 %.
f Brucella agar with laked horse blood 50 ml/l. and rifampicin 50 µg/ml.
[g] Hartemink *et al.*, 1996.
[h] Reinforced Clostridial agar with novobiocin 8 mg/l. and colistin 8 mg/l.

Molecular Methods of Microflora Analysis

Critical aspects of prebiotic evaluation, regardless of the model system used, are the reliable identification of bacterial groups and the accurate determination of the bacterial populations. Much of the data on the prebiotic potential of candidate prebiotic oligosaccharides has, to date, been performed using conventional microbiological techniques. This typically involves plating faecal samples or samples from gut model systems onto selective media designed to recover particular numerically predominant groups of bacteria (Wang and Gibson, 1993). The media used are, however, semi-selective at best, and are incapable of enumerating non-culturable bacteria (which may represent over 50% of the overall diversity). They also allow operator subjectivity in terms of phenotypic methods of microbial characterisation. Vast arrays of selective media are required in order to use this approach with the faecal microflora (Table 2).

An alternative approach is to apply molecular principles to more effectively characterise the microflora involved in selective fermentation studies (Collins and Gibson, 1999; Gibson and Collins, 1999; O'Sullivan, 1999; Kullen and Klaenhammer, 1999; Langendijk *et al.,* 1995; Wilson and Blitchington, 1996; McCartney *et al.,* 1996).

Genotypic Characterisations

The identity of gut bacteria can be determined by PCR-16S-rRNA gene restriction fragment length polymorphism and partial gene sequence analysis (Collins and Gibson, 1999). 16S rRNA molecules contain universally conserved, semi-conserved and non-conserved regions. The sequences of the "hypervariable" regions of the molecule are characteristic for different organisms and provide a rapid and reproducible means of identification. Full 16S rRNA genes can be amplified from single colonies utilising primers specific for conserved regions proximal to the 5' and 3' termini of the gene. The resulting rDNA is subjected to RFLP analysis involving digestion with restriction endonucleases followed by electrophoretic analysis. Each bacterial type is characterised by a simple, but highly specific, series of rDNA restriction patterns. A comprehensive rDNA signature database is established against which isolates can be compared and identified. rDNA products can then be sequenced directly using automated PCR-cycle sequencing. New sequences are compared with sequences in databases, and phylogenetic analyses performed to determine their identity (Suau *et al.* 1999).

Table 3. Molecular techniques applied to human faecal analysis (after McCartney and Gibson, 1998)

Restriction fragment length polymorphism RFLP	Compares banding patterns after digestion of chromosomal DNA with endonucleases
Pulsed field gel electrophoresis PFGE	Cutting restriction enzymes are used to reduce band numbers from RFLP
Ribotyping	A ribosomal DNA probe highlights bands within RFLP
Ribosomal DNA sequencing	Direct comparison of 16SrDNA gene sequences from sequencing of PCR amplified rDNA
Amplified fragment length polymorphism	RFLP amplified by polymerase chain reaction PCR.
Direct amplification	Culture independent PCR amplification of bacterial DNA in environmental samples
Genetic probes	Detection and/or identification of specific microbial groups within environmental samples by labelled hybridisation probes
Molecular marking	Use of a genetic tag that enables discrimination of target microorganisms within complex ecosystems

Probes

Microbiological changes occurring on fermentation of a test substrate can be monitored by gene probing procedures. These have the advantage that they do not involve culturing samples on agar plates. They thus enumerate total diversity including non-culturable species (Harmsen *et al.* 1999; 2000).

Fluorescent In Situ Hybridisation (FISH)

The most common approach involves the use of fluorescently labelled probes with microscopic counting. Hitherto, the best-validated probes are directed towards bifidobacteria, clostridia, bacteroides, lactobacilli and total bacteria. Genotypic probes targeting the predominant components of the gut microflora (bacteroides, bifidobacteria, clostridia and lactobacilli) are manufactured and 5' labelled with the fluorescent dye Cy3 such that changes in faecal bacterial populations can be measured. The most commonly used probes include

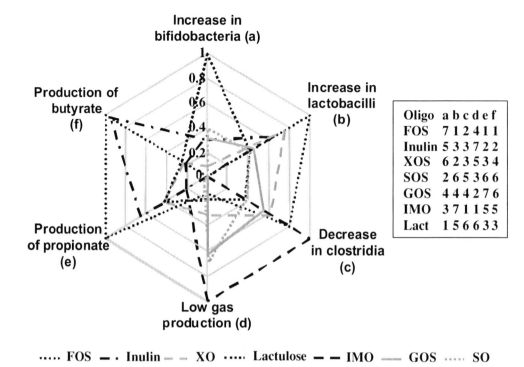

Oligo	a	b	c	d	e	f
FOS	7	1	2	4	1	1
Inulin	5	3	3	7	2	2
XOS	6	2	3	5	3	4
SOS	2	6	5	3	6	6
GOS	4	4	4	2	7	6
IMO	3	7	1	1	5	5
Lact	1	5	6	6	3	3

····· FOS — · Inulin ─ ─ XO ····· Lactulose — — IMO ▬▬ GOS ····· SO

Figure 1. Comparative properties of commercial prebiotics.

Bif164 specific for bifidobacteria (Langendijk *et al.*, 1996), Bac303 specific for bacteroides (Manz *et al.*, 1996), His150 specific for *Clostridium perfringens / histolyticum* sub.grp. (Franks *et al.*, 1998) and Lab158 specific for lactobacilli/enterococci (Harmsen *et al.* 1999). The total bacterial population is determined using the nucleic acid stain 4'6-diamidino-2-phenylindole (DAPI; Porter and Feig, 1980). Faecal samples are diluted and fixed in paraformaldehyde, allowing the storage of samples prior to counting. The various molecular techniques that have been used in the analysis of human faecal samples are listed in Table 3.

Comparative Analysis of Prebiotics in Batch Culture

Rycroft *et al.* (2001), performed faecal batch cultures on a range of oligosaccharide prebiotics. Microflora changes were determined and the evolution of gas and short chain fatty acids measured. The results are summarised in Figure 1.

Distinct differences were seen between the oligosaccharides tested although it must be borne in mind that the oligosaccharides used were commercial preparations of variable purity. Some of the prebiotics tested contain large quantities of carbohydrate that would, when ingested, be metabolised in the upper gut. In the *in vitro* batch culture the faecal bacteria ferment such sugars.

Structure-function Relationships

The list of recognised prebiotics represents a wide range of carbohydrate chemistry. Despite this, we still have very little understanding of the structure-function relationships in these molecules. Moreover, we do not have good comparative data on their fermentation properties (particularly at a microbial species level) and there is no predictive ability with respect to new carbohydrate structures isolated or synthesised. This is likely to hamper new product development based around the concept.

The prebiotic properties of carbohydrates are likely to be influenced by the following factors:

i) Monosaccharide composition. Recognised prebiotics are primarily built from glucose, galactose, xylose and fructose. The prebiotic potential of oligosaccharides composed of other monosaccharides such as arabinose, rhamnose, glucosamine and galacturonic acid is not known at the present time.

ii) Glycosidic linkage. The linkage between the monosaccharide residues is a crucial factor in determining both selectivity of fermentation and digestibility in the small intestine. The current paradigm for the selective fermentation of prebiotics is the cell-associated β-fructofuranosidase isolated from bifidobacteria. If this model holds true for other oligosaccharides and bacteria, then the linkage specificity of the glycosidases will be very important. Maltose is not recognised as prebiotic and is metabolised by the human intestinal brush border glycosidases. Isomaltose and isomalto-oligosaccharides (IMO), however, are prebiotic. Both are composed of α-glucosyl linkages but the 1-6 linkage in the IMO renders them partially resistant to metabolism in the small intestine and confers selectivity of fermentation in the colon.

iii) Molecular weight. Generally speaking polysaccharides are not prebiotic (Wang and Gibson, 1993). Conversely, all known prebiotics are low molecular weight. Inulin has the highest molecular weight, but most of the carbohydrate in inulin has a degree of polymerisation less than 25 with an average around DP 14 (De Leenheer, 1994). The effect of molecular weight on prebiotic properties can be seen from the fact that xylan is not selective whereas xylo-oligosaccharides are (Okazaki *et al.*, 1990; Jaskari *et al.*, 1998). Olano-Martin and co-workers have investigated this effect in more detail and found an increase in selectivity upon hydrolysis of dextran to IMO (Olano-Martin *et al.*, 2000) and upon hydrolysis of pectins to pectic oligosaccharides (Olano-Martin, 2001). The precise relationship between molecular weight and selectivity is not known at the present time for any polysaccharide/oligosaccharide system.

The 'Second Generation' of Prebiotics

Properties of Enhanced Prebiotics

Most of the prebiotics in industrial use are naturally occurring oligosaccharides or are the products of transfer reactions from sucrose and/ or lactose (Playne and Crittenden, 1996). As far as it is possible to discern, all of these prebiotics have arisen from a process of screening all molecules that it was possible to manufacture in reasonable quantity. There has been little or no attempt to deliberately design prebiotics rationally. This is due to the fundamental gaps in our knowledge about the biochemical basis of the effect discussed above, and a lack of manufacturing ability in the area of novel oligosaccharides.

Despite the current gaps in our knowledge, it is possible to produce a list of desirable attributes in a functionally enhanced prebiotic. It is also possible to identify certain properties that the oligosaccharide should possess in order to achieve these attributes (Table 4).

It must be recognised that it is likely to be impossible to achieve all of these attributes in a single oligosaccharide. Oligosaccharides with suites of properties suited to particular applications are, however, a reasonable target.

Persistence of the Prebiotic Effect to Distal Regions of the Colon

Many common diseases of the human large bowel arise in the distal colon, particularly colonic cancer (Rowland 1992). Prebiotics have been postulated to be protective against the development of colon cancer (Rowland and

Table 4. Design parameters for enhanced activity prebiotics

Desirable Attribute in Prebiotic	Properties of Oligosaccharides
Active at low dosage	Highly selectively and efficiently metabolised by *Bifidobacterium* and/or *Lactobacillus* sp.
Lack of side effects	Highly selectively metabolised by "beneficial" bacteria but not by gas producers, putrefactive organisms, etc.
Persistence through the colon	High molecular weight
Varying viscosity	Available in different molecular weights and linkages
Good storage and processing stability	Possess 1-6 linkages and pyranosyl sugar rings
Fine control of microflora modulation	Selectively metabolised by restricted species
Varying sweetness	Different monosaccharide composition
Inhibit adhesion of pathogens	Possess receptor sequence

Tanaka, 1993; Reddy *et al.*, 1997; Bouhnik *et al.*, 1996; Buddington *et al.*, 1996; Hylla *et al.*, 1998). If prebiotics are to have any protective value, however, they should preferably reach the distal colon. There are several ways that this might be achieved:

Increased Molecular Weight

Most current prebiotics are of relatively small DP, the exception being inulin. As discussed above, it is believed that the oligosaccharides must be hydrolysed by cell-associated bacterial glycosidases, prior to uptake of the resultant monosaccharides. It is, therefore, reasonable to assume that the longer the oligosaccharide, the slower the fermentation and hence the further the prebiotic effect will penetrate throughout the colon. For example, long chain inulin may exert a prebiotic effect in more distal colonic regions than the lower molecular weight FOS, which may be more quickly fermented in the saccharolytic proximal bowel. This approach has led to industrial forms of inulin/FOS mixtures with controlled chain length distributions ("Synergie II" manufactured by Orafti, Tienen, Belgium), which in theory should persist further in the hindgut. This approach might have great promise for making more persistent prebiotics. Manufacturing techniques for controlled chain length distribution oligosaccharides are also being developed (see below). As plant-derived inulin is intrinsically limited in its DP, microbial fructans such as laevan might be a better starting point for oligosaccharide generation. Laevan is a polysaccharide with a molecular weight measured in millions (Han, 1990); it is reasonable to assume that such a polysaccharide would take longer to be completely digested in the colon, increasing the chances of persistence through to the distal colon.

As discussed above, polysaccharides are generally not selectively fermented. There is likely to be a certain molecular weight for any given polysaccharide above which the selectivity is no longer great enough for the molecule to be considered as a prebiotic. Development of a persistent prebiotic might prove to be a compromise between persistence and prebiotic activity.

Chemical Modification

Modification of a proportion of the residues in an inulin or isomalto-oligosaccharide chain might conceivably lead to partial resistance to enzymic attack. The food industry has experience with chemically modified starches and celluloses. The modifying groups should be stable to stomach acid, but may well be removed in the colon by lipolytic enzymes and acyl esterases. The modification might also bring about desirable changes in physicochemical properties. Acylation of starch, for instance, results in a fatty mouthfeel. It would seem likely that this approach might have similar results with inulin.

Defined Health Outcomes

In the current regulatory climate in the EU specific health claims cannot be made for a food product (Gibson *et al.,* 2000). It is, however, possible to make a functional claim. It is our belief that modulation of the gut microflora composition using prebiotics is a functional aspect of intestinal microbiology. A recent consensus document (the gastrointestinal tract group of the EU Concerted Action on Functional Food Science in Europe, FUFOSE) summarised the established and possible effects of prebiotics (Salminen *et al.,* 1998). The established effects include non-digestibility and low energy value, a stool bulking effect and modulation of the gut microflora, promoting bifidobacteria and repressing clostridia.

SUBSTRATES   PRODUCTS   EFFECTS

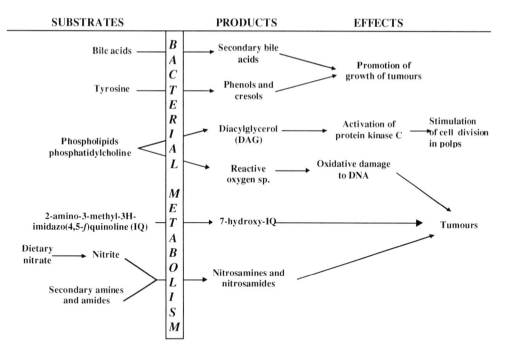

Figure 2. production of carcinogens and tumour promoters by the colonic microflora

Postulated areas for development in the future might include prevention of intestinal disorders (e.g. ulcerative colitis, irritable bowel syndrome) and gastrointestinal infections including diarrhoea, modulation of the immune response, prevention of colon carcinogenesis, reduction in serum triacylglycerols and cholesterol and the improvement of bioavailability of minerals such as calcium and magnesium.

Ways in which prebiotics might bring about some of these effects are discussed below. In many of these cases, the potential for development of prebiotics targeted at the health outcome is clear to see, however, all too often our lack of fundamental knowledge of the role of the colonic microflora in health is currently retarding progress.

Colonic Cancer

The second most prevalent cancer in humans is colon cancer (Gibson and Macfarlane, 1994), in addition it is thought that tumours arise 100 times more often in the large intestine compared to the small intestine (Morotomi *et al.* 1990). For this reason, many researchers believe that the colonic

Table 5. Genotoxic enzymes synthesised by colonic bacteria

Enzyme	Substrate
β-glucosidase	Plant glycosides e.g. rutin, cycasin
Azoreductase	Azo compounds e.g. benzidines
Nitroreductase	Nitro compounds e.g. nitrochrysene
β-glucuronidase	Biliary glucuronides e.g. benzidine
IQ hydratase-dehydrogenase	2-amino-3-methyl-3H-imidazo 4,5-*f*. quinoline IQ
Nitrate/nitrite reductase	Nitrate, nitrite

microflora has an important role to play in the development of bowel cancer (Rowland, 1988). It is known that several species of bacteria commonly found in the colon produce carcinogens and tumour promoters from food components that reach the colon. (Figure 2). Many also synthesise enzymes with genotoxic or toxic products (Table 5). Interest in a diet-mediated intervention in colon cancer arises due to the slow, progressive nature of the disease and the fact that we can influence colonic microbiology by diet. There have been several studies on the use of prebiotics in cancer prevention, mainly focussing on animal models.

It is thought that prebiotics may protect against development of colon cancer through at least two mechanisms.

i) Production of protective metabolites. Butyrate is a common fermentation end product and is known to stimulate apoptosis in colonic cancer cell lines and it is also the preferred fuel for healthy colonocytes (Prasad, 1980; Kim *et al.,* 1982). For these reasons, it is generally believed that it is desirable to increase the level of butyrate formed in the large gut. Some prebiotics are known to have this effect (Olano-Martin et al, 2000), although it must be borne in mind that lactobacilli and bifidobacteria do not produce butyrate. Known butyrate producers in the gut are clostridia and eubacteria (Cummings and Macfarlane, 1991). Development of prebiotics which stimulate eubacteria but not clostridia would be a desirable enhancement.

ii) Subversion of colonic metabolism away from protein and lipid metabolism. It is possible that prebiotics would induce a shift in bacterial metabolism in the colon towards more benign end products. An obvious target would be to shift the metabolism of clostridia and bacteroides away from proteolysis to saccharolysis.

Clostridia are also thought to be responsible for the bioconversion of dietary lipids into carcinogens (Morotomi *et al.,* 1990). These include diacylglycerol (DAG), formed from phosphatidylcholine in conjunction with bile salts. DAG is thought to activate protein kinase C, believed to be a potent stimulator of mucosal cell proliferation.

Lactic acid bacteria are believed to have inhibitory effects on several bacteria that produce carcinogenic enzymes (Table 5) and are themselves non-producers.

To date, few prebiotics have been evaluated in animal and human trials. Inulin, for instance has been shown to inhibit the formation of aberrant crypt foci in rats (Reddy *et al.*, 1997). Human studies are low in number and tend to focus on faecal markers of carcinogenesis rather than being epidemiological in nature. FOS, GOS and resistant starch have all been investigated in this regard. FOS has been found to reduce genotoxic enzymes concomitant with increasing bifidobacteria (Bouhnik *et al.*, 1996) and resistant starch has been found to reduce sterols, secondary bile acids and genotoxic enzymes, although no microbiological studies were performed (Hylla *et al.*, 1998).

A recent study on GOS however (Alles *et al.*, 1999), found no significant changes in bifidobacteria or in markers of carcingenesis. These results might, at first sight seem anomalous, as GOS are known prebiotics (Schotermann and Timmermans, 2000). However, the starting populations of bifidobacteria in the volunteers were rather high (9.2-9.4 log). It has been noted before (Rycroft *et al.*, 2001) that the magnitude of the response to prebiotics by bifidobacteria depends on the starting levels.

It is apparent that we currently have an inadequate knowledge of the effects of various prebiotics upon risk of colon cancer and more studies are needed to address this. Development of prebiotics with the goal of reducing biomarkers of cancer would, however, be desirable.

Immunological Effects

Lactic acid bacteria (LAB) have long been considered to be immuno-modulatory and there are several commercial products on the marketplace that have built upon this concept. LAB are thought to stimulate both non-specific host defence mechanisms and certain cell types involved in the specific immune response. The result is often increased phagocytic activity and/or elevated immunological molecules such as secretory IgA, which may affect pathogens such as salmonellae and rotavirus. Most attention in this respect has been directed towards the intake of probiotics (Schiffrin *et al.,* 1995; Perdigon and Alvarez, 1992) and interactions between cell wall components and immune cells. As prebiotics serve a similar end point to probiotics (i.e. improved gut microflora composition) similar effects may occur through their intake. A question arises whether increased immune function, even through non-pathogenic means, is a desirable trait. A more detailed understanding of the immunological responses to particular changes in the colonic microflora may allow the development of prebiotics with more desirable benefits.

Systemic Effects on Blood Lipids

It is widely believed that elevated cholesterol levels in the blood represent a risk factor for coronary heart disease, with low-density lipoproteins (LDL) being of most concern (Delzenne and Williams, 1999). There is also evidence that lactic acid bacteria may be able to reduce total and LDL cholesterol levels. The mechanisms by which lactic acid bacteria, and hence, indirectly, prebiotics, influence blood lipids are not clearly understood at the present time. It is possible that some LAB may be able to directly assimilate cholesterol. This has been hypothesised from some *in vitro* experiments, but is a source of contention in that the data are conflicting and precipitation of the cholesterol with bile salts at a low pH may occur giving misleading results. It has been suggested (Delzenne and Kok, 1999; Delzenne and Williams, 1999) that propionate produced by the bacterial fermentation of prebiotics inhibits the formation of serum LDL cholesterol. The difficulty with this hypothesis is that bacterial fermentation of prebiotics generally produces much more acetate than propionate – as the target LAB (lactobacilli, bifidobacteria) are not propionate producers. Moreover, acetate is a metabolic precursor of cholesterol and may therefore tend to increase, not decrease, serum levels. There is evidence that FOS decrease the *de novo* synthesis of triglycerides by the liver. The means by which this occurs is not fully understood but the effect appears be exerted at the transcriptional level. It is also possible that prebiotics (such as inulin) can modulate insulin-induced inhibition of triglyceride synthesis (Delzenne and Kok, 1999).

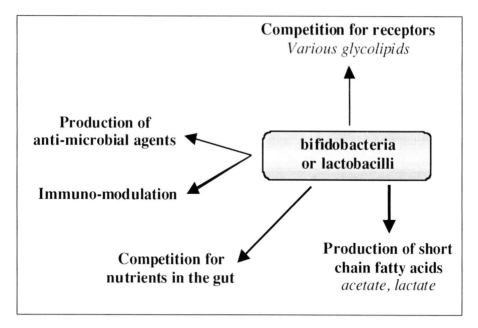

Figure 3. The barrier effect against GI pathogens

Human studies on the lipid lowering properties of prebiotics when consumed at a realistic (tolerable) dose are not clear-cut (Williams, 1999). Results are inconsistent and only FOS, inulin and GOS have been studied in this regard. It is possible that other prebiotics might lead to a more consistent effect, although until the mechanisms behind the effect are clarified, this will remain speculation.

The Barrier Effect Against GI Pathogens

One of the most mechanistically strong health benefits proposed for prebiotics is the barrier function against invading gastrointestinal pathogens. In particular, this may be a successful way of prophylactically addressing the burden of food safety through tackling problems such as campylobacters, salmonellae and *E. coli*.

There are several prospective mechanisms by which bifidobacteria and lactobacilli can inhibit pathogens (Figure 3). Fermentation of carbohydrates results in the production of short-chain fatty acids (SCFA) (Gibson and Roberfroid, 1995). These reduce lumenal pH in the colon to levels below those at which pathogens such as *E. coli* can effectively multiply. In addition, increased populations of bifidobacteria and lactobacilli can compete with

Table 6. Receptor saccharides for GI pathogens and toxins

Galα4Gal	*E. coli* P-piliated., Vero cytotoxin
Galβ4GlcNAcβ3Gal	*S. pneumoniae*
GalNAcβ4Gal	*P. aeruginosa, H. influenzae, S. aureus, K. pneumoniae*
Sialic acids	*E. coli* S-fimbriated.
Galα3Galβ4GlcNAc	*C. difficile* toxin A
Fucose	*V. cholerae*
GlcNAc	*E. coli, V. cholerae*
Mannose	*E. coli, K. aerogenes, Salmonella spp.*,Type 1-fimbriated.

other organisms for nutrients and receptors on the gut wall (Araya-Kojima *et al.*, 1995). Bifidobacteria and lactobacilli can also inhibit pathogens via a more direct mechanism. They are known to produce anti-microbial agents active against a range of pathogens (Gram positive and Gram negative) including salmonellae, campylobacters and *Escherichia coli* (Gibson and Wang, 1994b; Anand *et al.*, 1985; Araya-Kojima *et al.*, 1995). This anti-pathogen effect can be reproducibly seen in gut model systems and is likely to operate in the human gut.

Anti-adhesive Activities Against Pathogens and Toxins

The idea of combining prebiotic properties with anti-adhesive activities is currently under investigation. This would add major functionality to the approach of altering gut pathogenesis. Many intestinal pathogens utilise monosaccharides or short oligosaccharide sequences as receptors and knowledge of these receptor sites has relevance for biologically enhanced prebiotics (Table 6).

Binding of pathogens to these receptors is the first step in the colonisation process (Karlsson, 1989; Finlay and Falkow, 1989). There are currently several pharmaceutical preparations based upon such oligosaccharides in clinical trials. These agents are multivalent derivatives of the sugars and act as "blocking factors", dislodging the adherent pathogen (Heerze, *et al.*, 1994; Jararaman *et al.*, 1997). There is much potential for developing prebiotics, which incorporate such a receptor monosaccharide or oligosaccharide sequence. These molecules should have enough anti-adhesive activity to inhibit binding of low levels of pathogens. Whilst not a therapeutic option, such multifunctional prebiotics should increase host resistance to infection, reducing the likelihood of pathogen establishment and subsequent elaboration of virulence. They can thus be thought of as "decoy oligosaccharides" (Figure 4).

Targeted Prebiotics

The current concept of a prebiotic is an oligosaccharide that is selectively fermented by bifidobacteria and lactobacilli (Gibson and Roberfroid, 1995). Due to the difficulties of characterising the colonic microflora at the species level, virtually all of the data on prebiotic properties of oligosaccharides is on microflora changes at the genus level. It would, however, be highly desirable to develop prebiotics, which are targeted at particular species of *Bifidobacterium* and *Lactobacillus*. Such targeted prebiotics might be considered for several applications:

i) Synbiotics with defined health benefits. Many probiotic strains have been developed to have particular health benefits such as immune stimulation or anti-pathogen activity. In addition, commercial probiotic strains are selected for their survival characteristics such as resistance to acid and bile and their ability to be freeze-dried (Svensson, 1999; Lee *et al.*, 1999). Availability of prebiotics targeted at these strains would enable the development of synbiotic versions with enhanced survivability and colonisation in the gut.

ii) Infant formulae. It has long been known that the gut microflora of the breast-fed infant is dominated by bifidobacteria and that this is not the case for formula-fed infants (Cooperstock and Zedd, 1983; Benno *et al.*, 1984). This is thought to be one reason for the improved resistance to infection that the latter group experience. If prebiotics could be developed that have particular selectivity towards those bifidobacteria that are present in the gut of breast-fed infants, a new range of synbiotic formula foods could be envisaged.

iii) Functional foods for the elderly. Above the age of about 55-60 years, faecal bifidobacterial counts have been shown to markedly decrease compared to counts of younger people (Mitsuoka, 1990; Kleessen *et al.*, 1997). This decrease in bifidobacteria is a cause for concern as the natural elderly gut microflora may have become compromised through reduced bifidobacterial numbers resulting in a diminished ability to resist

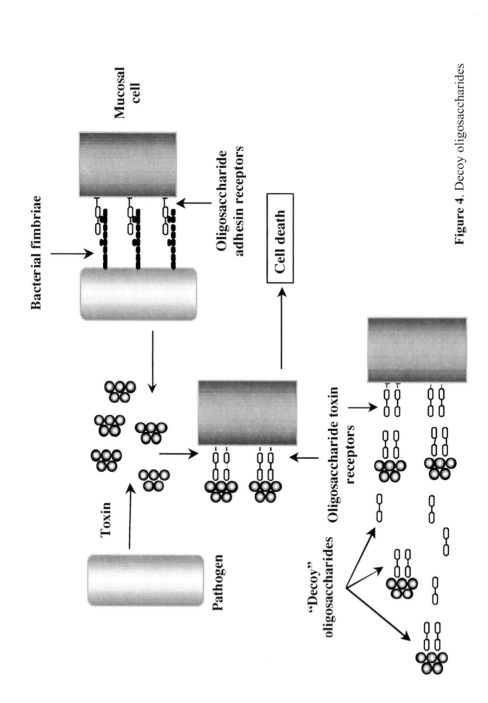

Figure 4. Decoy oligosaccharides

Table 7. Growth rates of selected gut bacteria on novel galacto-oligosaccharides

Probiotic	Growth rate h⁻¹, on oligosaccharide synthesised by						
	Oligomate	Bif. lactis	Bif. infantis	Bif. pseudolongum	Bif. adolescentis	Bif. angulatum	L. plantarum
L. acidophilus	0.69	1.16	0.93	0.57	1.30	0.72	0.54
L. plantarum	0.90	0.84	0.70	0.67	0.76	1.17	**0.83**
Bif. pseudolongum	0.56	0.69	0.66	**0.99**	0.64	0.69	0.44
Bif. longum	0.52	0.58	0.67	0.64	0.52	0.84	0.29
Bif. infantis	0.95	0.98	**1.20**	0.73	0.99	0.95	ND
Bif. lactis	0.95	**1.05**	0.76	0.79	0.73	1.05	0.59
Bif. angulatum	0.87	0.89	0.96	0.71	0.91	**1.27**	ND
Bif. adolescentis	1.02	0.48	0.48	0.81	0.83	1.01	0.46

Figures in bold indicate better growth on homologous oligosaccharide. ND: Not determined.

Table 8. Structures of prebiotic oligosaccharides

Oligosaccharide	Structure
Inulin	$Fru\beta2\rightarrow1Fru._n$, n = 1->60 $Fru\beta2\rightarrow1Fru._n\beta2\leftrightarrow1Glc$, n = 1->60
Lactulose	$Gal\beta1\rightarrow4Fru$
Galacto-oligosaccharides	Tri-to pentasaccharides with $Gal\beta1\rightarrow6Gal$ and $Gal\beta1\rightarrow3Gal$ linkages
Fructo-oligosaccharides	$Fru\beta2\rightarrow1Fru._n$, n = 1-5 $Fru\beta2\rightarrow1Fru._n\beta2\leftrightarrow1Glc$, n = 1-5
Isomalto-oligosaccharides	$Glc\alpha1\rightarrow6Glc._n$, n = 1-4
Soybean oligosaccharides	$Gal\alpha1\rightarrow6Glc\alpha1\leftrightarrow2\beta Fru$ $Gal\alpha1\rightarrow6Gal\alpha1\rightarrow6Glc\alpha1\leftrightarrow2\beta Fru$
Lactosucrose	$Gal\beta1\rightarrow4Glc\alpha1\leftrightarrow2\beta Fru$
Xylo-oligosaccharides	$Xyl\beta1\rightarrow4Xyl$

colonisation with invading pathogens. In this context it is noteworthy that the outbreak of *E. coli* O157:H7 in Lanarkshire, Scotland in 1996 resulted in 21 fatalities, all of them over 55 years of age. The development of functional foods for the elderly is currently the focus of the EU-funded project "Crownalife". This aims to characterise the gut microflora of elderly people across Europe, attempt to restore the youthful balance of the gut via a synbiotic intervention trial and also to develop prebiotics targeted at the bifidobacteria present in elderly people. The principle goal of the targeted prebiotic development is increased anti-pathogen activity.

Targeted prebiotics might be developed through one of two strategies. Screening an enhanced range of oligosaccharide structures for their prebiotic properties might identify prebiotics targeted toward particular species of probiotic. To this end, economical manufacturing technologies for complex oligosaccharides are required. These will allow the generation of wide structural diversity in candidate oligosaccharides for selectivity testing.

A more rational means of generating targeted prebiotics, however, depends upon the use of enzymes synthesised by the probiotic strains themselves. We expect that these enzymes will produce mixtures of oligosaccharide products, which are more readily hydrolysed by the cell-associated enzymes synthesised by the probiotic. In this way, it might be possible to produce oligosaccharide mixtures that are very selective, perhaps even for a single species of *Bifidobacterium* or *Lactobacillus*. Preliminary data using this approach are encouraging. Novel galacto-oligosaccharide mixtures have been synthesised using β-galactosidases from a range of probiotics (Rabiu *et al.*, 2001). Lactose was used as a glycosyl donor and acceptor in enzymatic synthesis reactions (see below) and growth of the probiotics tested on each synthetic mixture in pure culture (Table 7). Many probiotics had a higher growth rate on oligosaccharides synthesised by their own enzymes relative to other probiotics on the same substrate. The success of this approach in mixed culture will be tested when species-specific oligonucleotide probes become available.

Manufacture of Prebiotics

Current Technologies

There is a range of oligosaccharide structures that have found application as prebiotics (Table 8), although the prebiotic status of some of these has yet to be confirmed using rigorous molecular methodologies and in human volunteers.

These various molecules are manufactured using a range of extraction and chemical and enzymatic synthesis methods utilising cheap substrates.

Extraction

The simplest approach to the manufacture of prebiotics is to extract them from a biological material. This is currently performed on a commercial basis with inulin extracted from chicory (De Leenheer, 1994) and with raffinose and stachyose extracted from soy beans (Koga *et al.*, 1993b). Several companies exist in Europe, the largest manufacturer being Orafti in Tienen, Belgium, that extract inulin from chicory. The chicory extraction process is similar to the sugar beet process (De Leenheer, 1994). Chicory chips are extracted with hot water and the extracted chips are sold as animal feed. Proteins, peptide, anions, colloids and phosphates are then removed by liming and carbonatation at alkaline pH. The inulin is then demineralised by anion

$$\text{Gal}\beta1{\rightarrow}4\text{Glc} \xrightarrow{\ \ \ HO^-\ \ \ } \text{Gal}\beta1{\rightarrow}4\text{Fru}$$

Figure 5. Lobry de Bruyn-Alberda van Ekenstein isomerisation of lactose to lactulose

and cation exchange chromatography and decolourised by chromatography on activated carbon. The final product is sterilised, concentrated and spray-dried.

Soybean oligosaccharides are also extracted from their biological source with no further modification. They are the non-reducing α-galactosyl sucrose derivatives raffinose and stachyose. They are isolated from soybean whey (Koga *et al.*, 1993b) on a commercial basis by the Calpis Food Industry Co in Japan.

Chemical Approaches

Lactulose is currently the only prebiotic manufactured using a chemical rather than enzymatic approach. It is made by a base-catalysed Lobry de Bruyn-Alberda van Ekenstein isomerisation of the glucosyl moiety of the lactose to form fructose (Figure 5). The reaction results in both the pyranose and furanose forms of the lactulose. The reaction is catalysed by either sodium hydroxide or borate (Timmermans, 1994). Reaction rates are carefully controlled to minimise degradative side reactions. The Morinaga Milk Industry is the largest producer in Japan, while in the EU lactulose is manufactured by Solvay.

Enzyme Technology

The enzymatic manufacture of oligosaccharides proceeds via two general approaches, the hydrolysis of polysaccharides or synthesis of higher oligosaccharides from readily available disaccharides. Chemical methods of carbohydrate synthesis are generally inferior to enzyme-mediated approaches for industrial production. Chemical methods frequently suffer from side reactions resulting in unwanted colour and flavour compounds which must be removed in further refining steps. The field of food-grade oligosaccharide production is proving to be a fertile area for innovation in enzyme technology and there are likely to be new processes developed in the future (see below).

Polysaccharide Hydrolysis

There are currently two prebiotic oligosaccharides that are manufactured from their parent polysaccharides. Fructo-oligosaccharides can be manufactured from inulin and xylo-oligosaccharides from xylan. Fructo-oligosaccharides are manufactured in Europe by partial enzymatic hydrolysis of chicory inulin (De Leenheer, 1994). The fungal inulinase used gives low production of monosaccharides. As inulin contains chains terminating in either a reducing fructose residue or a non-reducing glucose residue, FOS produced from inulin have a degree of reducing activity.

Xylo-oligosaccharides are manufactured commercially in Japan by enzymatic hydrolysis of xylan from corn cobs (Koga *et al.*, 1993a). Other sources of xylan could be envisaged as a source of xylo-oligosaccharides, such as oat spelt xylan or wheat arabinoxylan, but these have not yet found commercial application. The xylan is extensively hydrolysed to its repeating unit, the disaccharide xylobiose although smaller quantities of higher oligosaccharides are also formed. Membrane processes to remove xylose and high molecular weight components can further purify the xylobiose.

Enzymatic Synthesis

Some prebiotics are manufactured commercially using enzymatic transfer reactions. These prebiotics include the isomalto-oligosaccharides, galacto-oligosaccharides, lactosucrose and some fructo-oligosaccharides (Playne and Crittenden, 1996). The reactions utilise cheap sugars such as sucrose and lactose as donors and acceptors.

Fructo-oligosaccharides are made commercially by Meiji Seika in Japan and in Europe by a collaboration between Meiji Seika and Eridania Beghin Say (Beghin-Meiji Industries). Fructosyltransferase from *Aureobasidium pullulans* or *Aspergillus niger* is used to build up higher fructo-oligosaccharides (Figure 6) from a 60 % (w/v) sucrose solution at 50-60°C in an immobilised cell-based reactor (Kono, 1993; Yun, 1996). All of the sucrose-derived FOS terminates in a non-reducing glucose residue. Higher product purities can be achieved by removal of the glucose and sucrose by ion-exchange chromatography (Kono, 1993).

Galacto-oligosaccharides are manufactured commercially from lactose using β-galactosidase (Matsumoto, 1993). This enzyme is a hydrolase enzyme

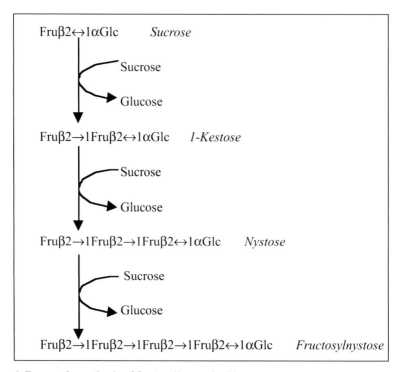

Figure 6. Enzymatic synthesis of fructo-oligosaccharides

and hence works by transferring galactose from lactose to water. Under conditions of high lactose concentration the enzyme utilises lactose as an alternative acceptor to water resulting in the formation of galacto-oligosaccharides (Timmermans, 1994). A variety of enzyme reactor configurations based upon free or immobilised β-galactosidases have been used to make galacto-oligosaccharides. A concentrated lactose solution is used and product oligosaccharide DP values range between 2 and 6 depending on the enzyme used. The products are complex mixtures principally containing $\beta1\rightarrow3$ and $\beta1\rightarrow6$ linkages. The largest manufacturers of transgalacto-oligosaccharides are Yakult Honsha and Snow Brand in Japan and Borculo Domo Ingredients in Europe.

A similar manufacturing technology is used to make lactosucrose (Kitahata and Fujita, 1993). In this case a mixture of sucrose and lactose is incubated with β-fructofuranosidase at high concentration. The fructosyl residue is transferred from the sucrose to the 1-position of the glucose moiety in the lactose, producing a non-reducing oligosaccharide (Figure 7). Lactosucrose is produced commercially by Hayashibara Shoji Inc. and Ensuiko Sugar refining Co. in Japan.

$$\text{Gal}\beta1\rightarrow4\text{Glc} + \text{Glc}\alpha1\leftrightarrow2\beta\text{Fru} \rightarrow \text{Gal}\beta1\rightarrow4\text{Glc}\alpha1\leftrightarrow2\beta\text{Fru} + \text{Glc}$$

β-Fructofuranosidase

Figure 7. Enzymatic synthesis of lactosucrose

Isomalto-oligosaccharides are also manufactured using enzymatic synthesis, albeit by a more complex process (Yatake, 1993). Firstly, starch is hydrolysed to malto-oligosaccharides by the combined action of α-amylase and pullulanase. This oligosaccharide mixture is then used as substrate in an enzymatic synthesis reaction. The α1→4 linked malto-oligosaccharides are converted into α1→6 linked isomalto-oligosaccharides by α-glucosidase. Glucose is removed from the mixture using chromatography to give the final IMO products. IMO is manufactured commercially by Showa Sangyu Co. in Japan. The commercial product consists of a mixture of IMO of differing molecular weight together with panose.

Emerging Technologies for Second Generation Prebiotics

Manufacturing technologies for functional carbohydrates are still developing. New technological approaches are likely to enable the development of the functionally enhanced second generation prebiotics discussed above.

Controlled Polysaccharide Hydrolysis

As discussed above, polysaccharide hydrolysis is used commercially to manufacture FOS and XOS as prebiotics. A hydrolysis process, which gave a higher degree of control over the molecular weight distribution of the products, however, would enable several of the functional enhancements discussed above. Control over the molecular weight distribution will influence the rheological properties and hence the technological application of the oligosaccharides and might lead to persistence of selective fermentation throughout the colon, i.e. towards distal regions - which would be a desirable attribute for a prebiotic (see above).

A promising approach to the controlled hydrolysis of polysaccharides is the use of endo-glycanases in enzyme membrane reactors (EMR, Figure 8). Such reactor systems have been established for the partial hydrolysis of dextran (Mountzouris *et al.,* 1998, 2001) and pectin (Olano-Martin *et al.,* 2001). By controlling factors such as residence time and the ratio of enzyme to substrate,

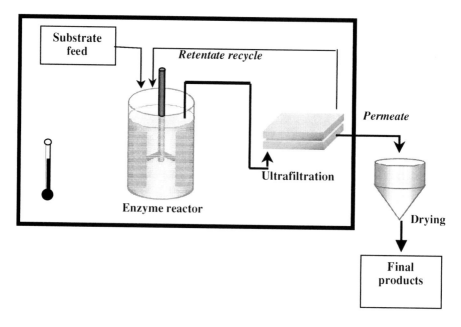

Figure 8. Enzyme membrane reactors for the controlled hydrolysis of polysaccharides

it was possible to convert dextran into different oligodextran preparations with average molecular weights varying from trisaccharide up to 12,000 Daltons. These have proven to be useful prebiotics in gut model systems (Olano-Martin *et.al.*, 2000) with selective fermentation extending through to the third vessel (i.e. mimicking the distal colon).

This approach can be extended to other polysaccharide systems. Of particular interest would be the use of bacterial extracellular polysaccharides (EPS). These polysaccharides display a huge degree of complexity and can often be manufactured economically by fermentation processes (Sutherland, 1975; Morin, 1998). They also contain a variety of rare sugars and a very wide range of glycosyl linkages. Suitable enzymes for the degradation of these polysaccharides can be isolated from bacteriophage. The use of phage-borne enzymes in EPS degradation has been used as an aid to the sequencing of polysaccharides (Stirm, 1994; Geyer *et al.*, 1983; Sutherland, 1999) but the approach has not been developed into a commercial manufacturing technology. The use of phage enzymes together with other glycosidases to "remodel" the resultant oligosaccharides could ultimately generate a huge diversity of oligosaccharide structures. Such a structure library could then be screened for enhanced selectivity towards probiotics at the specific level (see above).

Dextransucrase

Glcα1↔2βFru + Glcα1→4Glc   Glcα1→2Glcα1→6Glcα1→4Glc + Fru

Figure 9. Synthesis of novel hetero-oligosaccharides with dextransucrase

Enzymatic Synthesis of Oligosaccharides

There are several approaches to the enzymatic synthesis of oligosaccharides, each of which may potentially be developed into a practical manufacturing technology for novel prebiotics (Rastall and Bucke, 1992; Kren and Thiem, 1997; Crout and Vic, 1998). Many of the approaches that have been used at a laboratory scale are under investigation as to the feasibility of large-scale application.

Glycosyltransferases

Glycosyltransferases are divided into enzymes that utilise sugar nucleotides as donors, the so-called Leloir enzymes (Rastall and Bucke, 1992) and those utilising sugars such as sucrose as a glycosyl donor. These enzymes have not yet realised their full potential for the manufacture of novel prebiotics. Dextransucrase, for example has been used to generate novel selectively fermented oligosaccharides (Dols *et al.*, 1999). The enzyme normally produces dextran from sucrose by transfer of the glucosyl moiety. Under appropriate conditions, however, it can transfer glucose to an oligosaccharide acceptor such as maltose (Figure 9).

Such an approach has been used commercially to produce oligosaccharides that are selectively fermented by selected skin micro-organisms (Dols *et al.*, 1999). They have also been tested to some degree using gut micro-organisms (Djouzi *et al.*, 1995, 1997), however, the potential of these oligosaccharides as prebiotic food ingredients has not yet been fully established.

Glycosidases

Glycosidases are generally thought of as degradative enzymes. They can, however, be induced to act "in reverse" and bring about the synthesis of oligosaccharides (Rastall and Bucke, 1992; Kren and Thiem, 1997; Crout and Vic, 1998). There are actually two approaches to the manufacture of oligosaccharides by glycosidases.

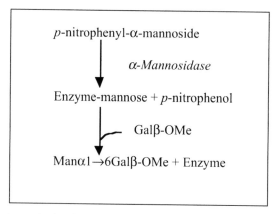

Figure 10. Kinetic synthesis of novel hetero-oligosaccharides

Kinetic synthesis

As discussed above for β-galactosidase, glycosidases catalyse hydrolysis of glycosides by transfer of a monosaccharide moiety from a reactive donor to water (McCarter and Withers, 1994). Under appropriate circumstances, they can be induced to transfer to another sugar to form an oligosaccharide (Figure 10). This most commonly involves the use of a reactive donor such as a *p*-nitrophenyl glycoside and a suitable acceptor (Rastall *et al.*, 1992).

The commercial application of kinetic synthesis to date has depended upon the use of cheap glycosyl donors such as lactose. The principal obstruction to the development of this approach is this need for cheap donors: this restricts the choice of types of glycosidase that can be used. Use of enzymes for which there is no readily available disaccharide donor would demand the use of a more expensive form such as a *p*-nitrophenyl glycoside. This expensive donor is used in greater than stoichiometric quantities and cannot be recycled. The approach may, however, be developed in the future for the synthesis of pathogen and toxin receptors for medical food applications; the trisaccharide receptor for toxin A from *Clostridium difficile* has been synthesised using this approach (Vic *et al.*, 1997).

Equilibrium synthesis

This is a potentially attractive means of manufacturing novel prebiotics and receptor-active oligosaccharides. The approach taken here is to use glycosidases in very highly concentrated sugar solutions at elevated

$$Man + Glc\alpha1\leftrightarrow2\beta Fru \rightleftharpoons Man\alpha1\rightarrow6Glc\alpha1\leftrightarrow2\beta Fru + H_2O$$

α-mannosidase/60°C/80 % w/w. carbohydrate

Figure 11. Equilibrium approach to enzymatic oligosaccharide synthesis

temperature. Under these conditions, the normal reaction equilibrium is shifted in the direction of condensation of saccharides to form higher oligosaccharides (Figure 11).

The equilibrium yields are around 10-25 % for hetero-oligosaccharides (Rastall *et al.*, 1992; Suwasono and Rastall, 1998a), but can be much higher at around 70% for homo-oligosaccharides (Johansson *et al.*, 1989). An important point, however is that the starting materials are not destroyed and can be recycled, resulting in much higher process yields. This approach, in principle, allows the synthesis of a wide range of hetero-oligosaccharides, as many enzymes will recognise a range of sugars as acceptors. Using, for example, α-mannosidase, it is possible to attach a mannose residue onto a large number of disaccharide acceptors to manufacture novel trisaccharides (Rastall *et al.*, 1992; Smith *et al.*, 1997; Suwasono and Rastall, 1998a). The biggest disadvantage with this approach is that the regioselectivity of the reaction is often low with the formation of mixtures of oligosaccharides containing many linkage isomers (Johansson *et al.*, 1989; Rastall *et al.*, 1992). Regioselectivity can, however, be regulated to some degree by a rational choice of enzyme and acceptor (Suwasono and Rastall, 1996; Suwasono and Rastall, 1998a). By this means it is possible to manufacture receptor active oligosaccharides such as the man$\alpha1\rightarrow2$man di- and trisaccharides (Suwasono and Rastall, 1996) which act as receptor for type-1 fimbriated enteric micro-organisms.

In order to realise the potential of such reactions, there are engineering challenges to be overcome. Some reaction engineering studies have been performed to date (Smith *et. al.*, 1997; Suwasono and Rastall 1998b), but more are required to design practical enzyme reactors for viscous sugar solutions. Isolation of the products is also a problem, although recent research into the use of nanofiltration as a means of fractionating oligosaccharides appears encouraging.

Fermentation and Metabolic Engineering

The production of prebiotic oligosaccharides by fermentation of genetically modified micro-organisms has tremendous potential. Although many micro-organisms produce EPS, very few are known to secrete oligosaccharides. One way of potentially achieving this is through the use of metabolic engineering. Globo-oligosaccharides (Galα1→4Galα1→4Gal) have been produced by co-culturing two metabolically engineered bacteria (Koizumi et al., 1998). Recombinant *E. coli* have been produced that overexpress the UDP-Gal biosynthetic genes. UDP-Gal is the glycosyl donor for galactosyltransferase and is produced in high levels by the recombinant *E. coli* in culture. Co-culture of this organism with another recombinant *E. coli*, expressing the α1,4-transferase genes from *Neisseria gonorrhoea*, resulted in manufacture of the globotriose from erotic acid, galactose and lactose. This approach is obviously expensive in terms of research and development, but the cost of the fermentation product is likely to be competitive.

Concluding Remarks

Prebiotic oligosaccharides are set to make a further impact on the area of functional foods. Current forms are efficient at stimulating bifidobacteria, however the next generation of prebiotics will entail multiple functionality. Biotechnological procedures are increasingly being applied to the area and developments with novel foods are proceeding quickly. It is clear that, for the further development of this field, reliable testing methods are used. One example is through the application of molecular procedures to improve our knowledge of the gut microflora, its interactions with the human host and its role in maintenance of health. Advances in these areas will ultimately provide us with very sophisticated dietary tools with which to manipulate this important ecosystem and help to improve public health.

References

Alles, M.S., Hartemink, R., Meyboom, S., Harryvan, J.L., van Laere, K.M.J.and Nagengast, F.M. 1999. Effect of transgalactooligosaccharides on the composition of the human intestinal microflora and on putative risk markers for colon cancer. Am. J. Clin. Nutr. 69: 980-91.

Anand, S.K., Srinivasan, R.A. and Rao, L.K. 1985. Antibacterial activity associated with *Bifidobacterium bifidum* - II. Cultured Dairy Products J. 2: 21-23.

Araya-Kojima, A., Yaeshima, T., Ishibashi, N., Shimamur, S. and Hayasawa, H. 1995. Inhibitory effects of *Bifidobacterium longum* BB536 on harmful intestinal bacteria. Bifidobacteria Microflora 14: 59-66.

Benno, Y., Sawada, K. and Mitsuoka, T. 1984. The intestinal microflora of infants: composition of fecal microflora in breast-fed and bottle-fed infants. Microbiology and Immunology 28: 975-986.

Bouhnik, Y., Flourie, B., Riottot, M., Bisetti, N., Gailing, M.F., and Guibert, A. 1996. Effects of fructo-oligosaccharides ingestion on faecal bifidobacteria and selected metabolic indexes of colon carcinogenesis in healthy humans. Nutr. Cancer 26: 21-29.

Buddington, R.K., Williams, C.H., Chen, S.-C. and Witherly, S.A. 1996. Dietary supplementation of neosugar alters the fecal microflora and decreases activities of some reductive enzymes in human subjects. Am. J. Clin. Nutr. 63: 709-716.

Collins, M.D. and Gibson, G.R. 1999.. Probiotics, prebiotics and synbiotics: Dietary approaches for the modulation of microbial ecology. American J. Clin. Nutrition 69: 1052-1057.

Cooperstock, M.S., and Zedd, A.J. 1983. Intestinal microflora of infants. In: Human Intestinal Microflora in Health and Disease. Hentges, D.J., ed.. Academic Press, London. p. 79-99.

Crittenden, R.G. 1999. Prebiotics. In: Probiotics: A Critical Review. Tannock, G., ed.. Horizon Scientific Press, Wymondham. p. 141-156.

Crittenden, R.G. and Playne, M.J. 1996. Production, properties and applications of food-grade oligosaccharides. Trends in Food Sci. Technol. 7: 353-361.

Crout, D.H.G., and Vic, G. 1998. Glycosidases and glycosyl transferases in glycoside and oligosaccharide synthesis. Curr. Opin. Chem. Biol. 2: 98-111.

Cummings, J.H., and Macfarlane, G.T. 1991. The control and consequences of bacterial fermentation in the human colon. J. Applied Bacteriol. 70: 443-459.

Cummings, J.H., Gibson, G.R. and Macfarlane, G.T. 1989. Quantitative estimates of fermentation in the hind gut of man. Acta Veterinaria Scandanavica 86: 76-82.

De Leenheer, L. 1994. Production and use if inulin: industrial reality with a promising future. In: Carbohydrates as Organic Raw Materials III. van Bekkum, H., Roper, H. and Voragen, A.G.J., eds. VCH, Weinheim. p. 67-92.

Delzenne, N.M. and Kok, N.N. 1999. Biochemical basis of oligofructose-induced hypolipidaemia in animal models. J. Nutr. 129:1467S-1470S.

Delzenne, N.M. and Williams, C.M. 1999. Actions of non-digestible carbohydrates on blood lipids in humans and animals. In: Colonic Microflora, Nutrition and Health. Gibson, G.R., and Roberfroid, M.B., eds. Dordrecht, Kluwer. p. 213-231.

Dols, M., Monchois, V., Remaud-Simeon, M., Willemot, R.-M. and Monsan, P.F. 1999. The production of α1→2-terminated Glucooligosaccharides. In: Carbohydrate Biotechnology Protocols. Bucke C., ed. Methods in Biotechnology 10: 129-139.

Djouzi, Z., Andrieux, C., Pelenc, V., Somarriba, S., Popot, F., Paul, F., Monsan, P. and Szylit, O. 1995. Degradation and fermentation of α-gluco-oligosaccharides by bacterial strains from human colon: in vitro and *in vivo* studies in gnotobiotic rats. J. Applied Bacteriol. 79: 117-127.

Djouzi, Z. and Andrieux, C. 1997. Compared effects of three oligosaccharides on metabolism of intestinal microflora in rats inoculated with a human faecal microflora. British J. Nutr. 78: 313-324.

Finlay, B.B. and Falkow, S. 1989. Common themes in microbial pathogenicity. Microbiol. Rev. 53: p. 210-230.

Franks, A.H., Harmsen, H.J.M., and Raangs, G.C. 1998. Variations of bacterial populations in human feces measured by fluorescent *in situ* hybridization with group-specific 16S rRNA-targeted oligonucleotide probes. Applied Environ. Microbiol. 64: 3336-3345.

Geyer, H., Himmelspach, K., Kwiatkowski, B., Schlecht, S. and Stirm, S. 1993. Degradation of bacterial surface carbohydrates by virus associated enzymes. Pure and Applied Chemistry 55: 637-653.

Gibson, G.R., Berry Ottaway, P. and Rastall, R.A. 2000. Prebiotics: New Developments in Functional Foods. Chandos Publishing Limited, Oxford.

Gibson, G.R. and Collins, M.D. 1999. Concept of balanced colonic microflora, prebiotics and synbiotics. In: Probiotics, Other Nutritional Factors and Intestinal Microflora. Hanson, L.A. and Yolken, R.H., eds.. Nestle Nutrition Workshop Series, Vol. 42, Nestec Ltd. Vevey/Lippincott-Raven Publishers, Philadelphia. p. 139-156.

Gibson, G.R. and Macfarlane, G.T. 1994. Intestinal bacteria and disease. In: Human Health - The Contribution of Microorganisms. S.A.W. Gibson., ed. Springer-Verlag, London. p. 53-62.

Gibson, G.R. and Roberfroid, M.B. 1995. Dietary modulation of the human colonic microflora: introducing the concept of prebiotics. J. Nutr. 125: 1401-1412.

Gibson, G.R. and Wang, X. 1994a. Enrichment of bifidobacteria from human gut contents by oligofructose using continuous culture. FEMS Microbiology Letts. 118: 121-128.

Gibson, G.R. and Wang, X. 1994b. Regulatory effects of bifidobacteria on the growth of other colonic bacteria. J. Applied Bacteriol. 77: 412-420.

Han, Y.W., 1990. Microbial Levan. In: Advances in Applied Microbiology 35. Neidleman, S.L., and Laskin, A.J., eds. Acadedmic Press, London. p. 171-194.

Harmsen, H.J.M., Gibson, G.R., Elfferich, P., Raangs, G.C., Wideboer-Veloo, A.C.M., Argaiz, A., Roberfroid, M.B. and Welling, G.W. 1999. Comparison of viable cell counts and fluorescent *in situ* hybridization using specific rRNA-based probes for the quantification of human fecal bacteria. FEMS Microbiology Ecol. 183: 125-129.

Harmsen, H.J.M., Wildeboer, A.C.M., Raangs, G.C., Wagendorp, A.A., Klijn, N., Bindels, J.C. and Welling, G.W. 2000. Analysis of intestinal microflora development in breast-fed and formula-fed infants by using molecular identification and detection methods. Journal of Pediatric Gastroenterol. Nutr. 30: 61-67.

Heerze, L.D., Kelm, M.A. and Talbot, J.A. 1994. Oligosaccharide sequences attached to an inert support SYNSORB. as potential therapy for antibiotic-associated diarrhoea and pseudomembranous colitis. J. Infect. Dis. 169: 1291-1296.

Hylla, S. Gostner, A., Dusel, G., Anger, H., Bartram, H.-P., Christl, S.U., *et al.* 1998. Effects of resistant starch on the colon in healthy volunteers: possible implications for cancer prevention. Am. J. Clin. Nutr. 67: 136-42.

Jayaraman, N., Nepogodiev, S.A. and Stoddart, J.F. 1997. Synthetic carbohydrate-containing dendimers. Chem, Eur. J. 3: 1193-1199.

Jaskari, J., Kontula, P., Siitonen, A., Jousimies-Somer, H., Mattila-Sandholm, T. and Poutanen,K. 1998. Oat β-glucan and xylan hydrolysates as selective substrates for *Bifidobacterium* and *Lactobacillus* strains. Applied Microbiol. Biotechnol. 49: 175-181.

Johansson, E., Hedbys, L., Mosbach. K., Larsson, P.-O., Gunnarsson, A. and Svensson, S. 1989. Studies on the reversed α-mannosidase reaction at high concentrations of mannose. Enzyme and Microbial Technol. 11: 347-352.

Karlsson, K.-A., 1989. Animal glycosphingolipids as membrane attachment sites for bacteria. Annual Rev. Biochem. 58: 309-350.

Kim, Y.S., Tsao, D., Morita, A. and Bella, A. 1982. Effect of sodium butyrate and three human colorectal adenocarcinoma cell lines in culture. Falk Symposium 31: 317-323.

Kitahata, S. and Fujita, K. 1993. Xylsucrose, isomaltosucrose and lactosucrose. In: Oligosaccharides. Production, Properties and Applications. Japanese Technol. Rev. 3: 158-174.

Kleessen, B., Sykura, B., Zunft, H-J. and Blaut, M. 1997. Effects of inulin and lactose on fecal microflora, microbial activity and bowel habit in elderly constipated persons. Am. J. Clin. Nutr. 65: 1397-1402.

Koga, K., and Fujikawa, S. 1993a. Soybean oligosaccharides. In: Oligosaccharides. Production, Properties and Applications. Japanese Technol. Rev. 3: 130-143.

Koga, Y., Shibuta, T. and O'Brien, R. 1993b. Soybean oligosaccharides. In: Oligosaccharides. Production, Properties and Applications. Japanese Technol. Rev. 3: 90-106.

Koga, T. 1993. Fructooligosaccharides. In: Oligosaccharides. Production, Properties and Applications. Japanese Technol. Rev. 3: 50-78.

Koizumi, S., Endo, T., Tabata, K. and Ozaki, A. 1998. Large-scale production of UDP-galactose and globotriose by coupling metabolically engineered bacteria. Nature Biotechnol. 16: 847-850.

Kren, V., and Thiem, J. 1997. Glycosylation emplying bio-systems: from enzymes to whole cells. Chem. Soc. Rev. 26: 463-473.

Kullen, M.J. and Klaenhammer, T. 1999. Genetic modification of lactobacilli and bifidobacteria. In: Probiotics: A Critical Review. Tannock, G. ed.. Horizon Scientific Press, Wymondham. p. 65-83.

Langendijk, P.S., Schut, F., Jansen, G.J., Raangs, G.C., Kamphuis, G.R., Wilkinson, M.H.F. and Welling, G.W. 1995. Quantitative fluorescence *in situ* hybridisation of *Bifidobacterium* spp. with genus-specific 16S rRNA-targeted probes and its application in fecal samples. Applied and Environ. Microbiol. 61: 3069-3075.

Lee, Y.-K., Nomoto, K., Salminen, S. and Gorbach, S.L. 1999. Handbook of Probiotics, Wiley, New York.

Macfarlane, G.T., Gibson G.R. and Cummings J.H 1992. Comparison of fermentation reactions in different regions of the colon. J. Appl. Bacteriol. 72: 57-64.

Macfarlane, G.T., Macfarlane, S. and Gibson, G.R. 1998. Validation of a three-stage compound continuous culture system for investigating the effect of retention time on the ecology and metabolism of bacteria in the human colonic microflora. Microbial Ecology 35: 180-187.

Manz, W., Amann, R. and Ludwig, W. 1996. Application of a suite of 16S rRNA-specific oligonucleotide probes designed to investigate bacteria of the phylum cytophaga-flavobacter-bacteroides in the natural environment. Microbiology 142: 1097-1106.

Matsumoto, K.1993. Galactooligosaccharides. Oligosaccharides. Production, properties and applications. Japanese Technol. Rev. 3: 90-106.

McBain, A.J. and Macfarlane, G.T. 1997. Investigations of bifidobacterial ecology and oligosaccharide metabolism in a three-stage compound continuous culture system. Scandinavian J. Gastroenterol. 32: 32-40.

McCarter, J.D. and Withers, S.G. 1994. Mechanisms of enzymatic glycoside hydrolysis. Curr. Opin. Structural Biol. 4: 885-897

McCartney, A.L., Wenzhi, W. and Tannock, G.W. 1996. Molecular analysis of the composition of the bifidobacterial and lactobacillus microflora of humans. Applied and Environ. Microbiol. 62: 4608-4613.

Mitsuoka, T. 1990. Bifidobacteria and their role in human health. J. Industrial Microbiol. 6: 263-268.

Molly, K., Vande Woestyne, M. and Verstraete, W. 1993. Development of a 5-step multi-chamber reactor as a simulation of the human intestinal microbial ecosystem. Applied Microbiol. Biotechnol. 139: 254-258.

Morin, A. 1998. Screening of polysaccharide producing organisms, factors influencing the production and recovery of microbial polysaccharides. In: Polysaccharides: structural diversity and functional versatility. Dimitiu, S., ed. Marcel Dekker Inc., New York.

Morotomi, M., Guillem, J.G., LoGerfo, P. and Weinsten, I.B. 1990. Production of diacylglycerol, an activator of protein kinase C by human intestinal microflora. Cancer Research 50: 3595-3599.

Mountzouris, K.C. Gilmour, S.G. Grandison, A.S. and Rastall, R.A. 1998. Modelling of oligodextran production in an ultrafiltration stirred cell membrane reactor. Enzyme and Microbial Technol. 24: 75-85.

Mountzouris, K.C. Gilmour, S.G. and Rastall, R.A. 2002. Continuous Production of Oligodextrans via controlled hydrolysis of dextran in an enzyme membrane reactor J. Food Sci. In Press.

Okazaki, M., Fujikawa, S. and Matsumoto, N. 1990. Effects of xylooligosaccharide on growth of bifidobacteria. J. Japanese Soc. Nutr. Food Sci. 43: 395-401.

Olano-Martin, E., Mountzouris, K.C., Gibson, G.R. and Rastall, R.A. 2000. *In vitro* fermentability of dextran, oligodextran and maltodextrin by human gut bacteria. Brit. J. Nutr. 83: 247-255.

Olano-Martin, E., Mountzouris, K.C., Gibson, G.R. and Rastall, R.A. 2001. Continuous production of oligosaccharides from pectin in an enzyme membrane reactor. J. Food Sci. 66 7: 966-971.

O'Sullivan, D.J. 1999. Methods for analysis of the intestinal microflora. In: Probiotics: A Critical Review. Tannock, G., ed.. Horizon Scientific Press, Wymondham. p. 23-44.

Perdigon, G. and Alvarez, S. 1992. Probiotics and the immune state, In: Probiotics: The Scientific Basis. Fuller, R., ed. Chapman and Hall, London. p. 146-180.

Playne, M.J. and Crittenden, R. 1996. Commercially available oligosaccharides, Bulletin of the Internat. Dairy Found. 313: 10-22.

Porter, K.G. and Feig, Y.S. 1980. The use of DAPI for identifying and counting aquatic microflora. Limnology and Oceanography 25: 943-948.

Prasad, K.N. 1980. Butyric acid: a small fatty acid with diverse biological functions. Life Sciences 27: 1351-1358.

Rabiu, B.A., Jay, A.J., Gibson, G.R. and Rastall, R.A. 2001. Synthesis and fermentation properties of novel galacto-oligosaccharides by beta-galactosidases from bifidobacterium species. Applied and Environ. Microbiol. 67: 2526-2530.

Rastall, R.A., and Bucke, C. 1992. Enzymatic synthesis of oligosaccharides. Biotechnol. Genetic Eng. Rev. 10: 253-281.

Rastall, R.A., Rees, N.H., Wait, R., Adlard, M.W., and Bucke, C. 1992. Alpha-mannosidase-catalysed synthesis of novel manno-, lyxo-, and heteromanno-oligosaccharides: A comparison of kinetically and thermodynamically mediated approaches. Enzyme and Microbial Technol. 14: 53-57.

Reddy, B.S., Hamid, R. and Rao, C.V. 1997. Effect of dietary oligofructose and inulin on colonic preneoplastic aberrant crypt foci inhibition. Carcinogenesis 18:1371-74.

Roberfroid, M.B., Van Loo, J.A.E., Gibson, G.R. 1998. The bifidogenic nature of inulin and its hydrolysis products. J. Nutr 128: 11-19.

Rowland, I.R., ed. 1988. Role of the Gut Microflora in Toxicity and Cancer. Academic Press, London.

Rowland, I.R. Metabolic interactions in the gut. 1992. In: Probiotics: the Scientific Basis. Fuller, R. ed. Chapman and Hall, Andover. p. 29-53.

Rowland, I.R. and Tanaka, R.1993. The effects of transgalactosylated oligosaccharides on gut microflora metabolism in rats associated with a human faecal microflora, J. Applied Bacteriol. 74: 667-674.

Rycroft, C.E., Jones, M.R., Gibson, G.R. and Rastall, R.A. 2001. A comparative *in vitro* evaluation of the fermentation properties of prebiotic oligosaccharides. J. Applied Microbiol. 91 (5): 878-887.

Salminen, S., Bouley, C., Boutron-Ruault, M.-C., Cummings, J.H., Franck, A., Gibson, G.R., Isolauri, E., Moreau, M.-C., Roberfroid, M. and Rowland, I.R. 1998. Functional food science and gastrointestinal physiology and function. British J. Nutr. 80: S147-S171.

Salminen, S., and Salminen, E. 1997. Lactulose, lactic acid bacteria, intestinal microecology and mucosal protection. Scand. J. Gastroenterol. 32: 45-48.

Suau, A., Bonnet, R., Sutren , M., Godon, J.J., Gibson, G.R., Collins, M.D. and Dore, J. 1999. Direct analysis of genes encoding 16S rRNA from complex communities reveals many novel molecular species within the human gut. Appl. Environ. Microbiol. 65: 4799-4807.

Schiffrin, E.J., Rochat, F., Link-Amster, H., Aeschlimann, J.M. and Donnet-Hughes, A. 1995. Immune system stimulation by probiotics. J. Dairy Sci. 78: 1597-1606.

Schoterman, H.C. and Timmermans, H.J.A.R. 2000. Galacto-oligosaccharides. In: Prebiotics and Probiotics. LFRA Ingredients

Handbook. G.R. Gibson and F. Angus, eds. Leatherhead Food RA Publishing. p. 19-46.

Smith, N.K., Gilmour, S.G. and Rastall, R.A. 1997.Statistical optimisation of enzymatic synthesis of derivatives of trehalose and sucrose. Enzyme and Microbial Technol. 21 (5): 349.

Stirm, S. 1994. Examination of the repeating units of bacterial exopolysaccharides. In: Methods in Carbohydrate Chemistry, Volume X. BeMiller, J.N., Manners, D.J. and Sturgeon, R.J., eds. John Wiley.

Sutherland, I. 1975. Surface carbohydrates of the prokaryotic cell. Academic Press, New York.

Sutherland, I. 1999. Polysaccharases for microbial exopolysaccharides. Carbohydrate Polymers 38: 319-328.

Suwasono, S. and Rastall, R.A. 1996. A highly regioselective synthesis of mannobiose and mannotriose by reverse hydrolysis using specific 1,2-alpha-mannosidase from *Aspergillus phoenicis*. Biotechnol. Letts. 18: 851-856.

Suwasono, S. and Rastall, R.A. 1998a. Enzymatic synthesis of manno- and heteromanno-oligosaccharides using α-mannosidases: a comparative study of linkage-specific and non-linkage-specific enzymes. J. Chem.Technol. Biotechnol. 73 (1): 37-42.

Suwasono, S. and Rastall, R.A. 1998b. Synthesis of oligosaccharides using immobilised 1,2-alpha- mannosidase from *Aspergillus phoenicis*: Immobilisation-dependent modulation of product spectrum. Biotechnol. Letters 20 (1). 15-17.

Svensson, U, 1999. Industrial Perspectives. In : Probiotics: A Critical Review. Tannock, G., ed. Horizon Scientific Press, Wymondham. p. 57-64.

Tannock, G.W. 1997. Influences of the normal microflora on the animal host. In: Gastrointestinal Microbiology. Vol. 2. Mackie, R.I., White, B.A. and Isaacson, R.E., eds. New York, Chapman and Hall. p. 537-587.

Timmermans, E., 1994. Lactose: its manufacture and physicochemical properties. In: Carbohydrates as Organic Raw materials III. van Bekkum, H., Roper, H. and Voragen, A.G.J., eds. VCH, Weinheim. p. 93-113.

Vic, G., Tran, C.H., Scigelova, M. and Crout, D.H.G. 1997. Glycosidase-catalysed synthesis of oligosaccharides: a one step synthesis of lactosamine and of the linear B type 2 trisaccharide α-D-Gal-1→3.-β-D-Gal-1→4.-β-D-GlcNAcSEt involved in the hyperacute rejection response in xenotransplantation from pigs to man and as the specific receptor for toxin A from *Clostridium difficile*. Chemical Commun. 1997 (2): 169-170.

Wang, X. and Gibson, G.R. 1993. Effects of the *in vitro* fermentation of oligofructose and inulin by bacteria growing in the human large intestine. J. Appl. Bacteriol. 75: 373-380.

Williams, C.M. 1999. Effects of inulin on lipid parameters in humans. J. Nutr. 129: 1471S-73S.

Wilson, K.H. and Blitchington, R.B. 1996. Human colonic bacteria studied by ribosomal DNA sequence analysis. Appl. and Environ, Microbiol, 62: p. 2273-2278.

Yatake, T. 1993. Anomalously linked oligosaccharides. In: Oligosaccharides. Production, Properties and Applications. Japanese Technol. Rev. 3: 79-89.

Yun, J. 1996. Fructooligosaccharides- occurrence, preparation and application. Enzyme and Microbial Technol. 19: 107-117.

From: *Probiotics and Prebiotics: Where Are We Going?*
Edited by: Gerald W. Tannock

Chapter 6

Prebiotics and Calcium Bioavailability

Kevin Cashman

Abstract

A prebiotic substance has been defined as a non-digestible food ingredient that beneficially affects the host by selectively stimulating the growth and/ or activity of one or a limited number of bacteria in the colon. Therefore, compared to probiotics, which introduce exogenous bacteria into the colonic microflora, a prebiotic aims at stimulating the growth of one or a limited number of the potentially health-promoting indigenous micro-organisms, thus modulating the composition of the natural ecosystem. In recent years, increasing attention has been focussed on the possible beneficial effects of prebiotics, such as enhanced resistance to invading pathogens, improved bowel function, anti-colon cancer properties, lipid lowering action, improved calcium bioavailability, amongst others. The objective of this chapter is to critically assess the available data on the effects of prebiotics on calcium bioavailability, and place it in the context of human physiology and, when possible, explain the underlying cellular and molecular mechanisms. The chapter will also try to highlight future areas of research that may help in the evaluation of prebiotics as potential ingredients for functional foods aimed at enhancing calcium bioavailability and protecting against osteoporosis.

Introduction

The maintenance of a community of bacteria which contains a predominance of beneficial species and minimal putrefactive (protein degrading) processes is believed to be important for maintaining intestinal health (Crittenden, 1999). Since specific components of the intestinal microflora have been associated with beneficial effects on the host, such as promotion of gut maturation and integrity, antagonisms against pathogens, and immune modulation, it would seem logical that the quantity of these components might be enhanced with dietary interventions (Brassart and Schriffin, 2000). Recently, Crittenden (1999) suggested that two separate approaches exist to increase the number of health-promoting organisms in the gastrointestinal tract. The first is the oral administration of live, beneficial microbes. This is the 'probiotic' approach, and is achieved most commonly by consumption of the probiotic bacteria, which have to date been selected mostly from lactic acid bacteria and bifidobacteria that form part of the normal intestinal microflora of humans, as milk-based products. However, since these organisms are indigenous to the colon, a second strategy to increasing their numbers is to supply those already present in the intestine with a selective carbon and energy source that provides them with a competitive advantage over other bacteria in the ecosystem, that is, to selectively modify the composition of the microflora using dietary components, the 'prebiotics'. These two kinds of dietary components have become the focus of great interest in the general population, the food industry, and the scientific community because of their potential for positively modifying biological and physiological processes and, thereby, possibly enhancing human health and well-being.

There is an impressive list of therapeutic and prophylactic attributes ascribed to the use of probiotics (see review by Tannock, 1999) and over the last two decades alone there has been a major international research effort to substantiate at least some of these health claims. In recent years, increasing attention has been focussed on the possible beneficial effects of prebiotics. The physiological importance and health benefits claimed for prebiotic substances are detailed in Table 1. Because of their putative beneficial effects, prebiotics (as well as probiotics) have been regarded as functional food ingredients. The working definition of a functional food is that "a food can be regarded as functional if it is satisfactorily demonstrated to affect beneficially one or more target functions in the body, beyond adequate nutritional effects, in a way which is relevant to either the state of well-being and health or the reduction of the risk of a disease" (Roberfroid, 2001).

Table 1. The physiological effects and putative health benefits claimed for prebiotic substances[1]

Physiological Effects	Possible Health Benefit
Selection of probiotic bacterial growth in large intestine (colonization resistance)	Enhanced resistance to invading pathogens
Increased stool frequency and stool weight	Improved bowel function/Laxative effects
Non-specific stimulation of immune function	Resistance to infection
Not hydrolysed by oral micro-organisms	Anticariogenic effect
Not glycaemic	Potentially useful for diabetes
Modulation of carcinogen metabolism	Anti-colon cancer properties
Reduced synthesis of VLDL cholesterol and serum triglycerides	Cardioprotective
Increased absorption of calcium and magnesium	Protection against osteoporosis

Prepared using information from Macfarlane and Cummings (1999); Crittenden (1999) and Van Loo *et al.* (1999) and Roberfroid (2001).
[1]Some of these benefits to health remain to be clearly established.

It is not the aim of the present article to extensively review the scientific base of the various health benefits of prebiotics, other than their putative beneficial effects on calcium absorption and bone health; the other health benefits have recently been overviewed in several excellent articles that form part of the proceedings of two recent International Symposia which focussed on the influence of prebiotics and probiotics on human health (Supplement to the February edition of the *American Journal of Clinical Nutrition*, 2001; and Supplement to the *British Journal of Nutrition*, in press). The objective of this chapter is to critically assess the available data on the effects of prebiotics on calcium bioavailability, and place it in the context of human physiology and, when possible, explain the underlying cellular and molecular mechanisms. The chapter will also try to highlight future areas of research that may help in the evaluation of prebiotics as potential ingredients for functional foods aimed at enhancing calcium bioavailability and protecting against osteoporosis. In the context of osteoporosis, the development of prebiotic-containing functional food products must be based on a detailed understanding of the influence of these dietary components on calcium bioavailability and bone health, and furthermore, must be supported by independent and appropriate scientific evidence to demonstrate efficacy with respect to the claimed health benefits. A good starting point before reviewing the putative beneficial effects of prebiotics on calcium bioavailability, is to begin with a definition of a prebiotic substance.

Definition of a Prebiotic

A prebiotic substance has been defined as "a non-digestible food ingredient that beneficially affects the host by selectively stimulating the growth and/ or activity of one or a limited number of bacteria in the colon" (Gibson and Roberfroid, 1995). Therefore, compared to probiotics, which introduce exogenous bacteria into the colonic microflora, a prebiotic aims at stimulating the growth of one or a limited number of the potentially health-promoting indigenous micro-organisms, thus modulating the composition of the natural ecosystem (Roberfroid, 2001). To be effective, prebiotics should escape digestion in the upper gut by pancreatic and brush-border enzymes, reach the large bowel (especially, the cecum), and be utilised selectively by a restricted group of micro-organisms that have clearly identified, health promoting properties, i.e., probiotic bacteria (usually bifidobacteria and lactobacilli) (Macfarlane and Cummings, 1999; Cummings *et al.,* 2001). Recently, Roberfroid outlined the following criteria that can be used to classify a food component as a prebiotic: resistance to digestion, hydrolysis and fermentation by colonic microflora, and most importantly, selective stimulation of growth of one or a limited number of bacteria in the faeces (*in vivo* in humans) (Roberfroid, 2001).

In practise a range of dietary non-starch polysaccharides, resistant starches, undigested sugars, oligosaccharides and proteins are fermented by the microflora. Of these, it is the non-digested dietary carbohydrates that provide the principal substrates for colonic bacterial growth. Cummings *et al.* (2001) suggests that these are short-chain carbohydrates (SCCs) that are non-digestible by human enzymes. These range from small sugar alcohols and disaccharides, to oligosaccharides, and large polysaccharides (Table 2), all with a variety of sugar composition and glycosidic linkages. Analysis of these substances indicates that although some are very pure, containing

Table 2. Types of candidate prebiotic substances

Type of Short-chain Carbohydrates	Example(s) of Candidate Prebiotic Substances
Disaccharides	Lactose derivatives such as lactulose and lactitol
Oligosaccharides[1] e.g.,	
Fructo-oligosaccharides	*Raftilose®*
Galacto-oligosaccharides	*Oligomate®*
Soybean oligosaccharides	Raffinose and stachyose
Other Non-digestible oligosaccharides	Xylo-oligosaccharides, isomalto-oligosaccharides, lactosucrose, palatinose polycondensates
Polysaccharides	Inulin[2]
Resistant starch	
Type I	The physically inaccessible starch granules (such as whole and partially milled grains)
Type II	Native starch granules (e.g., in potato, banana, high amylose maize)
Type III	Retrograded starch formed during starch processing
Type IV	Chemically modified starches altered by cross-linking, esterification, or etherification

Prepared using information from Macfarlane and Cummings (1999) and Crittenden (1999).
[1]Oligosaccharides are usually defined as glycosides that contain between three and ten sugar moieties.
[2]Inulin extracted from chicory contains both oligosaccharides as well as polysaccharides.

86-87% oligosaccharides, e.g., inulin and oligofructose, in others the oligosaccharides fraction is minor (about 20-30%), the rest being free monosaccharides, starch, and non-starch polysaccharides (Cummings *et al.*, 2001). Such substances have attracted considerable attention in the past decade for their physiological and health promoting properties and thus, for their potential as candidates for functional food ingredients (Fooks *et al.*, 1999). It has been suggested that of the various SCCs available, the non-digestible oligosaccharides (NDOs, oligosaccharides that resist hydrolysis and digestion in the upper gastrointestinal tract but are hydrolysed and fermented in the large bowel (Delzenne and Roberfroid, 1994)) are the only known components for which convincing evidence has been reported in favour of a prebiotic effect (Roberfroid, 2001). However, not all NDOs have prebiotic properties, and inulin, fructo-oligosaccharides, and (to a lesser degree) galacto-oligosaccharides dominate the published reports (Macfarlane and Cummings, 1999). The inulin/oligofructose-type products are the prebiotics that have been investigated most extensively for their nutritional properties (Roberfroid and Delzenne, 1998; Roberfroid, 1999). These low molecular weight carbohydrates occur naturally in artichokes, onions, chicory, garlic, leeks, and, to a lesser extent, in cereals. Other oligosaccharides such as raffinose and stachyose are the major carbohydrates in beans and peas. These simple molecules can also be produced industrially, and a number of new potential prebiotics are being commercially developed (see review by Cummings *et al.,* 2001). The evidence that such ingredients can positively influence calcium absorption, and possibly bone health, will be reviewed in the following sections. However, so that one can critically review the evidence, it is important firstly to briefly overview calcium absorption, calcium bioavailability and the various methodologies for assessing calcium bioavailability.

Bioavailability of Dietary Calcium

The terms 'bioavailability' and 'absorption' of a nutrient are sometimes used interchangeably in the literature; however, there is an important difference between them. The absorption of a nutrient describes the process by which the nutrient is transported from the gastrointestinal lumen, across the intestinal mucosa, to the serosa (see the section below dealing specifically with intestinal absorption of calcium). The bioavailability of a nutrient, on the other hand, defines that fraction of the ingested nutrient that is utilised for normal physiological functions or storage. This definition recognises that one of the major determinants of bioavailability is that proportion which is absorbed from the gastrointestinal tract, but that this is not the only factor influencing bioavailability since tissue utilization (or lack of utilization) of the absorbed nutrient may vary dramatically (Jackson, 1997). Specifically in the case of calcium, bioavailability may be defined as the amount of calcium in foods that can be absorbed and utilised by the body for normal metabolic functions.

Intestinal Calcium Absorption

Calcium in food occurs as salts or associated with other dietary constituents in the form of complexes of calcium ions (Ca^{2+}). Calcium must be released in a soluble, and probably ionised, form before it can be absorbed (i.e., its transfer from the intestinal lumen to the circulatory system). Once in a soluble form, calcium is absorbed by two routes, transcellular and paracellular transport (Bronner, 1987). The saturable, transcellular pathway is a multi-step process, involving the entry of luminal Ca^{2+} across the microvillar membrane into the enterocyte, then movement through the cytosol (i.e., translocation to the basolateral membrane), followed by active extrusion from the enterocyte into the lamina propria and, eventually, into the general circulation (see Figure 1). The intracellular Ca^{2+} diffusion is thought to be facilitated by a cytosolic calcium-binding protein, calbindin D_{9K}, whose biosynthesis is dependent on vitamin D. Calbindin D_{9K} facilitates the diffusion of Ca^{2+} across the cell by acting as an intracellular calcium ferry or a chaperone. The active extrusion of Ca^{2+} at the basolateral membrane takes place against an electrochemical gradient and is mainly mediated by Ca-ATPase. The entry of Ca^{2+} across the apical membrane of the enterocyte is strongly favoured electrochemically because the concentration of Ca^{2+} within the cell (10^{-7} to 10^{-6} M) is considerably lower than that in the intestinal lumen (10^{-3} M), and the cell is electronegative relative to the intestinal lumen (Fullmer, 1992). Therefore, the movement of Ca^{2+} across the apical membrane does not require the expenditure of energy. It has been controversial as to whether a transporter or a channel is responsible for this process. It was widely believed that, because of the impermeability of lipid membranes to Ca^{2+}, a Ca^{2+} channel or integral membrane transporter must reside in the brush border membrane. Evidence would now suggest that the recently cloned calcium transport protein (CaT1) is a good candidate for this putative Ca^{2+} channel (Peng *et al.*, 1999). While each step in the transcellular movement of Ca^{2+} has a vitamin D-dependent component, calbindin D_{9K} is believed to be the rate-limiting molecule in vitamin D-induced transcellular calcium transport.

The paracellular route of calcium absorption involves a passive calcium transport through the tight junctions between mucosal cells (see Figure 1); it is non-saturable, essentially independent of nutritional and physiological regulation, and is concentration dependent. Some debate still persists as to whether indeed the paracellular pathway is vitamin D-dependent (Chirayath *et al.*, 1998; Fleet and Wood, 1999). Most calcium absorption in humans occurs in the small intestine, but there is some evidence (Barger-Lux *et al.*,

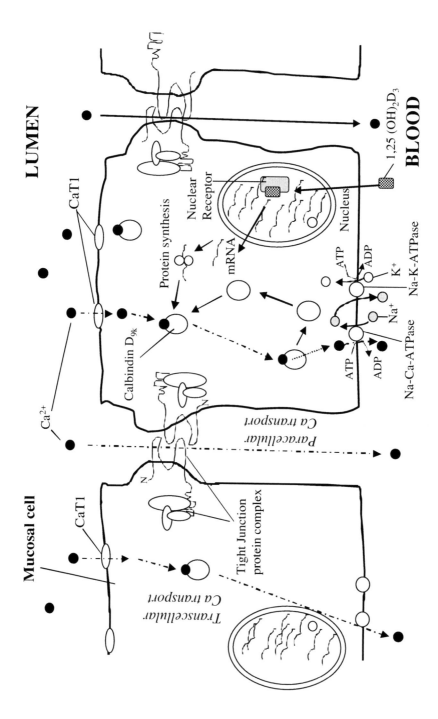

Figure 1. A schematic representation of transepithelial Ca^{2+} transport in the intestine. The central feature is that calcium absorption occurs by two independent processes, namely transcellular and paracellular transport of Ca^{2+} across an epithelium.

1989) for a small colonic component (typically believed to be no more than 10% of total calcium absorption). However, the large intestine may represent a site of increased importance for calcium absorption when acidic fermentation takes place (Younes *et al.,* 1996). This is important if one remembers that consumption of prebiotics will lead to acidic fermentation in the large intestine and this issue will be dealt with later in the section dealing with mechanisms of enhanced calcium absorption by prebiotics. When dietary calcium is abundant, the paracellular pathway is thought to be predominant. In contrast, when dietary calcium is limited, the active, vitamin D-dependent transcellular pathway plays a major role in calcium absorption. Transcellular calcium absorption responds to calcium needs, as reflected by changes in plasma Ca^{2+} concentration, by hormone-mediated up- or down-regulation of calbindin D_{9K} in mucosal cells; for example, reduced plasma Ca^{2+} evokes a parathyroid hormone mediated increase in plasma 1,25-dihydroxyvitamin D_3, which stimulates increased calbindin D_{9K} synthesis in the intestinal mucosa (Bronner, 1987).

Methods for Measuring Calcium Bioavailability

Greger (1992) suggests that the bioavailability of a nutrient reflects the sum of the effects of various factors (dietary and other) on the absorption, transport, cellular organisation, storage and excretion of the nutrient. Therefore, bioavailability is not definable by a single test or variable. Rather, the study of bioavailability is dependent on the strategic use of *in vitro,* subcellular and cellular systems, animal models and human subjects in an integrated manner.

Calcium status can be evaluated by measurements of the bone mass at different sites of the skeleton. Currently, dual energy X-ray absorptiometry (DEXA) is clearly the method of choice to assess bone mass, owing to its accuracy, precision and low radiation exposure. DEXA can be used to detect changes in the bone mass over time and thus to assess the effect of intervention measures aimed to prevent loss of bone mass or to increase this mass (Institute of Medicine, 1997). The measurement of bone mass is considered the best way to evaluate the long term effects on calcium status of factors which influence calcium metabolism or calcium absorption (Schaafsma, 1997). As already mentioned, calcium absorption is an important component of calcium bioavailability. The various methods of assessing intestinal calcium absorption have been reviewed by Schaafsma (1997). Assessment of intestinal absorption, based on measurements of the difference between calcium intake and calcium excretion (i.e., metabolic balance technique) has major inherent shortcomings, including errors in estimating intake, incompleteness of faecal collections, and the inability to measure true calcium absorption. The latter point is important, considering the magnitude of the endogenous faecal calcium excretion (i.e., calcium secreted into the gastrointestinal tract which is not re-absorbed) and therefore the large difference between true- (accounting for endogenous loss) and apparent calcium absorption (which does not account for endogenous calcium loss). Measurement of true calcium absorption can be performed with radioactive tracers of calcium, such as ^{45}Ca and ^{47}Ca. These tracers are relatively cheap, can be measured accurately and can be used in very small (tracer) amounts. A disadvantage, however, is the ionizing radiation. Therefore, for ethical reasons, nutrition studies with these tracers in human volunteers are not preferred. Application of stable isotopes (^{42}Ca, ^{44}Ca, ^{46}Ca and ^{48}Ca) to measure calcium absorption has become common place in spite of the high isotope costs. Principles are similar to tracer studies with radioisotopes. Various methods of mass spectrometry analysis of stable calcium isotope enrichement can be used with this approach. True calcium absorption from a particular food can be measured after labelling this food extrinsically or intrinsically with a stable calcium isotope. If at the same time of administration of this food another stable calcium isotope is administered intravenously, true calcium absorption can easily be measured from stable calcium enriched values in samples of serum and urine obtained 24-48 hour after administration. The timing of the urine collection can be modified to take account of predominantly the small intestinal component of calcium absorption (i.e., use of 24-hour urine collection) or to include the later colonic component (36-48 hour urine collection). A faecal collection method can also be used as a more laborious alternative for this dual labelling. It is also worth noting that dietary intervention trials in humans which employ such approaches, and which are aimed at assessing the influence of a dietary component on intestinal calcium absorption, are best carried out in randomized, double-blind, crossover design format, if possible. This type of study design, which is widely used in clinical, medical and pharmaceutical research, is considered a good approach for evaluating the efficacy of functional food ingredients.

While ideally calcium bioavailability should be measured by human studies, such studies are often very time consuming and expensive to run. As an alternative, experimental animal models, especially the laboratory rat, have been used quite extensively for assessing the impact of dietary and other factors on calcium absorption. The balance and isotope tracers approaches, mentioned above, can and have also been used in rats to determine calcium absorption. The adult rat, for example, has been shown to be a useful model for studies on calcium bioavailability since absorption mechanisms for calcium are similar in rats and in humans and a number of dietary and physiological factors affect calcium absorption similarly in the two species (Cashman and Flynn, 1996). However, while studies using laboratory animals are less expensive than studies in humans, they are somewhat limited by uncertainties with regard to differences in metabolism between animals and humans. More recently, intestinal cells in culture (particularly, the Caco-2 cells) have gained in popularity as an *in vitro* model of calcium absorption. When the human colon carcinoma cell line, Caco-2, is grown on microporous membranes in bicarmel chambers the cells differentiate spontaneously into bipolar enterocytes that exhibit many of the characteristics of normal epithelial cells, such as microvilli of the brush border membrane, tight intercellular junctions and the excretion of border associated enzymes. As these cells differentiate within the chambers, the apical pole extends into the upper chamber while the basal lateral pole, in contact with the porous membrane, is exposed to the lower chamber. The design of the bicarmel chambers permits the study of calcium uptake from the apical chamber, transport into the cell and vectoral secretion into the basal chamber. In particular, these cells have a functional vitamin D receptor and have calcium transport kinetics that suggest the presence of both a saturable and nonsaturable calcium transport pathway, similar to observations in human and animal intestine (Fleet and Wood, 1999). Furthermore, it has been shown that in these cells, 1, 25 dihyroxyvitamin D_3 treatment induces the saturable, but not diffusional, component of calcium transport and induces accumulation of calbindin D_{9K} mRNA (Fleet and Wood, 1999). Therefore, this relatively simple *in vitro* method appears to be a good model for predicting calcium bioavailability in humans under certain conditions.

A Stimulatory Effect of Prebiotics on Calcium Bioavailability – What is the Evidence?

This section will review the various lines of evidence for putative beneficial effects of prebiotics on calcium bioavailability. As already mentioned, an effect on calcium bioavailability can be regarded as an effect on intestinal calcium absorption and/or on bone status. The scientific data which is currently available on the effect of prebiotics on calcium bioavailability comes from animal studies as well as from a limited number of human studies.

Evidence of a Stimulatory Effect of Prebiotics on Calcium Absorption in Animals

Numerous studies have repeatedly shown that prebiotics, such as oligofructose (also known as oligoifructose) and inulin, galacto-oligosaccharides, resistant starches or lactulose effectively stimulate calcium absorption in the rat and these have been reviewed in recent articles (Franck, 1998; Van Loo *et al.,* 1999; Scholz-Ahrens *et al.,* 2001a; Scholz-Ahrens and Schrezenmeir, 2001). Most experiments on prebiotics lasted 3-4 weeks and were carried out in young growing rats or in models of disease or altered physiological status, such as, gastrectomized, ovariectomized, cecectomized, and magnesium, calcium or iron deficient rats (see review by Scholz-Ahrens *et al.,* 2001a; Scholz-Ahrens and Schrezenmeir, 2001). It is widely believed that the effect of prebiotic substances on calcium absorption in these animal models occurs at the level of the large intestine (Ohta *et al.,* 1994; Baba *et al.,* 1996), although Brommage *et al.*(1993) reported that lactulose stimulated calcium absorption to the same extent in cecectomized rats as in sham-operated control rats. The mechanisms by which these prebiotic substances enhance calcium absorption in rats are discussed in a later section in this chapter. Some interesting points arise from the findings of some of these studies and may, indeed, point the way in terms of future research that is needed, especially in human studies. For example, an interesting observation from some of these animal studies is that the effect of prebiotics on calcium absorption may be modulated by the amount of calcium in the diet. For example, Chonan and Watanuki (1996) reported a stimulatory effect of galacto-oligosaccharides on calcium absorption in intact (i.e., not surgically altered) rats when the diets of these animals contained 5 g Ca/kg diet, but not when the diet contained only 0.5 g Ca/kg diet. Similarly, Scholz-Ahrens *et al.* (2001a) reported that the effect of oligofructose on metabolic calcium balance in ovariectomised rats became more prominent when dietary calcium was high (10 g Ca/kg diet) as compared to when it was at the recommended level (5 g Ca/kg diet). Lactulose has also been shown to significantly increase (*P*<0.001) calcium absorption in young growing intact rats, but only if the diet contained 5 g Ca/kg diet or more; no stimulatory effect of lactulose was observed with a dietary calcium level of 2 g Ca/kg diet (Brommage *et al.,* 1993).

There is some evidence of a dose-dependent effect of prebiotic substances on calcium absorption in rats. For example, Levrat *et al.* (1991) found that dietary inulin (in the range 0 – 200 g/kg diet) stimulated intestinal calcium absorption in a dose-dependent manner. This coincided with a dose-

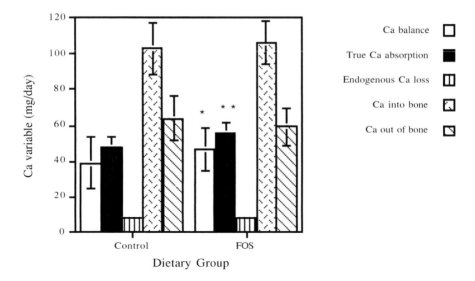

Figure 2. The effect of dietary fructo-oligosaccharides (FOS) (50 g/kg diet) on calcium balance, true calcium absorption, endogenous calcium loss, and calcium kinetics into and out of bone in young male rats. Values are means and standard deviations for 8 rats per group. *,**Significantly different from control group (*,** $P<0.05$, $P<0.01$). Graph constructed from data from Morohasi *et al.* (1998).

dependent decrease in cecal pH and a rise in cecum weight, cecal wall weight, and cecal pool of total short-chain fatty acids (SCFAs). Brommage *et al.* (1993) reported a near linear increase in intestinal calcium absorption in rats fed a diet containing 0, 50, and 100 g lactulose per kg diet; there was no further increase when the diet contained 150 g lactulose/kg diet.

The majority of the animal studies which demonstrate a positive effect of prebiotics on calcium absorption were based on using the calcium balance method (Morohashi *et al.*, 1998). However, as already mentioned, a simple calcium balance study can examine only apparent calcium absorption, urinary calcium excretion and calcium balance, and cannot be used to evaluate true intestinal calcium absorption, the excretion of calcium into the intestine or the kinetics of calcium movement into or out of bone. Moreover, apparent calcium absorption cannot explain whether an increase is due to an enhancement of calcium absorption or a reduction of endogenous calcium excretion into the intestine. Morohashi *et al.* (1998) recently addressed this issue by carrying out a rat study in which animals were supplemented with fructo-oligosaccharides and then calcium balance and [45]Ca kinetics were determined (see Figure 2). This allowed them to investigate the effect of fructo-oligosaccharides on calcium metabolism at the level of the intestine, kidney and bone. They found that dietary fructo-oligosaccharides increased true intestinal calcium absorption and had no effect on endogenously excreted calcium relative to the control diet (see Figure 2). Urinary calcium excretion was significantly ($P<0.01$) higher in rats fed fructo-oligosaccharides than in those fed the control diet. However, despite this, calcium balance was still significantly higher ($P<0.01$) in the rats fed fructo-oligosaccharides. They reported that calcium flow into and out of bone (i.e., bone formation and bone resorption, respectively) was unaffected by dietary fructo-oligosaccharides, despite the increased absorption and balance. This seems surprising considering the same group previously reported that fructo-oligosaccharides enhance bone mineral density and calcium content in rat bone (Ohta *et al.*, 1998a,b). An effect of prebiotics on bone arising from improved calcium absorption would presumably act through altered rates of bone turnover, which in turn would influence bone mass (Cashman and Flynn, 1999). However, Morohashi *et al.* (1998) suggest that because the effects on bone turnover were likely to be subtle, the techniques used in their study may not have been sensitive enough to detect such changes.

Another important consideration with respect to the stimulatory effect of prebiotics on calcium absorption is whether the enhanced calcium absorption is maintained over the longer term. The duration of many of the experiments with prebiotics was relatively short (14–28 d, see reviews by Scholz-Ahrens *et al.*, 2001a; Scholz-Ahrens and Schrezenmeir, 2001). The issue of adaptation of intestinal calcium absorption over time is highlighted by the findings of Brommage *et al.* (1993) which showed that dietary lactulose (50 g/kg diet), which stimulated calcium absorption in rats on the first day of the study, failed to enhance calcium absorption by the seventh day of the study. The authors hypothesized that this adaptive response of intestinal calcium absorption over time occurred by a down-regulation of the active, transcellular route of calcium absorption which counter balanced the lactulose-induced increase in passive, paracellular calcium absorption. However, it should be noted that the capacity for this adaptation is limited, and Brommage *et al.* (1993) suggested that providing higher levels of lactulose in their study could possibly have resulted in continued elevation of intestinal calcium absorption. This raises the issue of defining the minimum effective dose of dietary prebiotics for the prolonged stimulation of calcium absorption. There is some evidence from repeated balance studies in gastrectomised and intact rats that the stimulating effect of fructo-oligosaccharides on calcium absorption was maintained over several weeks (Ohta *et al.*, 1994, 1998a). Chonan and Watanuki (1996) found that calcium absorption was stimulated in intact rats after 8-10 days and also after 18-20 days when galacto-

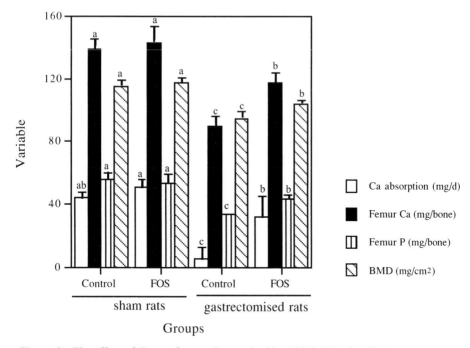

Figure 3. The effect of dietary fructo-oligosaccharides (FOS) (75 g/kg diet) on apparent calcium absorption, femur calcium content, femur phosphorus content, and femur BMD in male rats that underwent either a sham operation or a gastrectomy. Values are means and standard deviations for 7 rats per group. For a given parameter, values with different superscript letters are significantly different ($P<0.05$). Graph constructed from data from Ohta *et al.* (1998a).

oligosaccharides (50 g/kg diet) where included in the diet (containing 5 g Ca/kg diet).

It is also worth noting that certain prebiotic substances (such as, oligofructose, transgalacto-oligosaccharides, lactulose, resistant starch) have been shown to stimulate intestinal Mg absorption in various rat models (see reviews by Scholz-Ahrens *et al.,* 2001a; Scholz-Ahrens and Schrezenmeir, 2001) and this may be important in terms of a beneficial effect of these compounds on bone mineral density.

Evidence of a Beneficial Effect of Prebiotics on Bone Health in Animals

There is some evidence that the stimulatory effect of prebiotic substances on calcium absorption in rats can be translated into functional benefits for bone health, such as increased bone mineralization, bone density and improved bone structure. This topic has been recently reviewed in two excellent articles by Scholz-Ahrens *et al*. (2001a) and Scholz-Ahrens and Schrezenmeir (2001).

In addition to the stimulatory effect of fructo-oligosaccharides on calcium absorption in gastrectomised rats, they also prevented the changes indicative of post-gastrectomy-induced osteopenia, such as reduced bone calcium content and bone mineral density (Ohta *et al.,* 1998a, see Figure 3). Using a rat model of postmenopausal bone loss, namely the ovariectomised rat, Chonan *et al.* (1995) showed that dietary galacto-oligosaccharides stimulated intestinal calcium absorption relative to a control diet, and importantly, the bone ash weight and bone calcium content of the ovariectomised rats fed the galacto-oligosaccharide-containing diet were significantly higher than those of the animals fed the control diet. The positive effect of dietary galacto-oligosaccharides and oligofructose on bone status has also been demonstrated in healthy intact rats. For example, Chonan and Watanuki (1996) showed that supplementation of the diet (containing 5 g Ca/kg diet) with galacto-oligosaccharides (50 g/kg diet) stimulated calcium absorption relative to the control diet, and furthermore, bone calcium content was significantly higher in the animals fed the galacto-oligosaccharides-containing diet than those of the animals fed control diet. Recently, Takahara *et al.* (2000) reported that fructo-oligosaccharides (50 g/kg diet) stimulated calcium absorption and enhanced femoral bone volume and mineral concentrations in young growing intact rats.

As was noted with their effect on calcium absorption, the effect of certain prebiotic substances on bone status may be modulated by the amount of calcium in the diet. For example, Scholz-Ahrens *et al.* (2001b,c) reported that the positive effect of oligofructose on calcium content of bone and on the prevention of ovariectomy-induced loss of trabecular structure became more prominent when dietary calcium was high (10 g Ca/kg diet) compared to when it contained the recommended level (5 g Ca/kg diet). Similarly, Chonan and Watanuki (1996) reported a positive effect of galacto-oligosaccharides on bone mineralisation in intact rats when the diets of these animals contained 5 g Ca/kg diet, but not when the diet contained only 0.5 g Ca/kg diet.

An important consideration in interpreting the data from the above animal studies of the effect of prebiotic substances on calcium bioavailability is whether prebiotic substances have similar effects on the large intestine of humans as they do in rats. It is possible that the rat is particularly sensitive to prebiotic substances, in terms of their stimulatory effect on cecal fermentation and cecal enlargement. Therefore, a review of the evidence for a stimulatory effect of prebiotic substances on calcium absorption in human subjects is the most appropriate data when considering the efficacy of prebiotics for enhanced calcium bioavailability.

Evidence of a Beneficial Effect of Prebiotics on Calcium Absorption in Humans

According to the definition of a prebiotic, it must escape digestion in the small intestine in humans. While in healthy individuals with sufficient brush-border β–galactosidase activity, lactose is completely digested, in individuals with insufficient β–galactosidase activity, it may escape digestion, and thus, may be available for fermentation by the microflora in the colon. While numerous studies have shown a stimulatory effect of lactose on calcium absorption in the rat, studies on the effect of lactose on calcium absorption in humans have yielded inconsistent results. Miller, in a review of this area, concluded that it is likely that lactose enhances calcium absorption in human infants and in rats, while at levels normally present in milk, it does not have a significant effect on calcium absorption by healthy adults consuming normal diets (Miller, 1989). It is possible, however, that in subjects with low brush border β–galactosidase activity, lactose may stimulate calcium absorption because it reaches the terminal ileum and colon, where it can be fermented by the intestinal microflora (i.e., it may behave as a prebiotic). In this regard, Griessen *et al*. (1989) reported that calcium absorption was similar from milk (21.4%) and lactose-free milk (lactose replaced by glucose) (26.8%) in healthy adult subjects, but lactose increased calcium absorption in β–galactosidase-deficient subjects.

Lactulose is a synthetic disaccharide which does not exist in nature. It can be made on a large scale from lactose by alkaline isomerization and is often used in the treatment of constipation and chronic hepatic encephalopathy. Lactulose is not digested in the stomach or small intestine, but is fermented in the colon by indigenous microflora. For this reason, it has been regarded as a potential prebiotic substance (Table 2). Recently, van den Heuvel *et al*. (1999a) investigated the effect of 2 doses of lactulose (5 and 10 g/d compared with 0 g/d) on calcium absorption in a randomized, double-blind, crossover design study with 12 healthy postmenopausal women. Lactulose was given at breakfast for 9 d. True intestinal calcium absorption was measured by using the dual-stable calcium isotope-labeling (^{44}Ca and ^{48}Ca) technique, which allowed the measurement of late colonic calcium absorption. Calcium absorption during the three dietary periods, namely control (without lactulose) and 5 g and 10 g lactulose daily was (mean ± SD) 27.7 ± 7.7%, 30.0 ± 7.6%, and 32.2 ± 7.0%, respectively, with the difference in calcium absorption between the control and the 10 g dose being significant ($P<0.01$).

There is now increasing evidence that certain NDOs (such as, inulin, fructo- and galacto-oligosaccharides, see Table 2) can improve calcium absorption in adolescents and adults. For example, Coudray *et al.* (1997) investigated the effect of chicory inulin and sugar beet fibre on calcium absorption in a crossover design study. They fed 9 healthy young adult males (mean age, 25.5 years) a control diet (containing 18 g dietary fibre per day) or the same diet supplemented with 40 g per day of either chicory inulin or sugar beet fibre for a period of 28 days (2 days of control diet followed by 14 d of progressive increase in test fibre amount and then 12 days at 40 g/d) and determined the apparent absorption of calcium by using the classic balance approach. They found that upon inulin ingestion, apparent calcium absorption increased significantly ($P<0.01$) from 21.3% to 33.7% (an increase of 58%); ingestion of sugar beet fibre had no effect. In another randomized, double blind, crossover study, van den Heuvel *et al.* (1999b) fed 12 healthy male adolescents (aged 14-16 years) either orange juice supplemented with oligofructose (15 g/d) or sucrose (control treatment) three times daily for 9 days, after which time, they measured true fractional calcium absorption by a dual stable calcium isotope technique. An increase of 26% in true fractional calcium absorption (47.8% with placebo to 60.1% with oligofructose, $P<0.05$) was observed upon ingestion of the daily 15 g supplement of oligofructose. In an earlier randomized crossover study by the same group, a daily supplement of 15 g of oligofructose, inulin, or galacto-oligosaccharides for 21 d had no effect on true calcium absorption (measured by a dual stable isotope technique) in healthy adult men (aged 20 to 30 years) (van den Heuvel *et al.,* 1998). However, in that study the colonic component of calcium absorption (a putative target for enhancement by prebiotics; see later) was not included because the urine collection was limited to 24-hours after isotope administration. This is in contrast with the later study in which the dual-labeling technique was slightly modified (i.e., urine was collected for 36-hours after isotope administration) that allowed the measurement of late colonic effects on calcium absorption. In a third recent study by the same group, the effect of galacto-oligosaccharides, also referred to as transgalacto-oligosaccharides (TOS), on calcium absorption was re-evaluated, but in this case in postmenopausal women and importantly, using the modified dual-labeling technique (36-hour urine collection) for determining true calcium absorption (van den Heuvel *et al.,* 2000). This study was also a double-blind, randomized crossover study, consisting of two 9-day treatment periods in which the women drank yogurt drinks twice per day (at breakfast and lunch) containing either TOS (20 g/d) or a reference substance, sucrose. TOS significantly ($P<0.05$) increased true calcium absorption by 16% (20.6 ± 7.0% to 23.9 ± 6.9%) in these postmenopausal women.

In a very recent randomized, double-blind, crossover design study, 29 young adolescent girls (aged 10-14 years, consuming a relatively high calcium intake (1500 mg/d)) received either 8 g servings of an inulin-oligofructose mix or placebo (sucrose) in a calcium-fortified orange juice daily for 3 weeks (Abrams and Griffin, 2001). True calcium absorption was measured using a dual stable isotope method at the end of each three-week period. A 48-hour urine collection was carried out after isotope administration so as to detect any modulatory effect of the enriched inulin on the colonic component of calcium absorption. Consumption of the enriched inulin resulted in a 18% increase ($P<0.01$) in true fractional calcium absorption and in an absolute increase in calcium absorption of 90 mg/day (Abrams and Griffin, 2001).

An important consideration in terms of the effects of these prebiotic substances on calcium bioavailability is whether the additional calcium absorbed is retained within the body for use in the various physiological functions but particularly for use by skeletal tissue. Importantly, there was no significant ($P>0.05$) difference in urinary calcium excretion during the prebiotic dietary periods compared to control periods in the above studies, suggesting that the additional calcium that was absorbed was retained within the body.

Mechanistic Aspects of the Stimulatory Effect of Prebiotics on Calcium Bioavailabilty

Several theories have been proposed to explain the stimulatory effect of prebiotic substances on intestinal calcium absorption. These theories, which are outlined in Table 3, refer to effects on the two routes of calcium absorption, namely transcellular and paracellular calcium transport, in the small and/or large intestine.

Prebiotic substances that escape digestion in the small intestine are substrates for the formation of short-chain fatty acids (SCFAs, essentially acetate, propionate, and butyrate) and other organic acids (e.g., lactate) in the large

Table 3. Possible mechanisms by which prebiotics stimulate intestinal calcium absorption

---

1. Enhanced passive calcium absorption in the large intestine due to an increased solubility of calcium. The increased solubility occurs as a consequence of reduced luminal pH which, in turn, arises from the increased production of lactic acid and short chain fatty acids due to stimulated microbial growth and fermentation by the presence of prebiotics.

2. A direct effect of these short chain fatty acids on transcellular calcium absorption (by increasing the exchange of cellular H^+ for luminal Ca^{2+}).

3. Butyrate may (a) induce cell growth and thus increase the absorptive surface area of the gut, and/or (b) increase mucosal calbindin D_{9K} levels in the large intestine (and reduce levels in the small intestine).

4. The synthesis of polyamines, metabolites generated by several strains of microbes, may be increased by prebiotics, and these metabolites might, at least in part, (a) lead to cell growth and enlargement of the absorptive surface in the gut, and/or (b) stimulate gene expression, possibly including calbindin D_{9K}.

---

intestine by the intestinal microflora. These SCFAs contribute to a reduced luminal pH in the large intestine, which is associated with an increased amount of soluble calcium, especially in the cecum. Moreover, this increased solubility may lead to an increased paracellular transport of calcium in the distal part of the small intestine and beginning of the large intestine (van den Heuvel *et al.,* 1999a). It is also possible that SCFAs directly influence the transcellular route of calcium absorption by modifying the exchange of intracellular H^+ for Ca^{2+} present in the distal colon. The protonated SCFA molecule diffuses across the apical membrane of the intestinal epithelial cells and once within the cell it dissociates, resulting in an increased intracellular H^+, which is secreted from the cell in exchange for Ca^{2+} from the distal colon. Once outside the cell, H^+ becomes available to protonate another SCFA molecule. Therefore, there is an increased exchange of cellular H^+ for luminal Ca^{2+} (van den Heuvel *et al.,* 1999a). There is also some evidence from animal studies that certain prebiotics may influence transcellular calcium transport by altering the intracellular synthesis of the vitamin D receptor and calbindin D_{9K}. For example, in gastrectomised rats fed fructo-oligosaccharides, the amount of calbindin D_{9K} was increased in both cecal and colorectal segments and decreased in the proximal and distal small intestine. The overall effect, however, was an improved calcium absorption due to the prebiotic (Ohta *et al.,* 1998c). The mechanism(s) by which prebiotic substances modulate calbindin D_{9K} levels are not clear but may be due to increased synthesis of bioactive compounds such as butyrate and, possibly, certain polyamines. This issue has recently been extensively reviewed by Scholz-Ahrens and Schrezenmeir (2001).

Finally, in addition to possible effects of prebiotic substances on the paracellular and/or transcellular processes of calcium absorption, it may also be possible that such substances influence gut morphology and its absorptive surface, possibly via an increased production of butyrate and/or certain polyamines (see review by Scholz-Ahrens and Schrezenmeir, 2001). For example, the villus crypt height, number of epithelial cells per crypt, and cecal vein flow have all been reported to be increased by prebiotics (see review by Scholz-Ahrens *et al.,* 2001a).

Conclusions and Suggested Further Research Areas

Upon critical review of the available data on the effect of certain prebiotics on calcium absorption, I think it is fair to conclude that, in general, prebiotic substances, such as inulin, oligofructose (fructo- oligosaccharide) and galacto-oligosaccharide, and lactulose have been found to stimulate calcium absorption and retention in rats. Furthermore, at least in some animal studies, the enhanced calcium absorption appeared to lead to improved bone status. Therefore, there is relatively good evidence of a beneficial effect of prebiotics on calcium bioavailability in rats. While there have only been a few studies on the effect of prebiotics (such as, lactulose, inulin, and fructo- and galacto-oligosaccharides) on calcium absorption in humans, so far, there would appear to be a stimulatory effect by these prebiotics on true intestinal calcium absorption, at least in subgroups of the population which have increased calcium requirements (e.g., adolescents and postmenopausal women).

One of the difficulties in the communication of the benefits of functional foods is that the term 'health claim' is defined differently in different countries. A recent Consensus Document on scientific concepts in functional foods in Europe has proposed use of two types of health claims, Type A and Type B claims (Diplock *et al.*, 1999). Type A claims refer to 'enhanced function claims' while Type B claims refer to 'reduced risk of disease claims'. It was proposed that these claims should be based on evidence related to markers which are linked to clearly defined and measurable outcomes and are significantly and consistently modulated in rigorously controlled studies by the particular food component (Ashwell, 2001). Furthermore, enhanced function claims should be accompanied by evidence based on valid, reproducible, sensitive and specific markers relating to the target function or biological response, whereas reduction of disease risk claims would only be justified if the evidence is based on valid, reproducible, sensitive and specific markers relating to an intermediate endpoint of improved state of health and well-being and/or reduction of risk of disease (Diplock *et al.* 1999). Therefore, currently there is some, albeit limited, evidence to support a Type A health claim (i.e., an enhanced function [improved calcium absorption in the present context]) for these prebiotic substances. There is, however, essentially no evidence to support a Type B claim (disease risk reduction) for these prebiotics. Therefore, there are still many out-standing research questions that would need to be answered before prebiotics could be marketed as functional food ingredients aimed at improving calcium bioavailability (i.e., improved absorption and utilization by the body) and reducing risk of osteoporosis. The following are just some of the questions which would need to be addressed in future humans studies:

1. Human studies to show that the benefits of prebiotic substances to true calcium absorption persist after their prolonged use (at least one year).
2. Human studies to investigate whether certain prebiotics have more efficacy (in terms of promoting true calcium absorption) than others, and to determine the minimum effective doses of these compounds.
3. Human studies to investigate whether these prebiotics enhance true calcium absorption in population subgroups other than those who have increased calcium requirements namely, adolescents or postmenopausal women.
4. Human studies to investigate whether the habitual dietary calcium content modulates the stimulatory effect of prebiotics on true calcium absorption.
5. Human studies to investigate whether the beneficial effects of prebiotic substances on calcium absorption are translated into benefits to bone health. These may take the form of studies in which bone mineral density, bone mineral content, bone turnover and, indeed, bone structure and quality (all of which can act as surrogate markers of osteoporosis risk) are assessed in appropriately designed intervention trials.

Needless to say, further research in experimental animals and possibly even in cells in culture may also help in better understanding the mechanistic aspects of the effects of prebiotics on calcium bioavailability. Future studies in model systems (and possibly in biopsy samples from human subjects) should consider using the newer molecular biology tools, such as transcriptomics and proteomics, to help gain new insights into the effect of prebiotics on calcium and bone metabolism. Such an integrated research approach (i.e., human, animal and cellular research) to assess the functionality of new foods and food ingredients, including the prebiotics, would help address the growing concerns of regulatory bodies and health professionals, as well as consumer organization, with respect to the rapid appearance on the market place of more and more foods claiming to possess health-promoting properties, despite the fact that such products, to a large extent, having not being subjected to fundamental research to substantiate manufacturers' health claims.

References

Abrams, S.A., and Griffin, I.J. 2002. Inulin and oligofructose and calcium absorption: Human data. Br. J. Nutr. In press.

American Journal of Clinical Nutrition. 2001. Proceedings of the International Symposium on Probiotics and Prebiotics, Kiel, Germany, June 11-12, 1998. Am. J. Clin. Nutr. 73 (supplement): 361S-498S.

Ashwell, M. 2001. Functional foods: a simple scheme for establishing the scientific validity for all claims. Public Health Nutr. 4: 859-862.

Baba, S., Ohta, A., Ohtsuki, M., Takizawa, T., Adachi, T., and Hara, H. 1996. Fructooligosaccharides stimulate the absorption of magnesium from the hindgut in rats. Nutr. Res. 16: 657-666.

Barger-Lux, M.J., Heaney, R.P., and Recker, R.R. 1989. Time course of calcium absorption in humans: evidence for a colonic component. Calcif. Tissue Int. 44: 308-311.

Brassart, D., and Schriffin, E.J. 2000. Pro- and Prebiotics. In: Essential of Functional Foods. M.K. Schmidl and Labuza T.P., eds. Aspen Publishers, Inc., Maryland. p. 205-216.

British Journal of Nutrition. 2002. Proceedings of the 3rd ORAFTI Research Conference, London, February 8-9, 2001. Br. J. Nutr. In press.

Brommage, R., Binacua, C., Antille, S., and Carrié. A-L. 1993. Intestinal calcium absorption in rats is stimulated by dietary lactulose and other resistant sugars. J. Nutr. 123: 2186-2194.

Bronner F. 1987. Intestinal calcium absorption: mechanisms and applications. J. Nutr. 117: 1347-1352.

Cashman, K.D., and Flynn, A. 1996. Effect of dietary calcium intake and meal calcium content on calcium absorption in the rat. Br. J. Nutr. 76: 463-470.

Cashman, K.D., and Flynn, A. 1999. Optimal nutrition: calcium, magnesium and phosphorus. Proc. Nutr. Soc. 58: 477-487.

Chirayath, M.V., Gajdzik, L., Hulla, W., Graf, J., Cross, H.S., and Peterlik, M. 1998. Vitamin D increases tight-junction conductance and paracellular Ca^{2+} transport in Caco-2 cell cultures. Am. J. Physiol. 274: G389-G396.

Chonan, O., and Watanuki, M. 1996. The effect of 6'-galactooligosaccharides on bone mineralization of rats adapted to different levels of dietary calcium. Int. J. Vitam. Nutr. Res. 66: 244-249.

Chonan, O., Matsumoto, K., and Watanuki, M. 1995. Effect of galactooligosaccharides on calcium absorption and preventing bone loss in ovariectomized rats. Biosci. Biotechnol. Biochem. 59: 236-239.

Coudray, C., Bellanger, J., Castiglia-Delavaud, C., Rémésy, C., Vermorel, M., and Rayssignuier, Y. 1997. Effect of soluble or partly soluble dietary fibers supplementation on absorption and balance of calcium, magnesium, iron, and zinc in healthy young men. Eur. J. Clin. Nutr. 151: 375–380.

Crittenden, R.G. 1999. Prebiotics. In: Probiotics: A Critical Review. G.W. Tannock, ed. Horizon Scientific Press, Wymondham. p. 141-156.

Cummings, J.H., Macfarlane, G.T., and Englyst H.N. 2001. Prebiotic digestion and fermentation. Am. J. Clin. Nutr. 73: 415S-420S.

Delzenne, N., and Roberfroid, M.B. 1994. Physiological effects of nondigestible oligosaccharides. Z. Lebensm Wiss Technol. 27: 1-6.

Diplock, A.T., Aggett, P.J., Ashwell, M., Bornet, F., Fern, E.B., and Roberfroid, M.B. 1999. Scientific concepts of functional foods in Europe: Consensus Document. Br. J. Nutr. 81: S1-S27.

Fleet, J.C., and Wood, R.J. 1999. Specific $1,25(OH)_2D_3$-mediated regulation of transcellular calcium transport in Caco-2 cells. Am. J. Physiol. 276: G958-G964.

Fooks, L.J., Fuller, R., and Gibson, G.R. 1999. Prebiotics, probiotics and human gut microbiology. Int. Dairy J. 9: 53-61.

Franck, A. 1998 Prebiotics stimulate calcium absorption: a review. Milchwissenschaft 53: 427-429.

Fullmer, C.S. 1992. Intestinal calcium absorption: calcium entry. J. Nutr. 122: 644-650.

Gibson, G.R., and Roberfroid, M.B. 1995. Dietary modulation of the human colonic microflora: Introducing the concept of prebiotics. J. Nutr. 125: 1401-1412.

Greger, J.L. 1992. Using animals to assess bioavailability of minerals: implications for human nutrition. J. Nutr. 122: 2047-2052.

Griessen, M., Cochet, B., Infante, F., Jung, A., Bartholdi, P., Donath, A., Loizeau, E., and Courvoisier, B. 1989. Calcium absorption from milk in lactase-deficient subjects. Am. J. Clin. Nutr. 49: 377–384.

Institute of Medicine. 1997. *Dietary Reference Intakes: Calcium, Magnesium, Phophorus, Vitamin D, and Fluoride.* National Academy Press: Washington, D.C.: Food and Nutrition Board.

Jackson, M.J. 1997. The assessment of bioavailability of micronutrients: Introduction. Eur. J. Clin. Nutr. 51: S1-S2.

Levrat, M.A., Remesy, C., and Demigne, C. 1991. High propionic acid fermentations and mineral accumulation in the cecum of rats adapted to different levels of inulin. J. Nutr. 121: 1730-1737.

Macfarlane, G.T., and Cummings, J.H. 1999. Probiotics and prebiotics: can regulating the activities of intestinal bacteria benefit health? B.M.J. 318: 999-1003.

Miller, D.D. 1989. Calcium in the diet: food sources, recommended intakes, and nutritional bioavailability. Adv. Food Nutr. Res. 33: 103-156.

Morohashi, T., Sano, T., Ohta, A., and Yamada, S. 1998. True calcium absorption in the intestine is enhanced by fructooligosaccharide feeding in rats. J. Nutr. 128: 1815-1818

Ohta, A., Baba, S., Takizawa, T., and Adachi, T. 1994. Effects of fructooligosaccharides on the absorption of magnesium in the magnesium-deficient rat model. J. Nutr. Sci. Vitaminol. 40: 171-180.

Ohta, A., Ohtsuki, M., Hosono, A., Adachi, T., Hara, H., and Sakata, T. 1998a. Dietary fructooligosaccharides prevent osteopenia after gastrectomy in rats. J. Nutr. 128: 106-110.

Ohta, A., Ohtsuki, M., Uehara, M., Hosono, A., Hirayama, M., Adachi, T., and Hara, H. 1998b. Dietary fructooligosaccharides prevent postgastrectomy anemia and osteopenia in rats. J. Nutr. 128: 485-490.

Ohta, A., Motohashi, Y., Ohtsuki, M., Hirayama, M., Adachi, T., and Sakuma, K. 1998c. Dietary fructooligosaccharides change the intestinal mucosal concentration of calbindin-D9k in rats. J. Nutr. 128: 934-939.

Peng, J.B., Chen, X.Z., Berger, U.V., Vassilev, P.M., Tsukaguchi, H., Brown, E.M., and Hediger, M.A. 1999. Molecular cloning and characterization of a channel-like transporter mediating intestinal calcium absorption. J. Biol. Chem. 274: 22739-22746.

Roberfroid, M.B. 1999. Concepts in functional foods: the case of inulin and oligofructose. J. Nutr. 127: S1398-S1401.

Roberfroid, M.B. 2001. Prebiotics: preferential substrates for specific germs? Am. J. Clin. Nutr. 73: 406S-409S.

Roberfroid, M.B., and Delzenne, N. 1998. Dietary fructans. Annu. Rev. Nutr. 18: 117-143.

Schaafsma, G. 1997. Bioavailability of calcium and magnesium. Eur. J. Clin. Nutr. 51: S13-16.

Scholz-Ahrens, K.E., and Schrezenmeir, J. 2002 Inulin, oligofructose and mineral metabolism -Experimental data and mechanisms. Br. J. Nutr. In press.

Scholz-Ahrens, K.E., Schaafsma, G., van den Heuvel, E.G.H.M., and Schrezenmeir, J. 2001a. Effects of prebiotics on mineral metabolism. Am. J. Clin. Nutr. 73: 459S-464S.

Scholz-Ahrens, K.E., van Loo, J., and Schrezenmeir, J. 2001b. Effect of oligofructose on bone mineralization in ovariectomized rats is affected by dietary calcium. Am J Clin Nutr 73: 498S (abstr).

Scholz-Ahrens, K.E., van Loo, J., and Schrezenmeir, J. 2001c. Long-term effect of oligofructose on bone trabecular structure in ovariectomized rats. Am. J. Clin. Nutr. 73: 498S (abstr).

Takahara, S., Morohashi, T., Sano, T., Ohta, A., Yamada, S., and Sasa, R. 2000. Fructooligosaccharide consumption enhances femoral bone volume and mineral concentrations in rats. J. Nutr. 130: 1792-1795.

Tannock, G.W. 1999. Introduction. In: Probiotics: A Critical Review. G.W. Tannock, ed. Horizon Scientific Press, Wymondham. p. 1-4.

Van den Heuvel, E.G., Muijs, T., Van Dokkum, W., and Schaafsma, G. 1999a. Lactulose stimulates calcium absorption in postmenopausal women. J. Bone Miner. Res. 14: 1211-1216.

van den Heuvel, E.G.H.M., Muys, T., van Dokkum, W., and Schaafsma, G. 1999b. Oligofructose stimulates calcium absorption in adolescents. Am. J. Clin. Nutr. 69: 544–548.

van den Heuvel, E.G.H.M., Schaafsma, G., Muys, T., and van Dokkum, W. 1998. Nondigestible oligosaccharides do not interfere with calcium and nonheme iron absorption in young, healthy men. Am. J. Clin. Nutr. 67: 445–451.

van den Heuvel, E.G., Schoterman, M.H., and Muijs, T. 2000. Transgalactooligosaccharides stimulate calcium absorption in postmenopausal women. J. Nutr. 130: 2938-2942.

Van Loo, J., Cummings, J., Delzenne, N., Englyst, H., Franck, A., Hopkins, M., Kok, N., Macfarlane, G., Newton, D., Quigley, M., Roberfroid, M,, van Vliet, T., and van den Heuvel, E. 1999. Functional food properties of non-digestible oligosaccharides: a consensus report from the ENDO project (DGXII AIRII-CT94-1095). Br. J. Nutr. 81: 121–132.

Younes, H., Demigne, C., Remesy, C. 1996. Acidic fermentation in the caecum increases absorption of calcium and magnesium in the large intestine of the rat. Br. J. Nutr. 75: 301-314.

From: *Probiotics and Prebiotics: Where Are We Going?*
Edited by: Gerald W. Tannock

Chapter 7

The Possible Role of Probiotic Therapy in Inflammatory Bowel Disease

Michael Schultz and Heiko C. Rath

Abstract

Despite many years of extensive research, the role of the luminal bacterial microflora in the pathogenesis of chronic inflammatory bowel diseases has not been fully clarified. There is mounting evidence that a genetically determined immune response is reacting overly aggressively towards components of the intestinal microflora. Recent work has suggested that the course of the disease might be altered by manipulating the intestinal microflora with the use of antibiotics or probiotics. However few clinical trials have been conducted, and the results of *in vitro* experiments are contradictory regarding the effects on the human immune system.

This chapter summarizes recent *in vitro* and *in vivo* data regarding possible disease initiating and perpetuating microorganisms, and possible therapeutic mechanisms of probiotic bacteria relevant to inflammatory bowel diseases. Furthermore, we will review clinical trials examing the efficacy of probiotic microorganisms in inflammatory bowel diseases.

Introduction

Crohn's disease and ulcerative colitis, collectively referred to as inflammatory bowel diseases (IBD), are chronic aggressive disorders, which affect approximately 35-55/100,000 people in Western Europe. Incidence and prevalence varies substantially between regions of the world with regard to different standards in lifestyle and hygiene. The disease normally starts in childhood or youth with a peak between 20 and 30 years of age. Although there are many similarities concerning pathomechanisms and clinical course, the disorders have very distinct features. Ulcerative colitis is characterized by inflammation with superficial ulcerations limited to the mucosa of the colon. It normally starts in the rectum and continuously spreads throughout the large intestine. Crohn's disease, however, is characterized by a discontinuous pattern, potentially affecting the whole gastrointestinal tract. In contrast to ulcerative colitis, the inflammation is transmural with large ulcerations and occasional granuloma. Both entities are mainly located in areas with high bacterial concentrations, such as the terminal ileum and cecum in Crohn's disease and the rectum in ulcerative colitis. Here, bacterial concentrations are as high as 10^{12} per gram dry weight of luminal content. Modern categorizing techniques reveal 300 – 400 distinct bacterial organisms, of which only a few are well known, most are unculturable, unnamed and only described by shape and staining features.

It is the enormous bacterial concentration and the close contact to the gastrointestinal mucosa, the largest surface of the body, equipped with a highly sophisticated immune system, which always incriminated luminal bacteria in the pathogenesis of inflammatory bowel diseases. Since the first descriptions by Samuel Wilks in 1859 and Burrell B. Crohn in 1932, extensive studies have tried to identify a specific pathogen that could be of etiologic importance for ulcerative colitis and Crohn's disease. Research in this field was intensified by the fact that *Helicobacter pylori* was found to be the cause for peptic ulcer disease. On the other hand, overwhelming evidence accumulated, supporting the hypothesis that inflammatory bowel diseases are characterized by a genetically determined, overly aggressive, immune response to ubiquitous luminal antigens, including members of the gut microflora and their products.

However, not all bacterial subsets of the luminal microflora possess equal pro-inflammatory properties. It has been suggested that the microflora can be divided into three functional categories: neutral organisms (e.g. *E. coli*), health promoting organisms (probiotics, e.g. *Lactobacillus, Bifidobacterium,* etc.), and potentially pathogenic organisms (e.g. *Bacteroides sp.*) (Schultz *et al.*, 1997). Antibiotic trials in both IBD (Sutherland *et al.*, 1997) and experimental disease have demonstrated that mainly Crohn's disease can be attenuated with broad spectrum antibiotics or antimicrobial agents affecting primarily the anaerobic fraction of the microflora (Rath *et al.*, 2001; Onderdonk *et al.*, 1981; Rath *et al.*, 1995; Braat *et al.*, 1998). Also fecal stream diversion and bowel rest seem to be an effective tool to reduce the bacterial load in an affected area and promote wound healing (Scholmerich, 1998). Since any antibiotic therapy can be associated with major or minor side effects, the focus of ongoing research was recently shifted towards alternative methods to alter the intestinal microflora in patients with IBD. As early as 1907, Elias Metchnikoff suggested that intestinal bacteria, and especially lactic acid bacteria, play a beneficial role regarding the health of patients (Metchnikoff, 1907). These beneficial microorganisms are commonly called probiotics and are defined as living food supplements which beneficially affect the health status of the host. Preferably, they should be of human origin, nonpathogenic, resistant to technical processes as well as to gastric acid and bile salts, adhere to intestinal epithelial cells, persist for some time in the gut lumen, modulate the immune response and produce antimicrobial substances (Dunne *et al.*, 1999). Since then, probiotic activity has been associated with lactic acid bacteria, bifidobacteria, and other nonpathogenic strains, including certain enterococci and *E. coli* strain Nissle 1917, and non-bacterial organisms such as *Saccharomyces boulardii* (Shanahan, 2001). However, comprehensive comparison of probiotic performance has not yet been performed. Clinical observations including a few controlled trials, basic research, and animal studies have suggested a potential role for probiotic bacteria within the treatment regimens for IBD (Gorbach, 2000; Marteau *et al.*, 2001). So far, it is known that different probiotics act differently regarding the local effect on the intestinal microflora, and the modulation of the immune system. In that respect, few studies have been conducted using probiotic microorganisms in combination. While the clinical efficacy was well demonstrated, the mode of action of these organisms is still largely unclear and *in vitro* studies are inconclusive.

This chapter summarizes recent *in vitro* and *in vivo* data regarding the role of intestinal bacteria in the pathogenesis of chronic intestinal inflammation and possible therapeutic mechanisms of probiotic bacteria relevant to inflammatory bowel disease. Furthermore, we will review clinical trials examining the efficacy of probiotic microorganisms in IBD.

Is IBD the Result of a Persistent Infection?

Pathogenic Microbes in the Pathogenesis of IBD

Characteristic of IBD are the similarities with a variety of intestinal disorders caused by definite infective agents. Macroscopically and microscopically, Crohn's disease can barely be distinguished from ileocecal and rectal manifestations of abdominal tuberculosis, featuring ulcers, abscesses and fistula. *Salmonella*, *Yersinia enterocolitica*, *Shigella* and *Campylobacter* are other pathogens leading to intestinal inflammation resembling Crohn's disease or ulcerative colitis. Although no evidence exists that an infection with these pathogens can initiate IBD, clinical observations demonstrate that certain intestinal and extraintestinal bacterial infections sometimes precede or reactivate chronic intestinal inflammation. Possible disease modifying mechanisms of transient pathogens are the disruption of the mucosal barrier, allowing increased uptake of luminal antigens, mimicry of self-antigens and permanent modulation of the mucosal immune system via the activation of the NF-κB pathway. The few persistent pathogens under current investigation for their active etiological role in inflammatory bowel disease will be reviewed.

Possible Disease Initiating Pathogens

Mycobacterium paratuberculosis

In 1913 Dalziel reported similarities between ileocecal tuberculosis and granulomatous enterocolitis in ruminants (Johne's disease) and a human idiopathic granulomatous enterocolitis which was later described by Crohn *et al.* (Dalziel, 1913; Crohn *et al.*, 1932). He initially suggested a common etiology of these disorders. Later, B.B. Crohn rejected this hypothesis in view of the absence of acid–fast bacilli. The debate about a pathogenic role for *M. paratuberculosis* in IBD was resurrected following the detection of *M. paratuberculosis* in resected Crohn's disease tissues by direct culture and PCR as well as immunologic data and results from antibiotic trials.

Chiodini *et al.* (1984) first isolated a primarily unclassified mycobacterium from resected intestinal tissues of patients with Crohn's disease. This organism was later identified by the same group as *Mycobacterium paratuberculosis*. Subsequently they reported cell-wall-deficient (CWD) microorganisms after 18 and 30 months of incubation of media inoculated with resected intestinal tissues from patients with Crohn's disease (Chiodini *et al.*, 1986). This CWD form could account for the previously described inability to detect mycobacteria in affected intestinal tissues from patients with Crohn's disease. Ultrastructural examination revealed the presence of spheroplasts which did not stain with conventional dyes and failed to grow on hypertonic media. It was postulated that these cell wall-deficient forms may later transform into the bacillary form of *Mycobacterium* species. Spheroplasts were isolated from 61% of patients with Crohn's disease but not from tissues of patients with ulcerative colitis or with other intestinal inflammations. Unfortunately, later studies either failed to detect any mycobacteria or revealed comparable contaminations of a variety of mycobacteria in patients with Crohn's disease and controls (Graham *et al.*, 1987). Several shortcomings of direct culture methods were criticized which may account for the poorly reproducible results: the very low concentrations of the pathogen in the tissue, its high variability, the slow growth and therefore high risk of environmental contamination. Since staining with Ziehl-Neelsen was unable to recover CWD variants of mycobacteria, a further attempt to investigate unculturable microorganisms in tissue was immunohistochemistry. But low sensitivity and specifity accounted for low detection and discrimination rate. In an immunohistochemical search for mycobacteria in the intestinal tissues of patients with Crohn's disease with antibodies against a variety of typical and atypical mycobacteria, including cell-wall-defective forms, no mycobacteria in any of the 67 specimens from 30 patients were detected (Kobayashi *et al.*, 1989).

A promising step towards higher sensitivity and specifity was revealed by PCR technology, since it was able to include CWD forms, and discriminate more sophisticatedly between different strains of mycobacteria. Using IS900, a highly specific target region from the *Mycobacterium paratuberculosis* genome, several studies were performed. Beginning in 1992, Sanderson *et al.* detected *M. paratuberculosis* in 65% of patients with Crohn's disease, in 4% of ulcerative colitis and in 13% of controls. This was later confirmed and extended to the point that other *Mycobacterium* spp. were distributed more equally between the groups (Moss *et al.*, 1992) emphasizing the etiological role of *M. paratuberculosis*. The intensity of the IS900-PCR signal was very low and there was no relation to mycobacterial growth. Moreover, even using this highly sophisticated technique, subsequent studies revealed variable results and many failed to detect any mycobacterial DNA (Kanazawa *et al.*, 1999).

To determine the etiology of a possible infectious disease, the immunologic response to several specific antigens of mycobacteria has been investigated. Using ELISA techniques, Thayer *et al.* reported antibodies reactive with mycobacterial antigens in 23% of patients with Crohn's disease and none in controls or patients with ulcerative colitis (Thayer *et al.*, 1984). These results were later criticized for methodological insufficiency and could not be reproduced by others. Stainsby *et al.* (1993) investigated sera from 38 patients with Crohn's disease, 15 patients with ulcerative colitis, and 30 healthy controls using eight mycobacterial preparations. There was almost equal evidence for a high contact rate with environmental mycobacteria in all patients and controls. A subset of patients with active Crohn's disease had raised IgG concentrations to *M. paratuberculosis*, but this phenomenon was also observed in patients with celiac disease.

Kreuzpaintner *et al.* (1995) investigated species-specific antigens and detected antibodies directed against 45/48-kD antigens, which were significantly more prevalent in patients with Crohn's disease than in various controls. Antibodies were detected in 64.7% of patients with Crohn's disease, 10% of patients with ulcerative colitis, 5% of patients with carcinoma of the colon, and none in healthy controls. Interestingly, 180 days after intestinal resection, the antibody response was reduced, or no longer detectable. This concept might be promising for further evaluations, although it remains contradictory whether mycobacteria induce intestinal inflammation or more easily invade the ulcerated, highly inflamed mucosa. So far, no evidence for a cell mediated immune response towards mycobacteria has been gathered from patients with IBD.

One particular study illustrates the difficulty of detection of mycobacteria in patients with Crohn's disease: Collins *et al.* (2000) conducted a multi-diagnostic test in 439 IBD patients and 324 controls in the United States and Denmark. Using the IS900 PCR, specific to *Mycobacterium paratuberculosis*, 19% of CD patients and 26% of UC patients were positive, but only 6% of the controls. U.S. patients with Crohn's disease were significantly more positive than those from Denmark. This was explained by the higher incidence of BCG vaccination in Europe than in the U.S. which might be also be protective for *M. paratuberculosis* infection as suggested by the authors. Interestingly, the incidence of IBD did not vary to the same degree between the two regions. Culture of the same tissue, which was found positive for the IS900 PCR, failed to grow *Mycobacterium paratuberculosis* from patients and controls. In the U.S., patients with Crohn's disease had significantly more serological evidence of *M. paratuberculosis* than patients with ulcerative colitis or controls. However no difference was detected in patients from Denmark and controls. No patient in the entire study was positive for two tests.

Unfortunately, therapeutic approaches did not reduce the confusion regarding the etiopathogenic role of *M. paratuberculosis* in inflammatory bowel diseases. Several studies have been conducted, using classical antimycobacterial regimes in the treatment of active Crohn's disease. Most were case studies or open uncontrolled trials with little or no effect. Only three controlled trials are reported here. Twenty seven patients with active Crohn's disease were included in a two year randomized double blind, crossover, controlled trial with rifampicin plus ethambutol (Shaffer *et al.*, 1984). Fourteen patients completed the trial, while four patients were withdrawn due to adverse effects. There was no significant difference of CDAI or any clinical indicator of disease activity between the treatment groups. Moreover, beneficial effects were not detected in any of the subgroups of patients.

Swift *et al.* (1994) included 130 patients with active symptoms of Crohn's disease in a double blind, randomized, controlled trial with rifampicin, isoniazid, and ethambutol, or identical placebos over a period of two years. There was no difference in concomitant treatment between the two groups. Seventeen patients in the antibiotic group had side effects compared with three patients on placebo. All of the patients withdrawn early due to adverse reactions were in the active group. There was no significant difference between groups regarding radiological changes of the extent of the disease. In a follow-up investigation after 5 years there was no evidence of consistent benefit or disadvantage from the antimicrobial chemotherapy regarding the number of acute relapses, surgical episodes, hospital admissions, disease activity, blood tests, or medication required for Crohn's disease (Thomas *et al.*, 1998).

Using a very broad spectrum of antimycobacterial drugs, forty patients with refractory, steroid- dependent Crohn's disease were randomized to receive 2 months of tapering steroids plus either a 9-month regimen of ethambutol, clofazimine, dapsone and 1-day dose only of rifampicin, or identical placebo, in a double-blind, placebo-controlled, crossover trial (Prantera *et al.*, 1994). Three out of 19 patients on active medication relapsed during the study period, compared with 11/17 on placebo. Nine patients with relapse or persistent activity of the disease on placebo were crossed over to active drug; five achieved sustained remission, two failed, and two were withdrawn for side effects. There was no substantial endoscopical or radiological effect. Taking into account the very broad antibacterial spectrum of this combination, the positive influence on chronic active intestinal inflammation cannot be restricted to the antimycobacterial effect. In contrast, experimental data support a beneficial effect of very broad spectrum antibiotic regimes on established spontaneous and induced experimental colitis in various animal models (Rath *et al.*, 2001; Dieleman *et al.*, 2001b).

Data on clarithromycin, an antibiotic agent, effective against atypical mycobacteria such as *M. paratuberculosis*, in the treatment of Crohn's disease are rare and contradictory. In 1995, Graham performed a randomized, placebo controlled, crossover trial to test clarithromycin 500mg bid. over 3 months in 15 patients with active Crohn's disease with one year follow up (Graham *et al.*, 1995). Five out of 7 patients receiving clarithromycin achieved remission at the end of the treatment period compared with 1/8 patients in the placebo group. Crossover therapy revealed an effect in only 1/7 patients from the former placebo group now receiving clarithromycin. Once in remission, no relapse was reported within the one year follow-up. On the other hand, in 1999, the same group conducted a second controlled trial randomizing 31 patients into the treatment group with clarithromycin 500mg bid and ethambutol 15mg/kg/day or placebo over 3 months based on increased mucosal permeability assessed with the lactulose-mannitol test (Goodgame *et al.*, 1999). There was no difference between both groups at the end of the treatment period and the one year follow up. Both trials were never published in peer-reviewed journals. Since clarithromycin is the first line antibiotic in the treatment of atypical mycobacteria from the *M. avium* complex (including *M. paratuberculosis)* these conflicting data further question *M. paratuberculosis* as the primary etiopathogenic agent in Crohn's disease.

Also ambiguous are possible mechanisms of zoonotic transmission of *M. paratuberculosis*. There are no records of higher incidence of Crohn's disease in families on farms with *M. paratuberculosis*-infected animals. Attempts to experimentally induce granulomatous colitis by transmitting *M. paratuberculosis*-infected tissues in various animals or direct infection with viable bacteria were mainly negative (Bolton *et al.*, 1973). Although there are reports that *M. paratuberculosis* is found in breast milk of patients with IBD significantly more often than in controls (Naser *et al.*, 2000), no evidence of increased incidence of Crohn's disease in the children of these patients has been noted so far, compared with bottle-fed children from Crohn's disease patients.

In summary, it is still unclear whether *Mycobacteria paratuberculosis* plays a role in chronic IBD in a small group of patients, or, as an environmental pathogen, is a matter of superinfection of Crohn's disease-associated ulcers.

Measles Virus

The hypothesis of a possible measles virus (paramyxovirus)-induced IBD has been extensively studied by Wakefield *et al.* (1993), suggesting a granulomatous vasculitis, induced by persistent measles infection of mesenteric endothelial cells, leading to focal mesenteric ischemia as the underlying pathogenic abnormality in Crohn's disease. In this setting, the persistent infection with measles is a result of tolerance towards virus antigens and the reactivated inflammation is caused by the breakdown of this tolerance induced by other pathogens or different measles virus strains. Clinical evidence for this hypothesis was provided by the occurrence of intestinal aphthous ulcers (Hudson *et al.*, 1992) and vasculitis in acute measles infection and the fact that persistent measles infection is occasionally associated with subacute sclerosing panencephalitis. Other investigators however, reported, that granulomas in Crohn's disease were preferentially associated with lymphatic channels rather than with blood vessels (Talbot *et al.*, 1992). In further studies Wakefield *et al.* (1993) detected measles-like virus particles in gut tissues, especially in the granulomas of Crohn's disease patients by immunohistochemical staining, *in situ* hybridization, and electron microscopy. Other investigators failed to confirm these results using very sensitive polymerase chain reaction primers (Iizuka *et al.*, 1995).

Much attention has been given to epidemiological studies in the context of measles infections and IBD, but again, there are conflicting data. In one study, patients born 3 months following the peak of a measles epidemic had a significantly higher risk of acquiring Crohn's disease than expected (Ekbom *et al.*, 1994). In a case control study, however, no association between measles epidemics and the incidence of Crohn's disease could be documented (Thompson *et al.*, 1995a). A Swedish study has shown an increased risk of developing Crohn's disease in newborns from mothers with measles infection during pregnancy (41) but, on the other hand, a danish study could not link any cases of Crohn's disease due to maternal measles infection during pregnancy (Nielsen *et al.*, 1998). Thompson *et al.* (1995b) reported an increased risk of Crohn's disease and ulcerative colitis in 3000 vaccinated people of a 1964 measles vaccine trial in the United Kingdom, evoking a flood of letters criticizing the study design and the inappropriate statistical methods used (Thompson *et al.*, 1995b). Moreover, others failed to support these reports (Feeney *et al.*, 1997).

Montgomery *et al.* (1998) later presented data, suggesting, that the concurrent infection with measles and mumps in early childhood increased the risk for developing Crohn's disease but not measles infections alone. Using the Bradford-Hill criteria to analyze the relationship between measles virus and Crohn's disease, Robertson and Sandler did not find any scientific support for an etiological role of measles virus in IBD (Robertson *et al.*, 2001).

In summary, whether measles virus are pathogenic for Crohn's disease is still a very controversial issue and no recommendation can be given with regard to measles vaccination.

Other Possible Primary Pathogens

Liu *et al.* (1995) reported antigens for *Listeria monocytogenes* in macrophages and giant cells next to typical gut lesions, within the lamina propria, and in mesenteric lymph nodes in 75% of patients with Crohn's disease, compared with 13% of patients with ulcerative colitis and 0% of controls. But *Listeria* are ubiquitous organisms and the simultaneous presence of *E. coli* and streptococcal antigens in over 80% of the positive Crohn's disease patients suggest a non-specific secondary invasion rather than a primary, causative infection. Other investigators have failed to detect *Listeria* in colonic tissues of Crohn's disease patients (Chiba *et al.*, 1998a), but superinfection with this ubiquitous pathogen may lead to exacerbation of disease (Chiba *et al.*, 1998b). Recently, a novel bacterial genomic sequence was detected in 43% of affected tissues from Crohn's disease but in less than 10% in ulcerative colitis and controls (Sutton *et al.*, 2000). Although this sequence later was sourced to *Pseudomonas fluorescens* (Wei *et al.*, 2001) possible consequences regarding the pathogenic importance and therapeutic implication are still under investigation.

Adherent *Escherichia coli* have been linked with ulcerative colitis (Burke *et al.*, 1988) and to a lesser extend with Crohn's disease (Giaffer *et al.*, 1992) but, so far, other investigators could not confirm these results. However, enteroadherent *E. coli* may be associated with early recurrence of active Crohn's disease after surgery (Darfeuillie-Michaud *et al.*, 1998). The importance of these findings for IBD remains to be further investigated.

Possible Disease Perpetuating Pathogens

IBD is characterized as a chronic relapsing disorder, and some observations suggest, that the relapse might be triggered by a transient gastrointestinal or extraintestinal infection. Kangro *et al.* (1990) reported that in children, up to 60% had respiratory infections and Weber *et al.*, 1992) reported that 18% had enteric infections preceding a relapse of Crohn's disease or in 13% of ulcerative colitis. Further viral (e.g. cytomegalovirus, rotavirus, respiratory syncytial virus), bacterial (e.g. *Clostridium difficile, Salmonella, Shigella, Yersinia* and *Campylobacter*) and protozoal pathogens (e.g. *Entamoeba histolytica* and *Gardia lamblia*) have been associated with an aggravation of quiescent intestinal inflammation. Although this association is controversially discussed, and extensive diagnostic testing in most cases not pursued, a co-infection with a potentially perpetuating pathogen in an acute relapse situation needs to be carefully considered.

Reports of infections with typical enteric pathogens such as *Salmonella, Yersinia, Shigella* and *Campylobacter* in patients with an acute relapse of IBD are inconclusive and vary from 0 to 60% depending on the techniques used to detect these organisms (Weber *et al.*, 1992; Kallinowski *et al.*, 1998). This leads to the recommendation to restrict stool cultures to patients with a complicated flare up of their known Crohn's disease, with no or inadequate response to steroid treatment.

Infection with *Clostridium difficile* has been linked to hospitalized IBD patients with a severe exacerbation (Meyers *et al.*, 1981). Uncomplicated flare ups in outpatients seem not to be driven by this pathogen. The importance of a preceding antibiotic therapy is controversially discussed. However, the prognostic impact of *C. difficile*-driven exacerbations, leading to live threatening complications such as toxic megacolon in ulcerative colitis, and the positive outcome following treatment with metronidazole underlines the importance to include the possibility of an infection with *C. difficile* of hospitalized patients with severe relapse in the diagnostic workup (Bolton *et al.*, 1980).

Although an infection with cytomegalovirus (CMV) is clinically relevant only in immunocompromised patients, such as patients, infected with HIV and low $CD4^+$ T-cell counts, or patients undergoing chemotherapy, there is evidence that CMV infections may account for at least some relapses of IBD with a fulminant course and improvement after antiviral therapy (Cottone *et al.*, 2001). In cases of steroid refractory courses of disease with lymphopenia, fever, myalgia, and mildly elevated liver enzymes, diagnostic

CMV serology may be performed. In some cases of pouchitis, CMV infection mimics the relapse of IBD, with small ulcers in the inflamed mucosa, otherwise seen in ulcerative colitis. Wakefield suggested a potential role for co-infections with several herpes virus types (Wakefield *et al.*, 1992).

In summary, steroid resistance, or any unusual aggravated course of inflammatory bowel disease should lead to serious consideration of the possibility of a co-infection with various pathogens. Typical clinical symptoms, and evidence of antibiotic therapy prior the onset of relapse, can narrow the possible candidates. Bacterial culture or serology should complete the diagnostic work-up and specific antimicrobial or antiviral therapy should be applied.

Mechanism of Pro- and Anti-inflammatory Capabilities of Commensal Bacteria

Altered Bacterial Function in IBD

Several mechanisms of altered functional specifications of resident luminal bacteria may account for increased proinflammatory capabilities in ulcerative colitis and Crohn's disease. *E. coli* strains with abnormal invasion, chemotaxis, secretion of toxins, or epithelial adhesion, isolated from patients with inflammatory bowel disease are under investigation by several groups (Sartor, 1999). The median index of adhesion to buccal epithelial cells for *E. coli* from patients with ulcerative colitis in relapse was significantly higher (43%) than that for healthy controls (5%) or patients with infectious diarrhea (14%). Of specific note, the index was not significantly different among isolates from patients with active ulcerative colitis, Crohn's disease (53%), or ulcerative colitis in remission (30%) (Burke *et al.*, 1988).

Abnormal enterotoxin secretion of commensal bacteria may result in increased uptake of enteric bacteria and their products. Confluent HT-29 enterocytes were incubated with *Bacteroides fragilis* enterotoxin, followed by incubation with enteric bacteria, such as *Salmonella typhimurium, Listeria monocytogenes, Proteus mirabilis, Escherichia coli*, and *Enterococcus fecalis*. While *B. fragilis* enterotoxin did not affect enterocyte viability, it increased internalization of the other bacterial strains (Wells *et al.*, 1996).

Butyrate and other short chain fatty acids (SCFA) are produced by luminal anaerobic bacteria from food components such as non-digestible carbohydrates and proteins. SCFA are an essential fuel for colonic enterocytes and have anti-inflammatory capabilities (Segain *et al.*, 2000). It has been demonstrated, that butyrate decreased TNF production and proinflammatory cytokine mRNA expression in lamina propria mononuclear cells from Crohn's disease patients. It furthermore abolished LPS induced cytokine expression in peripheral blood mononuclear cells and activation of NFκB while inhibitory IκB remained stable.

Ulcerative colitis has been described as a "starving disease", due to the correlation of the inflammation with decreased luminal concentrations of mainly butyrate (Roediger, 1980). Clinical effects of butyrate enemas however, are inconclusive (Steinhart *et al.*, 1996; Scheppach *et al.*, 1992).

Hydrogen sulfides in the gut lumen are potential candidates to form persulphides with butyryl-CoA, which would inhibit cellular short-chain acyl-CoA dehydrogenase and oxidation, and to induce an energy-deficient state in colonocytes and consecutively lead to mucosal inflammation (Roediger *et al.*, 1993). It has been demonstrated that hydrogen sulfide concentrations are elevated in the luminal content of patients with ulcerative colitis. Sulfate-reducing bacteria, especially rapidly growing strains, such as *Desulfovibrio* spp. are a major source for luminal hydrogen sulfide, but concentrations in faeces of patients with ulcerative colitis are not significantly different from those in controls (Pitcher *et al.*, 2000).

IBD as the Result of an Aberrant Intestinal Microflora?

In recent decades abundant data have been generated from *in vitro* studies, experiments using animal models for intestinal inflammation and clinical trials suggesting a crucial role for normal luminal bacteria in the pathogenesis of IBD. The human body harbors 10 times more microorganisms than human cells. In a healthy body, the intestinal microflora forms an immunologically balanced symbiosis with the mucosal immune system. However, chronic intestinal inflammation, such as IBD, normally occurs in areas of highest bacterial concentrations – the terminal ileum, cecum and rectum. In these parts of the gastrointestinal tract, bacterial concentrations can reach up to 10^{12}/gram dry stool weight with a predominantly anaerobic composition (obligate anaerobes, such as *Bacteroides* spp. can outnumber facultative aerobic bacteria by 10^3). This delicate balance between the host defense and the potentially aggressive luminal content is disrupted in patients with inflammatory bowel disease.

The bacterial microflora develops sequentially during childhood and can be divided into a large and stable autochthonous and small transient portion. Data from different groups suggest, that the bacterial profile might be genetically determined and is relatively stable throughout life (van de Merwe *et al.*, 1988). There is evidence for the possibility that the fecal microflora differs in patients with active IBD compared to quiescent disease or healthy volunteers. This was demonstrated by Giaffer *et al.* (1991): while the total concentration of anaerobic organisms in IBD was not different compared to normal controls, *Bifidobacterium* seemed to be decreased, whereas *Bacteroides vulgatus* and *Bacteroides fragilis* were increased. The intestinal microflora in IBD was studied in more detail by other investigators, who in contrast demonstrated a significant decrease in the number of anaerobic bacteria and *Lactobacillus* in patients with active but not inactive ulcerative colitis (Fabia *et al.*, 1993; Hartley *et al.*, 1992). Furthermore, a decrease in fecal concentrations of *Bifidobacterium* in Crohn's disease and in patients with active pouchitis has been reported (Ruseler-van Embden *et al.*, 1983). Alteration of this balanced composition may result in chronic stimulation of the mucosal immune system.

The hypothesis of bacterial influence on the pathogenesis of IBD is strongly supported by the attenuation or complete absence of colitis in a germfree environment in a variety of animal models of chronic intestinal inflammation (Sartor, 1999). Germfree and disease-free HLA-B27 transgenic rats as well as IL-10 knockout (IL-10$^{-/-}$) mice, associated with a specific-pathogen free-microflora develop colitis and gastritis correlating with the time of association (Rath *et al.*, 1996; Sellon *et al.*, 1998). However, not all bacteria have equal proinflammatory capabilities as demonstrated by selective colonization in a gnotobiotic environment. HLA-B27 transgenic rats, raised under germfree conditions, develop only moderate colitis and gastritis if colonized with a cocktail of 5 different bacterial strains including *Bacteroides vulgatus*, and neither colonic nor gastric inflammation if colonized with the same cocktail excluding *B. vulgatus* (Rath *et al.*, 1996). Interestingly, monoassociation with *B. vulgatus* leads to colonic inflammation to the same degree as the whole cocktail including *B. vulgatus*, but not to gastritic inflammation indicating that some bacteria, although not involved in the induction of disease at the primary site, may account for secondary inflammation at remote sites of the gut (Rath *et al.*, 1999). Reports of *B. vulgatus*-induced colitis also exist from carrageenan-fed guinea pigs (Onderdonk *et al.*, 1981) and TCRα$^{-/-}$ mice (Kishi *et al.*, 2000).

The response to selective subsets of the gut microflora may be genetically determined and therefore host dependent. Gnotobiotic IL-10$^{-/-}$ mice, colonized with the cocktail as described above including *B. vulgatus* developed only mild colitis (Sellon *et al.*, 1998), while mice, monoassociated with *Enterococcus faecalis* exhibited moderate colitis and a selective *in vitro* T-cell response (Kim *et al.*, 2001). The more aggressive colitis in animal models colonized with a specific-pathogen-free microflora suggests synergistic effects of different strains. This is supported by the failure of metronidazole, which is selectively active against anaerobic bacteria, to attenuate established intestinal inflammation but to prevent the onset in different animal models for experimental colitis (Rath *et al.*, 2001; Dieleman *et al.*, 2001). Treatment with broad spectrum antibiotics (vancomycin/ imipenem) reduced established intestinal inflammation to an almost normal degree (Rath *et al.*, 2001).

The patho-mechanism by which intestinal bacteria exert their proinflammatory influence is not completely understood. It is still controversial whether gut bacteria and their products induce chronic intestinal inflammation and extraintestinal manifestation through antigens, such as peptidoglycan-polysaccharides (PG-PS), lipopolysaccharides (LPS), or bacterial DNA motifs (CpGs), which abnormally interact with the mucosal immune system or whether bacteria simply secondarily invade ulcers in inflamed areas and perpetuate established disease.

Liu *et al.* (1995) have detected *E. coli*, *Listeria* and *Streptococcus* antigen in Crohn's disease ulcers using immunohistochemistry. Macrophages and giant cells immunolabeled for this antigen were distributed underneath ulcers, along fissures, around abscesses, within the lamina propria, in granulomas, and in the germinal centers of mesenteric lymph nodes. More recently, quantitative competitive polymerase chain reaction (PCR) of paraffin-embedded intestinal specimens from 212 patients showed that *Pseudomonas fluorescens* DNA was present in 43% of colonic lesions in Crohn's disease but only in 9% in lesions of ulcerative colitis and 5% of non-inflammatory bowel disease specimens (Sutton *et al.*, 2000). This was prevalent regardless of disease status. Enzyme-linked immunosorbent assay analysis (ELISA) showed IgA seroreactivity in 54% Crohn's disease patients, but only in 4% of normal controls. There is also strong evidence, that bacteria from the distal intestine produce proinflammatory agents which activate the mucosal immune system. Bacterial cell wall polymers, such as LPS and PG-PS can induce experimental colitis in susceptible animal models (Sartor, 1999). PG-PS, injected subserosaly in the caecum of rats, can induce chronic, spontaneously relapsing, granulomatous disease, with local enterocolitis and associated extraintestinal manifestations such as distal arthritis and granulomatous hepatitis (Stimpson *et al.*, 1987). PG-PS may also account for hepatobiliary manifestations induced by experimental bacterial overgrowth in jejunal self-filling blind loops in genetically susceptible Lewis rats (Lichtman *et al.*, 1991). Formylated oligopeptides, such as N-formyl-methionyl-leucyl-phenylalanine (FMLP), synthesized by colonic bacteria could be important in the pathophysiology of colonic inflammation and is frequently associated with hepatobiliary complications (Hobson *et al.*, 1988). An enterohepatic circulation of synthetic FMLP has been demonstrated in the rat. Following colonic instillation of FMLP, the mean biliary excretion was almost 10-fold higher in rats with colitis compared with non-inflamed controls and could be attributed to the increased mucosal permeability (Ritter *et al.*, 1988).

Recently, some attention has been drawn to CpG motifs of bacterial DNA which stimulate a variety of immune competent cells to a T_{H1}-dominated response. Using the dextran sodium sulfate (DSS) mouse model, Obermeier could demonstrate that CpG´s aggravated the inflammation if given in the acute or chronic phase of intestinal inflammation, but pretreatment with CpG´s resulted in reduced inflammation suggesting induction of tolerance to bacterial antigens (Obermeier *et al.*, 2001).

Impact on the Luminal Microflora by Probiotic Therapy

Whether the intestinal microflora can be drastically altered by oral administration of high concentrations of probiotic microorganisms is still unclear. This would depend in part on the resistance of the administered strains to gastric acid, bile salts, and pancreatic enzymes, but the fate of the probiotic organism within the intestinal tract is known for only a few strains. *Lactobacillus casei* subsp. *rhamnosus* GG (*Lactobacillus* GG) for example can be detected in stool samples as viable bacteria for up to one week after the oral administration has ceased (Goldin *et al.*, 1992; Alander *et al.*, 1999) and *Lactobacillus acidophilus* (*Lactobacillus johnsonii* strain La1) and *Bifidobacterium sp.* can be found in ileal fluid following oral administration (Marteau *et al.*, 1992; Pochart *et al.*, 1992). Survival in gastric acid is also known for most other lactic acid bacteria, including *Lactobacillus reuteri* (various strains), *Lactobacillus gasseri* strain ADH, *Lactobacillus plantarum* 299, with the exception of *Lactococcus lactis*, which is largely inactivated in the duodenum by bile salts (Drouault *et al.*, 1999). The authors suggest, that this organism could therefore be used as a vector to deliver substances specifically into the duodenum, for example to treat pancreatic deficiencies. In a later study however, this *Lactococcus lactis* was engineered to secrete biologically active murine IL-10 in the colon. The anti-inflammatory effect was documented in the treatment of murine chronic colitis induced by DSS and in prevention of spontaneous colitis in IL-10$^{-/-}$ mice (Steidler *et al.*, 2000).

The ability to adhere to human intestinal epithelial cells might also be important for a suggested impact on the composition of the luminal microflora. Most studies have used *in vitro* settings with immortalized cell lines of human origin (HT-29 and Caco-2) to address this question, however, Alander *et al.* (1997) were able to demonstrate adherence of *Lactobacillus* GG to human colonocytes *in vivo*. The test subjects consumed a whey drink fermented with *Lactobacillus* GG for 12 days. The presence of *Lactobacillus* GG was then checked both in the fecal samples and in the colonic biopsies obtained from various locations in the large intestine following evacuation of the colon using laxatives. In all patients, *Lactobacillus* GG was the dominant fecal lactic acid bacterium, and in 4/5 patients, *Lactobacillus* GG could also be recovered from the biopsies, while in one patient, who suffered from ulcerative colitis, no *Lactobacillus* GG was isolated from biopsies. All other volunteers had no intestinal disorders. Furthermore, it has been shown that *Lactobacillus acidophilus* strains LB, LA1 and *Lactobacillus* GG bind to enterocyte-like Caco-2 cell lines (Chauviere *et al.*, 1992; Bernet *et al.*, 1994; Elo *et al.*, 1991), while *Lactobacillus bulgaricus* and several other *Bifidobacterium* and *Lactobacillus* strains showed none or very weak adhesive properties. Additionally, it was shown that *Lactobacillus acidophilus* LA1 inhibited cell attachment and invasion by enterotoxigenic and enteropathogenic *E. coli*, and *Salmonella typhimurium* and cell invasion by *Yersinia pseudotuberculosis* (Berent *et al.*, 1994). This was also documented for *E. coli* strain Nissle. *In vivo* experiments have suggested a protective effect of *E. coli* strain Nissle against *Salmonella typhimurium* infection. It was shown, using the human epithelial cell line INT407, that it directly inhibited the invasion of this pathogen and of adherent-invasive *E. coli*, (Boudeau *et al.*, 1999; Ölschläger *et al.*, 2001).

Several other lactic acid bacteria and *E. coli* strain Nissle 1917 have also been demonstrated to effectively inhibit the *in vitro* growth of many enteric pathogens including *Salmonella typhimurium*, *Staphylococcus aureus*, enteropathogenic *E. coli*, *Clostridium perfringens*, *Clostridium difficile*, *Listeria monocytogenes*, and *Candida albicans* (Silva *et al.*, 1987; Meurman *et al.*, 1995; Hockertz, 1997). Different mechanisms have been described as being responsible for this antimicrobial effect that could strengthen the position of a probiotic organism in a highly competitive environment such as the intestinal microflora. For *Lactobacillus rhamnosus* GG, the secretion of a microcin has been documented (Silva *et al.*, 1987) active against a wide range of other intestinal bacterial inhabitants, such as anaerobic bacteria (*Clostridium* spp., *Bacteroides* spp., *Bifidobacterium* spp.) and members of the family *Enterobacteriaceae*, *Pseudomonas* spp., *Staphylococcus* spp., and *Streptococcus* spp. as demonstrated by a microbiological assay. It did not inhibit other lactobacilli. *Lactobacillus plantarum* 299v enhances intestinal mucin gene expression (MUC2, MUC3) to inhibit the adherence of enteropathogenic *E. coli* to HT-29 cells (Mack *et al.*, 1999), but other mechanisms to enhance survival in a highly populated environment are also suspected (Hockertz, 1997).

The large bowel of humans and animal models is inhibited by a complex microbial community (Simon *et al.*, 1995). The collection of bacteria detected in faeces only reflects the bacteria present in the distal large bowel, whereas microorganisms, adherent to the gut wall, or present in more proximal segments of the intestinal tract are not taken into account. Due to feasibility aspects, most studies of the human intestinal microflora usually involve analyses of the bacterial composition in fecal samples. Therefore, studies on the impact of oral administration of probiotic microorganisms on the intestinal microflora are limited. Tannock *et al.* (2000) recently analyzed the fecal microflora of human subjects consuming *Lactobacillus rhamnosus* strain DR20. *Lactobacillus* and enterococcal contents of the intestinal microflora were only transiently altered, without markedly affecting biochemical and other bacteriological factors. Similarly, Venturi *et al.* (1999) demonstrated significant changes in the composition of the luminal microflora but just affecting a small fraction. Patients with ulcerative colitis in remission were treated with a probiotic preparation (VSL#3) p.o. for 12 months, containing large numbers of seven different bacterial strains (4 strains of lactobacilli, 3 strains of bifidobacteria, and 1 strain of *Streptococcus salivarius* subsp. *thermophilus*). Throughout the treatment period, increased fecal concentrations of the administered strains (*Streptococcus salivarius* subsp. *thermophilus*, bifidobacteria and lactobacilli) were found in all patients, but returned to basal levels only 15 days after administration had ceased. Concentrations of *Bacteroides*, clostridia, coliforms, total aerobic and anaerobic bacteria did not change significantly during treatment.

Alteration of the Gastrointestinal Mucosal Barrier

A major task of the intestine is to form a defensive barrier to prevent absorption of damaging substances from the external environment. There is evidence, that this permeability is increased in most patients with Crohn's disease and even in 10-20% of their clinically healthy relatives (Hilsden *et al.*, 1996). If this precedes the onset of the gastrointestinal manifestation of IBD, or is simply an unspecific epi-phenomenon, is still a subject for discussion (Teahon *et al.*, 1992).

The abnormal leakiness of the mucosa in Crohn's patients and their relatives can be amplified by oral administration of non-steroidal anti-inflammatory drugs (NSAIDs) and aspirin (Hollander, 1999). A broken mucosal barrier may then result in increased uptake of bacteria and their products (Hobson *et al.*, 1988) and lead to chronic inflammation and a vicious cycle when bacterial products, such as FMLP, will undergo enterohepatic circulation and by itself reduce the mucosal barrier even further (von Ritter *et al.*, 1988). Important contributors to the mucosal barrier are epithelial tight junctions and the superficial mucus. Defective tight junctions in dominant negative N-cadherin transgenic mice lead to focal chronic intestinal inflammation and adenomas (Hermiston *et al.*, 1995). It has furthermore been shown, in experimental models, that NSAIDs may disrupt the homeostasis of the intestinal microflora and induce overgrowth of some bacterial species which exacerbate NSAID-induced mucosal injury (Kent *et al.*, 1969; Reuter *et al.*, 1997; Satoh *et al.* 1983). Elliott *et al.* (1998) studied the effect of bacterial colonization on ulcer healing in rats. Gastric ulcers were induced by serosal injection of acetic acid. Within 6-12h of their induction, gastric ulcers were colonized by a variety of bacteria. Suppression of colonization with streptomycin and penicillin markedly accelerated healing. The induction of *Lactobacillus* colonization with oral administration of lactulose also accelerated the healing process, suggesting a protective effect by these bacterial species. Administration of an antibiotic-resistant *E. coli* strain reversed the beneficial effect of antibiotics.

The positive effect of *Lactobacillus* GG ingestion on the gastrointestinal mucosal barrier was studied in detail by several investigators. The intestinal permeability of 14 day old suckling rats was greatly enhanced by daily gavage with cow's milk, as documented at 21 days by increased absorption of intact horseradish peroxidase, but the addition of *Lactobacillus* GG to cow's milk counteracted this permeability disorder significantly (Isolauri *et al.*, 1993). Gotteland *et al.* (2001) showed that heat-killed *Lactobacillus* GG did not modify the indomethacin-induced increase of gastrointestinal permeability in human volunteers, however, live bacteria significantly reduced the alteration of gastric but not intestinal permeability. Oral administration of *Lactobacillus plantarum* 299v to an animal model of spontaneous bacterial peritonitis due to portal hypertension did not inhibit the bacterial translocation of intestinal *E. coli* and therefore did not change the course of the disease but other experimental settings showed contradictionary results (R. Wiest, personal communication).

Recent pilot studies have demonstrated that the new probiotic compound, VSL#3, is efficacious as maintenance therapy in pouchitis and ulcerative colitis (Venturi *et al.*, 1999; Gionchetti *et al.*, 2000) but the mechanism of action remains unclear. The aim of a further study was therefore to determine the efficacy of VSL#3 as a primary therapy in the treatment of colitis in the IL-10$^{-/-}$ mouse model, while mechanisms of action were investigated in T-84 monolayers *in vitro*. Treatment of IL-10$^{-/-}$ mice with VSL#3 resulted in normalization of colonic barrier integrity in conjunction with a reduction in mucosal secretion of TNF-α and IFN-γ and an improvement in histologic disease. *In vitro* studies showed that epithelial barrier function and resistance to *Salmonella* invasion could be enhanced by exposure to a protein-like soluble factor secreted by the bacteria found in the VSL#3 compound. Since VSL#3 is a compound of different bacteria, it remains unclear which member of the preparation is responsible for the observed effects (Madsen *et al.*, 2001).

Immune Effects of Commensal Bacteria

It is still unknown how the immune system is able to distinguish between self and foreign, good and bad, health-promoting and pathogenic luminal contents. Since the impact on the intestinal microflora by probiotics is limited, and the clinical relevance of the increased gut permeability in IBD is unclear, interest recently shifted towards the influence of probiotics on the host's immune system. Animal experiments document that the microflora is the major stimulus to the gut immune system and is also a potent regulator of the innate immune system, as seen in experiments with germfree animals (Wold *et al.*, 2000). The current hypothesis of the interaction between the intestinal microflora and the mucosal immune system suggests a central role for M cells located in the area of Peyer's patches to take up antigenic material (Kraehenbuhl *et al.*, 2000). The small intestine, where Peyer's patches are situated, is populated by a relatively small number of microorganisms (10^{3-8} CFU/g content), compared to the large intestine (10^{11-14} CFU/g content). It seems possible, that the oral intake of approximately 10^{10} probiotic microorganisms daily can influence the gut immune system by uptake through Peyer's patches (Wold, 2001). Further evidence of host-dependent immunologic response to residual bacterial microflora was provided by

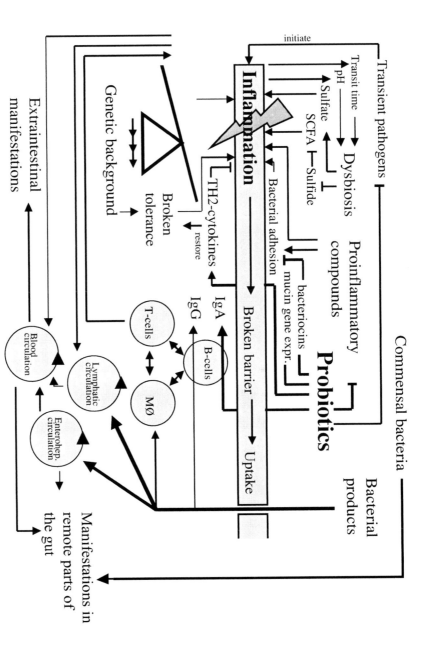

Figure 1. Pathogenetic mechanism of chronic inflammatory bowel disease. Commensal bacteria and their products are a constant challenge to an overly aggressive immune system in a genetically susceptible host. Probiotics may have beneficial influence in this proinflammatory cascade.

Duchmann *et al.* (1996). Mononuclear cells from spleen, small and large bowel of mice did not proliferate when exposed to autologous bacterial sonicates, but proliferated when exposed to bacterial sonicates derived from the heterologous intestine. Moreover, it was shown that both local and systemic tolerance to autologous microflora were broken in TNBS induced colitis. Tolerance to autologous microflora could be restored and colitis abrogated in mice systemically treated with IL-10 or antibodies to IL-12 suggesting a T_{H1}-type of immunoresponse, but the treatment did not influence the proliferation induced by heterologous microflora. It was furthermore shown, that LPMC isolated from inflamed areas of the intestine of patients with active IBD strongly proliferated after co-culture with sonicated autologous microflora in contrast to LPMC from non-inflamed areas from the same patients, and PBMC from patients with IBD or controls (Duchmann *et al.*, 1995). These cells from controls, however, strongly proliferated after co-culture with bacterial sonicates from heterologous intestinal contents.

T-cells play an important role in IBD. In particular, T_{H1}-mediated cytokines such as IL-2 and IFN-γ have been detected in the mucosa of Crohn's disease patients while in ulcerative colitis the profile of mucosal cytokines seems to belong to the T_{H2}-type with increased production of IL-4 and IL-10 (Sartor, 1994). This has largely been confirmed by animal models with experimental colitis (Sartor, 1996b). The T-cell response in experimental colitis is characterized by a T_{H1}-profile, as demonstrated by the induction of spontaneous chronic colitis in SCID-mice after transfer of CD45RB[high] cells, derived from BALB/c donors (Powrie *et al.*, 1990). This was confirmed in IL-10[-/-] mice which develop colitis and express high IL-12 and IFN-γ levels in the presence of conventional microflora. The disease is attenuated in specific pathogen free environment and completely absent under germfree conditions (Sellon *et al.*, 1998). This T_{H1}-response is directed against luminal bacteria, as demonstrated by IFNγ secretion of T-cells of $CD_3ε$ transgenic mice substituted with bone marrow cells after *in vitro* stimulation with sonicates of normal luminal microflora (Veltkamp *et al.*, 2001). The importance of T-cells was further illustrated by experiments with HLA-B27 transgenic rats. Creation of a self-filling blind loop of the caecum leads to highly aggravated inflammation with mainly anaerobic bacterial overgrowth. Exclusion of the cecum from the fecal stream, with subsequent decrease of total bacterial counts, does not only result in complete resolution of the local inflammation but also of the gastritis (Rath *et al.*, 1999), suggesting primary response of T-cells, located in the lymphoid aggregates of the caecal tip. These findings are further supported by attenuated colitis in T-cell receptor α[-/-] mice post appendectomy (Mizoguchi *et al.*, 1996). When

TCR-$\alpha^{-/-}$ mice underwent appendectomy at 3-5 weeks, the number of mesenteric lymph nodes cells at 6-7 months were significantly less than in the sham-operated TCR-$\alpha^{-/-}$ controls. Furthermore, appendectomy at 1 month of age suppressed the development of colitis at 6-7 months of age, leaving only 3% of knockout mice with intestinal inflammation, compared with 80% of controls. These results supported the concept that lymphoid follicles located in the appendix may play an important role in the development of chronic colitis.

There is a large body of evidence for the modulation of cytokine production by probiotic bacteria. While it could be demonstrated in both human IBD and animal models with experimental colitis that levels of certain proinflammatory cytokines (IFN-γ, TNF-α) decrease following the administration of probiotic preparations (Ulisse *et al.*, 2001; Helwig *et al.*, 1999; Schultz *et al.*, 1998), it remains unclear whether this effect is due to the attenuation of the disease, or is directly influenced by the probiotic microorganism. To overcome this problem, human intestinal cell lines (Haller *et al.*, 2000a) and peripheral blood mononuclear cells (Haller *et al.*, 2000b) were stimulated with nonpathogenic and pathogenic bacteria, and cytokine responses as well as cytokine gene transcription was measured. According to recent studies by Miettinen *et al.* (1996; 1998), lactic acid bacteria induce the production of several cytokines, including IL-12, IL-18, IFN-γ, TNF-α, IL-6 and IL-10 in human monocytes and additionally activate NF-κB and STAT signalling pathways (Mietinen *et al.*, 2000), but this does not result in a clear pro- or anti-inflammatory profile. We were able to demonstrate a significant induction of NF-κB-mediated IL-8 secretion of HT-29 cells upon stimulation with the probiotic microorganisms *E. coli* strain Nissle, but not with *Lactobacillus* GG (unpublished data). There was no effect by repeated oral exposure to viable or non-viable *Lactobacillus acidophilus*, *Lactobacillus* GG, or *Streptococcus thermophilus* on basal cytokine mRNA expression in Peyer's patches, spleen or mesenteric lymph nodes of mice (Tejada-Simon *et al.*, 1999). Recently, the probiotic compound VSL#3 has been studied in *in vivo* and *in vitro* settings regarding the impact on the cytokine profile in pouchitis. For this purpose, pouch biopsy samples were obtained from 7 patients with chronic active pouchitis before and after antibiotic (ciprofloxacin and rifaximin) and probiotic treatment. Tissue samples from 5 patients with a normal pouch have been used as control. Tissue levels of TNF-α were increased ($p<0.01$) in active pouchitis compared to the uninflamed pouch and reduced after antibiotic and probiotic treatment. Also IFN-γ and IL-10 tissue levels were augmented in pouchitis and also decreased following treatment. Levels of IL-4 and IL-10 were unchanged in inflamed pouches, and unaffected by antibiotic treatment. However, IL-10 increased following probiotic treatment. Moreover, inflamed pouches had higher levels of inducible nitric oxide synthase and gelatinase activities which decreased following treatment (Ulisse *et al.*, 2001).

Regarding humoral immune responses within the intestinal tract, it is noteworthy, that there is evidence of a local overproduction of IgG with a relative mucosal IgA deficiency in patients with Crohn's disease (Brandzaeg *et al.*, 1989). There have been several reports describing the effects of probiotics on sIgA, and especially several *Lactobacillus* sp. were shown to enhance IgA immune responses (Perdigon *et al.*, 1991; 1995). Malin and collegues studied the effect of oral bacteriotherapy with human *Lactobacillus* GG in children with Crohn's disease (n = 14) and juvenile arthritis (n = 9). The mean number of antibody secreting cells in the IgA class to β-lactoglobulin and casein increased significantly following 10 days of treatment. This indicates that orally administered *Lactobacillus* GG has the potential to increase the gut immune response and thereby promotes the gut immunological barrier (Malin *et al.*, 1996). Furthermore, Kaila *et al.* (1992) could demonstrate a significantly enhanced non-specific humoral immune response during the acute phase of rotavirus infection in children, reflected by increased numbers of IgG, IgA, and IgM secreting cells due to treatment with *Lactobacillus* GG. At convalescence, 90% of the study group versus 46% of the placebo group had developed an IgA specific antibody-secreting cell response to rotavirus.

While it is difficult to assess innate immunity, several studies have suggested an enhancement with probiotic bacteria. Non-specific immunity was measured by determining phagocytosis as well as a modified cytokine production (Delneste *et al.*, 1998; DeSimone *et al.*, 1993; McNab *et al.*, 1995). We were able to demonstrate a modulated cytokine response by peripheral blood mononuclear cells stimulated by intestinal bacteria in volunteers following 5 weeks of daily oral *Lactobacillus* GG intake with an increase of IL-10 secretion and a significant decrease of the proinflammatory cytokines TNF-α and IL-6, while IFN-γ showed a trend towards decreased secretion (Schultz *et al.*, 2000a). Schiffrin *et al.* (1995) demonstrated enhanced leukocyte phagocytosis of *E. coli* following the administration of *Lactobacillus johnsonii* strain La1 or *Bifidobacterium bifidum* strain Bb12 for 3 weeks in healthy volunteers.

Most of the described effects on the human immune response are mediated by viable bacteria only, while the administration of non-viable bacteria or bacterial components had no effect. However, the underlying mechanisms have not yet been identified. Recently Neish *et al.* (2000) demonstrated, that a non-pathogenic strain of *Salmonella* was able to abrogate synthesis of inflammatory cytokines by gut epithelial cells. The bacteria accomplished this by blocking degradation of IκB, an inhibitor of the master transcription factor NF-κB. It has been concluded that the normal gut microflora is therefore able to induce a distinctive form of tolerance in gut epithelial cells. However, whether this is the mode of action for probiotic bacteria to mediate the clinically observed anti-inflammatory effects intestinal diseases remains unclear.

Genetic Susceptibility Towards Intestinal Bacteria

There is strong evidence, that IBD depends on a specific genetic background, as demonstrated by the different susceptibility of various animal strains to experimental colitis, increasing frequency of inflammatory bowel diseases in affected families, higher concordance between monozygotic twins compared with dizygotic twins, and chromosomal alterations in Crohn's disease patients. Host response to commensal bacteria is influenced by genetic background. Lewis rats, injected with purified PG-PS into subserosal lymphoid aggregates, as described earlier in this chapter, spontaneously develop a relapse of chronic granulomatous enterocolitis with extraintestinal liver and joint manifestations, after an initial acute phase and consecutive remission, but Buffalo and Fischer rats display only mild local inflammation, and no extraintestinal manifestations (Sartor *et al.*, 1996a). Lewis rats with predominantly anaerobic bacterial overgrowth in a jejunal self-filling blind loop develop spontaneous hepatobiliary inflammation but Fischer and Buffalo rats do not (Lichtman *et al.*, 1990). Wistar rats exhibit a delayed inflammatory response. Lewis rats, subcutaneously injected with indomethacin develop chronic ulcerative small intestinal inflammation with hepatic manifestation, whereas this injury in Fischer rats spontaneously resolves and is only restricted to the mid small bowel in contrast to the above rat strains (Sartor, 1997). Lewis rats have a defective neuroimmunologic driven anti-inflammatory response, with low corticosterone levels in the mucosal tissue (Sternberg *et al.*, 1989), low activation of plasma kallikrein and abnormal high bradykinin liberation compared with Buffalo and Fischer rats (Sartor *et al.*, 1996a). Recently various inbred strains of mice (C3H/HeJ, C3H/HeJBir, C57BL/6J, DBA/2J, NOD/LtJ, NOD/LtSz- Prkdc(scid)/Prkdc(scid), 129/SvPas, NON/LtJ, and NON.NOD-H2g7) were screened for genetically determined differences in susceptibility to DSS-induced colitis. This study demonstrated major differences in genetic susceptibility to DSS-induced colitis among inbred strains of mice (Mahler *et al.*, 1998). C3H/HeJ mice developed severe colitis only after challenge with DSS, whereas the C3H/HeJBir mouse strain developed spontaneous colitis, however, it is not clear wheather this goes together with defects or deletions of the Toll-like receptor 4 locus (Sundberg *et al.*, 1994). Most recently, a polymorphism in the NOD2 gene was described as a member of a super family of apoptosis regulators, which is expressed in monocytes, and activates nuclear factor NFκB. It was concluded that this activation is regulated by an intracellular receptor for components of microbial pathogens, however, this topic is still under investigation. To date, it is not entirely clear, how the NOD2 mutation creates the IBD phenotype (Hugot *et al.*, 2001; Beutler *et al.*, 2001).

Clinical Rationale for Probiotic Therapy

Clinical data support the experimental evidence of the crucial role of luminal bacteria in the pathogenesis of IBD. Since recurrent aphthous lesions in the neo-terminal ileum of Crohn's disease patients are observed within a few months after curative resection of the distal ileum, fecal components were incriminated in the pathogenesis. Following curative ileal resection and ileocolonic anastomosis in 5 patients with Crohn's disease, a diverting terminal ileostomy was constructed 25-35 cm proximal to the anastomosis. Six months following exclusion, none of the 5 patients had endoscopic lesions in the neo-terminal ileum, but six months after re-anastomosis all patients had significant recurrence of disease (Rutgeerts *et al.*, 1991).

To further investigate the requirement of intestinal components in the course of mucosal inflammation, D'Haens *et al.* (1998) re-infused autologous intestinal luminal contents into the excluded ileum of 3 patients with Crohn's disease following ileocolonic resection and creation of a proximal loop ileostomy, and determined inflammatory changes by histology and electron microscopy. Eight days after infusion focal infiltration of mononuclear cells, eosinophils, and neutrophils into the lamina propria, small vessels, and epithelium was observed. Additionally, markers for epithelial transformation and lymphocyte recruitment were elevated, further emphasizing that intestinal components may trigger postoperative recurrence of Crohn's disease.

Although previously the requirement of viable bacteria was suggested, there is evidence that humoral and, more likely, cellular immune responses to bacterial components are the main mediators of chronic intestinal inflammation (Figure 1). Duchmann *et al.* (1996) performed an analysis of

T-cell clones from peripheral blood, and non-inflamed and inflamed intestinal tissue from IBD patients and control individuals. They demonstrated that all T-cell clones reacted selectively to bacterial stimuli and only < 3% were cross-reactive. These specific T-cell clones were more frequently isolated from inflamed tissue than from peripheral blood or non-inflamed intestinal tissue. They could further demonstrate that broken tolerance towards autologous bacteria may be a central feature of patients with inflammatory bowel disease as outlined in detail earlier in this chapter (Duchmann *et al.*, 1995).

Mucosal immunoglobulins were isolated from endoscopic washings from Crohn's disease patients, patients with ulcerative colitis, inflammatory and non-inflammatory controls. Total mucosal IgG was significantly higher in active Crohn's disease and ulcerative colitis compared with non-inflamed controls. However, no difference was detected in patients with non-specific inflammation, but specifity to proteins of a range of non-pathogenic commensal fecal bacteria was significantly higher in active CD than in UC and non-IBD inflammation or non-inflammatory controls (Macpherson *et al.*, 1996). These results further supported the hypothesis of a breakdown of tolerance to the normal microflora of the gut in patients with active inflammatory bowel disease, especially Crohn's disease. The etiopathogenic role of B-cell response in chronic intestinal inflammation remains unclear, since it does not play a significant role in chronic experimental colitis. Therefore it cannot be excluded, that elevated immunoglobulins are a secondary phenomenon, due to a broken intestinal barrier in active disease.

The discussion about a potential role of pathogenic or commensal bacteria in the pathogenesis of inflammatory bowel disease has led to multiple trials with therapeutic approaches using a variety of antimicrobial or probiotic agents. Most studies were performed under uncontrolled conditions. Few trials, however, met the rigorous criteria of evidence-based medicine. These trials represent bridges from bench results to bedside application.

Antibiotics Against Intestinal Bacteria

Metronidazole

Since there is strong evidence of a microbial influence in the pathogenesis of Crohn's disease, antibiotic therapy has a long history (Table 1). In 1978, Blichfeldt *et al.* reported, in a placebo controlled, double blind, crossover study, significant improvement of symptoms and laboratory values with metronidazole 1g daily over 4 months in patients with active Crohn's disease, restricted to the colon. All patients were additionally treated with salazosulfapyridin or prednisone. In a Swedish randomized, double-blind, crossover trial 78 patients with active Crohn's disease were investigated for two 4-month periods (Ursing *et al.*, 1982). To compare the efficacy of metronidazole with sulphasalazine the Crohn's Disease Activity Index (CDAI) by Best *et al.* (1976) and plasma levels of orosomucoid were evaluated. In the first four months no difference in CDAI and erythrocyte sedimentation rate was found between the treatment groups. However the plasma orosomucoid level was significantly reduced in the metronidazole group. In the patients who had active disease throughout the first period, a significant reduction of the CDAI could be observed in those who switched to metronidazole in the second four months, but not for those who switched to sulphasalazine. Those patients who had already responded during the first period remained stable in the second phase regardless of the treatment procedure. However, the plasma concentration of orosomucoid increased significantly in patients treated with sulphasalazine but not in the metronidazole group. These data suggest, that the antimicrobial capabilities of metronidazole are slightly more effective than sulphasalazine in the treatment of Crohn's disease.

Sutherland *et al.* (1991) included 105 patients with Crohn's disease in a double blind randomized placebo controlled study investigating the efficacy of metronidazole in two doses (20 mg/kg, 10 mg/kg) over a 14-week period. There was a significant improvement in CDAI, C-reactive protein concentrations, and serum orosomucoid in both metronidazole groups compared to placebo. Patients receiving metronidazole 20 mg/kg/day had a greater improvement in disease activity than those receiving 10 mg/kg/day. However, this was not significant, probably due to the small sample size. Seventeen patients were withdrawn due to adverse events, but there was no difference between the groups. Subgroup analysis revealed a higher efficacy in patients with involvement of the large intestine or affecting both small and large bowel than in those with small bowel disease only.

More recently, Rutgeerts *et al.* (1995) included sixty patients after curative ileal resection and primary anastomosis within 1 week after surgery in a double-blind, placebo controlled trial using metronidazole to prevent recurrence of Crohn's disease in the neoterminal ileum. During a treatment period of three months thirty patients received metronidazole (20 mg/kg) and the remaining patients placebo. After treatment was discontinued, 75% of patients in the placebo group had recurrent lesions in the neo-terminal ileum, compared with 52% of patients treated with metronidazole. The severity of relapse was significantly reduced in the metronidazole treated patients. Although discontinued after 12 weeks, metronidazole therapy statistically significantly reduced the clinical recurrence rates at 1 year, while the relapse rates at 2 years and 3 years were not significantly different although sharing a similar trend. Nine patients dropped out due to side effects, seven in the metronidazole group and two in the placebo arm. This study revealed the most compelling evidence of efficacy of metronidazole in Crohn's disease patients. It demonstrated the prophylactic capabilities of metronidazole to prevent early relapse in patients in remission. This is consistent with experiments in three different animal models of spontaneous and inducible colitis, where metronidazole prevented the onset of chronic colitis, but was unable to treat established intestinal inflammation (Rath *et al.*, 2001; Dieleman *et al.*, 2001), incriminating a more narrow spectrum of predominantly anaerobic bacteria in the induction of relapse in Crohn's disease than in the perpetuation of acute inflammation.

Wide spread concern exists regarding severe permanent adverse events, especially peripheral neuropathy by the use of metronidazole. The specific bactericidal effect of metronidazole is probably due to the reduction of the 5-nitro group to a hydroxylamine group, which is able to bind to DNA, and thereby inhibits further cell division. This mechanism is restricted to an environment only present in anaerobic micro-organisms. There is evidence that metabolic products of metronidazole also bind to neuronal RNA causing peripheral axonal degeneration (Bradley *et al.*, 1977), but reports on peripheral neuropathy are inconsistent in the literature. No symptoms have been reported in the use of a cumulative dose of less than 30g. Most symptoms were mild and transient and related to daily doses higher than 20 mg/kg/day (Duffy *et al.*, 1985). Discontinuation or dose reduction resulted in most cases in the resolution of symptoms. Due to these concerns, ornidazole was tested for the prevention of postoperative recurrence of Crohn's disease. In the interim analysis, the authors reported on a significantly increased rate of severe endoscopic lesions in patients on placebo in comparison with ornidazole-treated patients at the 3 and 12 months intervals, but no data on tolerability and clinical outcome was given (Rutgeerts *et al.*, 1999). In an open trial, Triantafillidis *et al.* studied the clinical effect of ornidazole in 25 patients with active Crohn's disease (CDAI > 150 points). At the end of the four-week treatment period, 75% of the patients were clinically in remission, while only minimal side effects were reported (Triantafillidis *et al.*, 1996).

Although clinical experience in ulcerative colitis reveals dramatic benefit of broad spectrum antibiotic regimen adjunctive to conventional steroid therapy,

Table 1. Antibiotic therapy in inflammatory bowel disease

Antibiotic Therapy	No. pts.	Type of Study	Aim of Clinical Trial	Outcome	Reference
I) Active Crohn's Disease					
Metronidazole	n = 22	randomized, double-blind, cross-over	Effect on symptoms and laboratory values in patients with active Crohn's disease	Significant improvement only in the 6/22 patients with colonic involvement	Blichfeldt, 1978
	n = 78	randomized, double-blind, cross-over	Cross-over trial to compare the efficacy of metronidazole with sulphasalazine in active Crohn's disease	Significant reduction of CDAI in patients who had active disease in the first 4 months and were switched to metronidazole	Ursing, 1982
	n = 105	randomized, double-blind, placebo-controlled	Comparison of different doses in Crohn's disease	Greater improvement in disease activity receiving 20mg/kg/day compared to 10mg/kg/day (NS), but 21 were withdrawn for deterioration of symptoms, 17 for adverse experiences, and 11 for protocol violation	Sutherland, 1991
Ciprofloxacin	n = 40	randomized	Efficacy of ciprofloxacin compared with mesalazine in active Crohn's disease	This study suggests that ciprofloxacin is as effective as mesalazine in treating mild to moderate flare-up of Crohn's disease	Colombel, 1999
Metronidazole/ Ciprofloxacin	n = 41	randomized, partially masked	Efficacy and safety of the combination of metronidazole and ciprofloxacin compared with methylprednisolone in active Crohn's disease	45.5% in the antibiotic group and 63% receiving steroids went into clinical remission at the end of the 12-week treatment period (NS)	Prantera, 1996
	n = 233	retrospective	Efficacy of metronidazole and/or ciprofloxacin in the treatment of acute Crohn's disease	70.6% of the pts. treated with the antibiotic combination achieved clinical remission, 72.8% with metronidazole, 69.0% with ciprofloxacin. Remission lasted 1 year in each group. Severe side effects, were observed in 20% of patients.	Prantera, 1998

Drug(s)	n	Study type	Aim	Results	Reference
	n = 72	open	Efficacy of the combination of ciprofloxacin and metronidazole in active Crohn's disease	Clinical response in 84% of pts. with or without ileal involvement in comparison to 64% of pts. with ileal disease alone and in 86% of pts. without resection compared to 61% of those with previous resection	Greenbloom, 1998
Rifampicin/ Ethambutol	n = 27	randomized, double-blind, cross-over	Efficacy of antimycobacterial therapy in active Crohn's disease	No significant difference of CDAI or any clinical indicator of disease activity between the treatment groups	Shaffer, 1984
Rifampicin/ Isoniazid/ Ethambutol	n = 130	randomized, double-blind	Efficacy of triple-antimycobacterial therapy in active Crohn's disease	No difference between the two groups regarding radiological changes. In a follow-up after 5 years there was no evidence of consistent benefit or disadvantage from the antibiotic therapy regarding the number of acute relapses, surgical episodes, hospital admissions, disease activity, blood tests, or medication required.	Swift, 1994 Thomas, 1998
Ethambutol/ Clofazimine/ Dapsone/ Rifampicin	n = 40	randomized, double-blind, cross-over	Efficacy of antimycobacterial therapy in refractory, steroid- dependent Crohn's disease	3/19 pts. on medication relapsed during the study period, compared with 11/17 on placebo. There was no substantial endoscopical or radiological effect.	Prantera, 1994
Clarithromycin	n = 15	randomized, double-blind, cross-over	Efficacy of antimycobacterial therapy in active Crohn's disease	5/7 pts. receiving clarithromycin achieved remission compared with 1/8 pts. in the placebo group. Cross-over revealed an effect only in 1/7 pts. from the former placebo group now receiving clarithromycin.	Graham, 1995
Clarithromycin/ Ethambutol	n = 31	placebo-controlled	Effect on mucosal permeability	No difference between at the end of the treatment period and the one year follow up.	Goodgame, 1999
Ornidazole	n = 25	open	Clinical effect in active disease	75% of pts. were clinically in remission after four weeks of treatment	Triantafillidis, 1996

Table 1 continued

II) Postoperative Crohn's Disease

Metronidazole	n = 60	randomized, double-blind, placebo-controlled	Prevention of recurrence of Crohn's disease in the neoterminal ileum	75% pts. in the placebo group, compared with 52% pts. in the metronidazole group at 3 month had recurrent lesions in the neoterminal ileum. Antibiotic therapy significantly reduced the clinical recurrence rates at 1 year.	Rutgeerts, 1995
Ornidazole	n = 71	double-blind, placebo-controlled	Prevention of postoperative recurrence of Crohn's disease	Significantly increased rate of severe endoscopic lesions in pts. on placebo in comparison with ornidazole treated pts. at 3 and 12 months, but no data on tolerability and clinical outcome	Rutgeerts, 1999

III) Chronic Pouchitis

Metronidazole	n = 13	double-blind, placebo-controlled, cross-over	Effect on diarrhea in chronic pouchitis	Significant decrease in stool frequency, but no effect on histological or serological marker	Madden, 1994
Rifaximin/ Ciprofloxacin	n = 18	open	Clinical effect on chronic active, treatment resistant pouchitis	88.8% pts. either significantly improved or went into remission	Gionchetti, 1999

IV) Ulcerative Colitis

Ciprofloxacin	n = 83	randomized, placebo-controlled	Long term effect of ciprofloxacin to induce and maintain remission of ulcerative colitis	Clinical relapse rate 21% in the ciprofloxacin -treated group and 44% in the placebo group. Endoscopic and histological findings demonstrated differences only at 3 months but not at 6 months.	Turunen, 1998
	n = 83	randomized, placebo-controlled	To assess the therapeutic role of ciprofloxacin as an adjunct to corticosteroids in acute severe ulcerative colitis	A short course of intravenous ciprofloxacin does not seem to augment the effect of corticosteroids for patients with acute, severe ulcerative colitis	Mantzaris, 2001

in some patients with severe disease, the efficacy of antibiotic therapy, especially metronidazole is somehow discouraging. Patients with ulcerative pouchitis, usually benefit from antibiotic therapy. Metronidazole 400 mg three times daily for seven days was given to patients with chronic pouchitis in a double-blind, placebo-controlled, crossover study (Madden *et al.*, 1994). At entry all patients had symptomatic pouchitis. Metronidazole decreased the number of bowel movements whereas placebo increased stool frequency. However, there was no significant change in the macroscopical or histological grade of inflammation or serum C-reactive protein level. Non-severe side effects were observed in 52% of the patients. In an uncontrolled observation in 52 patients with pouchitis, clinical response to antibiotic treatment with metronidazole was 96% (Hurst *et al.*, 1996). Some small uncontrolled trials report benefit of local instillation with 40 mg metronidazole suspension without detectable systemic concentrations and side effects.

Ciprofloxacin

An alternative to metronidazole is ciprofloxacin. Although it is not effective against facultative Gram negative anaerobes such as *Bacteroides* spp., it covers a broad spectrum of intestinal bacteria and in contrast to metronidazole has some proven beneficial effect in patients with ulcerative colitis. In a controlled trial, Turunen *et al.* (1998) evaluated the role of long term effect of ciprofloxacin to induce and maintain remission of ulcerative colitis in patients previously poorly responding to conventional therapy with steroids and mesalazine. The patients were initially treated with high doses of prednisone on a tapering regime and maintenance treatment with mesalazine was commenced. Ciprofloxacin or placebo were administered for 6 months. During the treatment period, the relapse rate was 21% in the ciprofloxacin-treated group and 44% in the placebo group. This difference was detected using predominantly clinical criteria. Endoscopic and histological findings demonstrated differences only at 3 months but not at 6 months. These modest results could not be confirmed in short term treatment of acute ulcerative colitis (Mantzaris *et al.*, 2001).

In another randomized controlled trial, Colombel *et al.* (1999) investigated the efficacy of ciprofloxacin 1 g/day compared with mesalazine 4 g/day for 6 weeks in the treatment of active Crohn's disease (Table 1). Complete remission was observed in 56% of patients treated with ciprofloxacin and 55% of patients treated with mesalazine, suggesting ciprofloxacin to be equally effective as mesalazine in the treatment of mild to moderate active Crohn's disease.

Prantera *et al.* (1996) conducted several trials using antibiotics in the treatment of active Crohn's disease (Table 1). They investigated the efficacy and safety of the combination of metronidazole and ciprofloxacin compared with methylprednisolone in a randomized, partially masked trial including 41 consecutive patients with active Crohn's disease. Patients were randomized to receive either ciprofloxacin 500 mg twice daily plus metronidazole 250 mg four times daily for 12 weeks, or methylprednisolone 0.7-1 mg/kg/day, with variable tapering to 4 mg daily. 45.5% in the antibiotic group and 63% receiving steroids went into clinical remission at the end of the 12-week treatment period. The difference did not reach statistical significance, probably due to the small sample size. Five patients in both groups were considered treatment failures. Side effects were observed in 27.3% of antibiotic treated patients and 10.6% on steroids. More recently, the same group retrospectively evaluated the efficacy of metronidazole and/or ciprofloxacin in the treatment of acute Crohn's disease (Prantera *et al.*, 1998). The clinical records of 233 inpatients treated for active Crohn's disease with metronidazole and/or ciprofloxacin (1 g/day each) during the period 1984-1996 were reviewed (Table 1). 70.6% of the patients treated with the antibiotic combination achieved clinical remission, 72.8% with metronidazole, 69.0% with ciprofloxacin. Remission lasted about one year in each group. Side effects, requiring discontinuation of therapy, were observed in 20% of patients.

The efficacy of the combination of ciprofloxacin and metronidazole in active Crohn's disease was further studied in a open trial by Greenbloom *et al.* (1998) (Table 1). At the end of the 10-week treatment period, 49/72 patients reached clinical remission (68%) and 55/72 patients showed a clinical response (76%). The effect was even greater in patients with colonic disease with or without ileal involvement (84%) in comparison to patients with ileal disease alone (64%) and in patients without resection (86%) compared with those with previous resection (61%).

In view of these data, it can be concluded, that the combination of metronidazole and ciprofloxacin has some beneficial effect, but the antimicrobial effect is not broad and strong enough to reduce the bacterial load to the degree needed to reliably inhibit the inflammatory process.

Other Antibiotics

Local therapy with rifaximin together with oral ciprofloxacin might be an acceptable alternative in chronic active, treatment-resistant pouchitis. Gionchetti and colleagues (1999) treated eighteen patients with rifaximin 1 g *bid* and ciprofloxacin 500 mg *bid* for 15 days (Table 1). Eighty eight per cent of patients either significantly improved or went into remission. No side-effects could be observed. The fact that rifaximin plasma levels and urinary excretion were not measurable suggested a primarily topical effect. A significant reduction of total anaerobes and aerobes was observed. There are also reports of beneficial effect of amoxycillin/clavulanic acid, tetracycline and erythromycin.

Based on these data, it is obvious that some patients, especially those with large bowel involvement, may benefit from a reduction of the bacterial load. Therefore, differential antibiotic strategies are required, taking into account whether the patient is in remission or in active disease. Metronidazole alone should be restricted to maintenance strategies of remission, especially after surgery. Broad spectrum combinations, such as ciprofloxacin and metronidazole may be of benefit as adjunctive therapy to standard protocols in the acute phase of inflammation. In some cases of colitis refractory to standard procedures including immunosuppressive drugs, accompanied with severe systemic disease, a bowel decontamination as described by Prantera *et al.* (1994) may be considered.

Probiotic Therapy in Rodent Models for Experimental Colitis

Beside *in vitro* experiments, the effects of probiotic microorganisms are often tested in animal models for experimental colitis. There are several models resembling different aspects of human IBD and underlining the importance of intestinal bacteria in the pathogenesis of IBD. In several animal models, intestinal inflammation is absent (HLA-B27 transgenic rats, IL-10$^{-/-}$ mice) or greatly attenuated (IL-2$^{-/-}$ mice) if raised in a sterile environment, whereas spontaneous colitis develops in a specific-pathogen-free environment, leading to severe bloody diarrhea and occasional rectal prolapse (Rath *et al.*, 1996; Sadlack *et al.*, 1993; Schultz *et al.*, 1999; Kuhn *et al.*, 1993; Sellon *et al.*, 1998; Elson *et al.*, 1995).

IL-10$^{-/-}$ mice raised in a specifc-pathogen-free environment spontaneously develop a chronic colitis, resembling human Crohn's disease (Kuhn *et al.*, 1993). Madsen *et al.* (1999) documented a decreased concentration of colonic *Lactobacillus* sp., and an increase in colonic mucosal adherent bacteria in neonatal IL-10$^{-/-}$ mice. Restoring *Lactobacillus* sp. to normal levels of control animals by oral administration of lactulose and rectal swabbing with *Lactobacillus reuteri* reduced colonic mucosal adherent bacteria and prevented the development of colitis. Other investigators confirmed these beneficial effects by *Lactobacillus* sp. in the prevention of experimental colitis. Healthy IL-10$^{-/-}$ mice, when transferred from a germfree environment into specifc-pathogen-free conditions will develop a rapidly progressing intestinal inflammation within four weeks (Sellon *et al.*, 1998). We were able to demonstrate, that the disease can be significantly attenuated, if the mice are colonized with *Lactobacillus plantarum* 299v 2 weeks prior to the transfer. This effect was accompanied by a significant reduction of IFN-γ secreted by stimulated mesenteric lymph node cells and colonic IL-12. Administration of *L. plantarum* at the time of transfer had no effect (Schultz *et al.*, 1998). Furthermore, Dieleman *et al.* showed that the recurrence of colitis in HLA-B27 transgenic rats after antibiotic treatment with vancomycin and imipenem can be prevented by administration of *Lactobacillus* GG as documented by significantly reduced histologic scores, caecal myeloperoxidase and IL-10 levels (Dieleman *et al.*, 2001a).

Another aspect of probiotic therapy is the question whether the use of prebiotics together with probiotics enhances the anti-inflammatory effects. Therefore, prebiotic oatbase is often administered together with lactobacilli. Recently, we were able to demonstrate an effect of oral administration of a new pre- and pro-biotic preparation on the treatment of established experimental colitis in HLA-B27 rats. The preparation consists of inulin as the prebiotic compound, and a combination of the probiotic microorganisms *L. acidophilus* La-5, *L. delbrückii* subsp. *bulgaricus*, *Bifidobacterium* Bb-12, and *Streptococcus thermophilus* with a total bacterial concentration of 8x10^{7}CFU/ml. At 4 months of age, following two months of treatment, untreated rats had severe colitis in all colonic segments, but the inflammation was significantly diminished in the colon of rats treated with the pre- and probiotic preparation. This effect was enhanced by combination with metronidazole, suggesting a synergistic effect of the combination of anti- and probiotics in the treatment of experimental colitis (unpublished data). The role of the prebiotic inulin remains to be studied in detail.

To study the effect of fermentation, acute colitis in rats was induced by methotrexate, resembling a major side effect of many chemotherapeutic agents. Administration of lactobacilli, but not oatbase, decreased the intestinal myeloperoxidase level, and re-established intestinal microecology. Both lactobacilli and oatbase reduced plasma endotoxin levels, but the effects of lactobacilli were greater with fermentation than without fermentation or oatbase alone, while *Lactobacillus plantarum* was more effective in reducing intestinal pathogens than *Lactobacillus reuteri* (Mao *et al.*, 1996). The effect on acute experimental colitis was confirmed by Holma *et al.*, but significant strain differences were documented, with *Lactobacillus reuteri* R2LC being superior to *Lactobacillus rhamnosus* GG in reducing the severity of acetic acid-induced colitis in rats (Holma *et al.*, 2001).

Various *in vitro* and *in vivo* studies have suggested possible mechanisms of probiotic bacteria that might counteract underlying pathophysiological causes of human IBD. Most intriguing is the fact that IBD seems to be due to an intolerance of the immune system towards ubiquitous intestinal bacteria. Therefore it appears logical to intervene by modulation of this intestinal microflora. Results of *in vivo* experiments using animal models for experimental colitis also suggested a potential use of probiotics to treat Crohn's disease and ulcerative colitis. It is evident that the tested organisms varied in efficacy, depending on the experimental model. Clinical studies will show whether a combination of various probiotic microorganisms will lead to better results than the use of single strains.

Probiotic Therapy of Inflammatory Bowel Diseases

Despite many years of extensive research, the mechanisms responsible for initiation and perpetuation of IBD remain unclear. The main theory is that IBD might result from a lack of tolerance to members of the intestinal microflora leading to an overreactive immune response. A specific therapy is still not available. Accumulating research suggests that probiotics might offer an alternative by alteration of the intestinal microflora and modulation of the immune response (Schultz *et al.*, 2000b; Shanahan, 2000a; Shananhan, 2000b; Dunne, 2001).

Single-strain Therapy

The treatment with well defined single strains offers the advantage of better handling and control, and the resulting therapeutic effects can be clearly attributed to the administered substance. However, as outlined above, the various health promoting aspects of probiotic therapy are not unique among all microorganisms. Therefore each strain might be useful for specific indications only, which have to be defined by clinical trials. Among the best studied organisms are lactic acid bacteria, however, there is evidence of successful use of *E. coli* and non-bacterial microorganisms such as *Saccharomyces boulardii*.

Saccharomyces boulardii

Saccharomyces boulardii is a non-pathogenic yeast that has been successfully used as a biotherapeutic agent to prevent antibiotic-associated diarrhea (Surawicz *et al.*, 1989 and 2000) and to treat other types of infection-related diarrhea.

Pein and Hotz (1993) performed a pilot, double-blind, controlled study of the efficacy of *Saccharomyces boulardii* on symptoms of Crohn's disease (Table 2). Twenty patients with active, moderate Crohn's disease were randomly assigned to receive either *Saccharomyces boulardii* or a placebo for 7 weeks in addition to the standard treatment. A significant reduction in the frequency of bowel movements and in disease activity was observed in the group receiving the probiotic agent, but not in the placebo group. In a later study, Guslandi *et al.* (2000) included thirty-two patients with Crohn's disease in clinical remission (CDAI < 150). They were randomly treated for six months with either mesalamine 1 g three times a day or mesalamine 1 g two times a day plus a preparation of *Saccharomyces boulardii* 1 g daily. Clinical relapses as assessed by CDAI values were observed in 37.5% of patients receiving mesalamine alone and in 6.25% of patients in the group treated with mesalamine plus the probiotic agent (Guslandi, 2000) (Table 2). The results of these two studies suggest, that *Saccharomyces boulardii* may represent a useful tool in the maintenance treatment of Crohn's disease.

Lactobacillus GG

Lactobacillus GG is probably the best studied probiotic bacterium (Saxelin, 1997) since it was first described by Gorbach and Goldin (1987). Its clinical potential has been documented for diarrheal disorders such as traveller's diarrhea (Oksanen *et al.*, 1990; Hilton *et al.*, 1997), treatment of recurrent infection with *Clostridium difficile* (Biller *et al.*, 1995) and rotavirus infection in children (Juntunen *et al.*, 2001; Isolauri, 2000), but few studies have been conducted to evaluate its potential in IBD. While effects have been documented only in children with Crohn's disease (Vanderhoof, 2000), several clinical trials have been initiated to study the effect in adult disease.

The effect of oral human *Lactobacillus* GG was investigated in 14 children with Crohn's disease in remission. Following 10 days of oral administration, the gut immune response was measured by determining the number of IgA-secreting cells in peripheral blood samples (Table 2). While no report was given regarding clinical effects or side effects, the results indicate that orally administered *Lactobacillus* GG has the potential to increase the gut IgA immune response and thereby to promote the gut immunological barrier (Malin *et al.*, 1996).

More recently, an open-label study by Gupta *et al.* (2000) examined the effect of *Lactobacillus* GG in four children with mildly to moderately active Crohn's disease (Table 2). Changes in intestinal permeability were measured by a double sugar permeability test and clinical activity was determined by measuring the pediatric Crohn's disease activity index. A significant improvement in clinical activity 1 week after starting *Lactobacillus* GG could be documented, which was sustained throughout the study period. Median pediatric Crohn's disease activity index scores at 4 weeks were 73% lower than baseline and intestinal permeability improved in an almost parallel fashion (Gupta *et al.*, 2000).

In a further open-label study, Friedman *et al.* (2000) reported that administration of *Lactobacillus* GG twice daily and fructooligosaccharide as a prebiotic for one month induced remission in 10 patients with chronic pouchitis as documented by complete suppression of symptoms and reversal of macroscopic endoscopic alterations (Table 2).

E. coli strain Nissle 1917

Infectious diarrheal diseases were a severe problem for soldiers during World War I. In 1917 Alfred Nissle isolated a specific *Escherichia coli*, later named strain Nissle, from the faeces of a soldier who seemed to be immune against this spreading disease (Nissle, 1966). In the following years, various *in vivo* and *in vitro* studies confirmed the probiotic potential of this microorganism.

In a clinical trial by Kruis *et al.* (1997), a total of 120 patients with inactive ulcerative colitis were included in a double-blind, double-dummy fashion to compare mesalazine 500 mg *tds* to an oral preparation of viable *E. coli* strain

Table 2. Table of clinical trials with probiotic microorganisms in inflammatory bowel disease

Probiotic Microorganism	No. pts.	Type of Study	Aim of Clinical Trial	Outcome	Reference
Saccharomyces boulardii	n = 20	randomized, double-blind, placebo-controlled	Effect on diarrhea in Crohn's disease in remission	Significant reduction in the frequency of bowel movements	Pein and Hotz, 1993
	n = 32	randomized, double-blind	Comparison of mesalamin to mesalamine plus probiotic in maintenance of remission in Crohn's disease	Rate of relapse significantly reduced	Guslandi, 2000
Lactobacillus GG	n = 14	open	Effect on gut immune response in children with Crohn's disease	Significant increase of gut IgA-secreting cells	Malin, 1996
	n = 4	open	Effect on intestinal permeability and clinical activity in children with mildly active Crohn's disease	Significant improvement of clinical activity and intestinal permeability	Gupta, 2000
	n = 10	open	Effect on chronic pouchitis	Complete suppression of symptoms and reversal of macroscopic endoscopic alterations	Friedman, 2000
E. coli strain Nissle 1917	n = 120	Randomized, double-blind	Comparison to mesalazine to maintain remission in ulcerative colitis	Comparable relapse-free period on either medication	Kruis, 1997
	n = 116	Randomized, double-blind	Comparison to mesalazine to maintain remission in ulcerative colitis	Comparable relapse-free period on either medication	Rembacken, 1999

	n	Study design	Objective	Results	Reference
	n = 327	Randomized, double-blind	Comparison to mesalazine to maintain remission in ulcerative colitis	Equivalence of the maintenance effects of E. coli strain Nissle 1917 and mesalamine	Kruis, 2001
	n = 28	Randomized, double-blind	Maintenance of remission in Crohn's disease after medical induction	No difference in the number of pts. reaching remission but fewer relapses on probiotic medication	Malchow, 1997
VSL#3	n = 40	Randomized, double-blind, placebo-controlled	Maintenance of remission in chronic pouchitis	Remission was maintained in 85% on probiotic treatment vs. 0% on placebo	Gionchetti, 2000
	n = 20	open	Maintenance of remission in ulcerative colitis	15/20 pts. maintained remission on probiotic treatment	Venturi, 1999
	n = 40	double-blind, randomized	Prevention of post-operative recurrence in Crohn's disease with pro- and antibiotic combination	4/20 relapsed on probiotic treatment compared to 8/20 on mesalamine	Campieri, 2000

Nissle (Serotype 06: K5: H1) for 12 weeks with regard to the efficacy in preventing a relapse of the disease (Table 2). Life table analysis showed a relapse-free time of 103 ± 4 days for mesalazine and 106 ± 5 days for *E. coli* Nissle 1917 (non-significant) (Kruis *et al.*, 1997).

These promising results were later confirmed by Rembacken *et al.* (1999) and finalized by Kruis *et al.* (2001) in a multicentric, randomised, double-blind study including 327 patients with ulcerative colitis in remission (Table 2). All patients received either mesalazine (500mg *tid*) or *E. coli* Nissle 1917 (200mg) and were followed up for a maximum of 12 months. After remission was achieved, patients were maintained on either mesalazine or *E. coli* and followed up for a maximum of 12 months. Relapse rates were 45.1% under *E. coli* Nissle 1917 and 36.4% under mesalazine. The statistical analysis showed significant equivalence ($p<0.02$) of the maintenance effects of *E. coli* Nissle 1917 and mesalamine. It was therefore concluded that *E. coli* strain Nissle offers an alternative to mesalazine to prevent relapse in ulcerative colitis, especially to those patients allergic or unresponsive to mesalazine (208). Malchow *et al.* (1997) included 28 patients with active Crohn's disease (CDAI >150) (Table 2). The disease had to be limited to the large intestine. Following a one-year treatment period, combining a tapering regimen of steroids with *E. coli* strain Nissle or placebo, there was no difference in the number of patients reaching remission. However, in the *E. coli* Nissle group, fewer patients relapsed (33.3% vs. 63.6%).

Other Single Strain Therapies

Despite the fact that plenty of *in vitro* work was done to evaluate the probiotic properties of various *Lactobacillus acidophilus* strains, no clinical trial has been conducted yet regarding the effects on patients with IBD. There is one report on a clinical observation that the orally administered prebiotic substrate lactulose is split by bacteria in the large intestine, leading to a reduction in fecal pH and creating intestinal conditions beneficial to *Lactobacillus acidophilus* and inhibitory to coliform bacteria, *Bacteroides* sp., *Salmonella* and *Shigella*. The authors conclude that lactulose may be used for treatment of IBD as bacteria and bacterial endotoxin might have an important role in the pathogenesis of this disease (Liao *et al.*, 1994).

Multi-strain Therapy

While single strain therapies might be easier to handle, and the observed clinical effect can be clearly attributed to the administered probiotic preparation, mixtures of probiotic strains might have the advantage of combining several probiotic properties and lead to better clinical results.

VSL#3

VSL#3 is the first product combining different probiotic microorganisms studied in a scientific manner. While the selection criteria are not available, VSL#3 has been successfully used to maintain remission in ulcerative colitis, Crohn's disease and pouchitis. Furthermore, Gionchetti *et al.* (2000) has combined clinical trials with mechanistic studies to explain the observed effects.

In a first trial, the efficacy of VSL#3 was evaluated in maintenance of remission of chronic pouchitis compared to placebo. Forty patients with pouchitis in clinical and endoscopic remission were randomized to receive either VSL#3, 6 g/day, or placebo for 9 months (Table 2). All patients received one month of antibiotic treatment (1g ciprofloxacin and 2g rifaximin daily) prior to probiotic/placebo treatment). Patients were assessed clinically every month and endoscopically and histologically every 2 months or in the case of a relapse. At the end of the study period, three patients (15%) in the VSL#3 group had relapses within the 9-month follow-up period, compared with 20 (100%) in the placebo group (P< 0.001) (Gioncheti *et al.*, 2000).

In an uncontrolled clinical trial a similarly beneficial effect on ulcerative colitis was documented by the same group. To evaluate the effects on intestinal microflora and the clinical efficacy of VSL#3, twenty patients with ulcerative colitis in remission and intolerant or allergic to 5-ASA, have been treated for 12 months (Table 2). Fecal samples for stool culture were obtained from the patients at the beginning of the trial and after 10, 20, 40, 60, 75, 90 days, 12 months and at 15 days after the end of the treatment. As a result, 15/20 treated patients remained in remission during the study, one patient was lost to follow up, while the remaining relapsed (Venturi *et al.*, 1999).

In a third study, the authors compared the efficacy of VSL#3 combined with an antibiotic treatment to mesalamine in prevention of post-operative recurrence of Crohn's disease. Forty patients were randomized to receive either rifaximin 1.8g/d for 3 months followed by VSL#3 or mesalamine 4g/d for 12 months (Table 2). Endoscopic examination was performed after 3 and 12 months. After one year, 4 patients in the antibiotic/probiotic group had a severe endoscopic recurrence compared to 8 patients in the mesalamine group, while no side effects were reported in the antibiotic/probiotic group (Campieri *et al.*, 2000).

Other Probiotic Combinations

No other combination of probiotic microorganisms has been clinically studied up-to-date, but there is evidence, that the combination of pre- and probiotics might have an advantage (Madsen *et al.*, 1999; Friedman *et al.*, 2000) over single- or multi-strain combinations.

Another promising therapeutic concept is the use of prebiotics alone, and by that the enhancement of the endogenous probiotic microflora. This has been demonstrated by the use of germinated barley foodstuff (GBF) containing glutamine-rich protein and the hemicellulose-rich fiber which was made from brewer's spent grain by physical isolation (milling and sieving). Both *in vivo* (Kanauchi *et al.*, 1998) and *in vitro* (Kanauchi *et al.*, 2000) studies demonstrated that the fiber fraction of GBF supports maintenance of epithelial cell populations, facilitates epithelial repair, and suppresses epithelial NF-κB-DNA binding activity through generating increased short-chain fatty acid (especially butyrate) production by luminal microflora thereby preventing experimental colonic injury. Based on these observations, clinical studies were initiated in patients with mild to moderate active ulcerative colitis. The patients who had been unresponsive or intolerant to standard treatment received 30g of GBF feeding daily in a nonrandomized, open-label fashion. At 4 weeks, this treatment resulted in a significant clinical and endoscopic improvement independent of disease extent. The improvement was associated with an increase in stool butyrate concentrations and in luminal *Bifidobacterium* and *Eubacterium* levels (Mitsuyama *et al.*, 1998).

Summary and Concluding Remarks

In summary, despite the fact that probiotic properties of ubiquitous luminal bacteria have been suspected for almost a century, it was not until recently that the scientific background was investigated. While probiotics were first used in veterinary medicine to fight concommitant infections in industrialized animal farms, clinical observations soon followed. It was a long way from single case reports to the first clinical trials conducted in a randomized, placebo-controlled, double-blind fashion to evaluate the clinical effects of probiotic preparations in comparison to standard medication.

Inflammatory bowel diseases seem to be an ideal indication to use probiotic preparations for treatment. Since there is abundant evidence for the important role of luminal bacteria in the pathogenesis of IBD, the alteration of the intestinal microflora by probiotic microorganisms should lead to clinical improvement of the condition. Furthermore, immunomodulatory effects of probiotics, as documented in various *in vitro* studies, as well as the stabilization of the intestinal barrier should be able to counteract some suspected pathomechanisms in IBD.

While most data regarding the role of intestinal bacteria in the pathogenesis of IBD were aquired in animal models of experimental intestinal inflammation, few experiments were performed to study the effect of probiotics in treatment or prevention. It was documented that oral administration of different strains was able to prevent or at least significantly attenuate the disease, however, treatment of established colitis seemed difficult. Further studies showed that combination with antibiotic therapy enhanced the effect.

The immunomodulatory effects of probiotic preparations were documented on human intestinal cell lines or peripheral blood mononuclear cells. While a definite effect on cytokine production was obvious, no clear picture regarding a T_{H1}- or T_{H2}-mediated cytokine response was seen. Promising *in vivo* and *in vitro* results led to the first clinical trials. The few trials published to date document the possible use of probiotics in inflammatory bowel disease. So far, *E. coli* strain Nissle 1917 and *Saccharomyces boulardii* have shown promising clinical effects in the maintenance of remission in IBD, and can be recommended at least for patients allergic or intolerant to mesalamine. Most intriguing is the fact that the combination of different probiotic microorganisms as in VSL#3 produced exciting results. More clinical trials are currently under way to evaluate the effect in various conditions such as post-operative maintenance of remission and the treatment of acute IBD. In the near future, it will be necessary to learn more about the interaction of probiotic microorganisms with the cells of the host (e.g. epithelial and other immunocompetent cells). Furthermore, the effects of combinations of various probiotics should be evaluated, to design more powerful preparations, using the different probiotic potentials of single strains to enhance the clinical effect.

Acknowledgements

Heiko C. Rath is supported by the Deutsche Forschungsgemeinschaft (German Research Council) RA 671/4-1 and RA 671/6-1, the Deutsche Crohn und Colitis Vereinigung (Crohn's and Colitis Foundation of Germany) and the Wihelm Sander Stiftung (Wilhelm Sander Foundation) 98.078.1. Michael Schultz is supported by the Deutsche Crohn und Colitis Vereinigung (Crohn's and Colitis Foundation of Germany).

Reference List

Alander, M., Korpela, R., Saxelin, M., Vilpponen-Salmela, T., Mattila-Sandholm, T., and von Wright, A. 1997. Recovery of *Lactobacillus rhamnosus* GG from human colonic biopsies. Lett. Appl. Microbiol. 24: 361-364.

Alander, M., Satokari, R., Korpela, R., Saxelin, M., Vilpponen-Salmela, T., Mattila-Sandholm, T., and von Wright, A. 1999. Persistence of colonization of human colonic mucosa by a probiotic strain, *Lactobacillus rhamnosus* GG, after oral consumption. Appl. Environ. Microbiol. 65: 351-354.

Bernet, M.F., Brassart, D., Neeser, J.R., and Servin, A.L. 1994. *Lactobacillus acidophilus* LA1 binds to cultured human intestinal cell lines and inhibits cell attachement and cell invasion by enterovirulent bacteria. Gut 35: 483-489.

Best, W.R., Becktel, J.M., Singleton, J.W., and Kern, F. 1976. Development of a Crohn's disease activity index. Gastroenterology 70: 139-144.

Beutler, B. 2001. Autoimmunitiy and apoptosis: the Crohn's connection. Immunity 15: 5-14.

Biller, J.A., Katz, A.J., Flores, A.F., Buie, T.M., and Gorbach, S.L. 1995. Treatment of recurrent *Clostridium difficile* colitis with *Lactobacillus* GG. J. Pediatr. Gastroenterol. Nutr. 21: 224-226.

Blichfeldt, P., Blomhoff, J.P., Myhre, E., and Gjone, E. 1978. Metronidazole in Crohn's disease. A double blind cross-over clinical trial. Scand. J. Gastroenterol. 13: 123-127.

Bolton, P.M., Owen, E., Heatley, R.V., Williams, W.J., and Hughes, L.E. 1973. Negative findings in laboratory animals for a transmissible agent in Crohn's disease. Lancet 2: 1122-1124.

Bolton, R.P., Sherriff, R.J., and Read, A.E. 1980. *Clostridium difficile* associated diarrhoea: a role in inflammatory bowel disease? Lancet 1: 383-384.

Boudeau, J., Glasser, A.L., Masseret, E., Joly, B., and Darfeuille-Michaud, A. 1999. Invasive ability of an *Escherichia coli* strain isolated from the ileal mucosa of a patient with Crohn's disease. Infect. Immun. 67: 4499-4509.

Braat, H., Dieleman, L.A., Sellon, R.K., Schultz, M., and Sartor, R.B. 1998. Effects of antibiotics on the initiation and perpetuation of colitis in the IL-10 KO mice. Gastroenterology 114: G3853.

Bradley, W.G., Karlsson, I.J., and Rassol, C.G. 1977. Metronidazole neuropathy. B. M. J. 2: 610-611.

Brandzaeg, P., Halstensen, T.S., Kett, K., Krajci, P., Kvale, D., Rognum, T.O., Scott, H., and Sollid, L.M. 1989. Immunobiology and immunopathology of human gut mucosa: Humoral immunity and intraepithelial lymphocytes. Gastroenterology 97: 1562-1584.

Burke, D.A., and Axon, A.T. 1988. Adhesive *Escherichia coli* in inflammatory bowel disease and infective diarrhoea. B.M.J. 297: 102-104.

Campieri, M., Rizzello, F., Venturi, A., Poggioli, G., Ugolini, F., Helwig, U., Amasini, C., Romboli, E., and Gionchetti, P. 2000. Combination of antibiotic and probiotic treatment is efficacious in prophylaxis of post-operative recurrence of Crohn's disease: a randomized controlled study vs. mesalamine. Gastroenterology 118: A4179.

Chauviére, G., Coconnier, M.-H., Kernéis, S., Fourniat, J., and Sevrin, A.L. 1992. Adhesion of *Lactobacillus acidophilus* strain LB to human enterocyte-like Caco-2 cells. J. Gen. Microbiol. 138: 1689-1696.

Chiba, M., Fukushima, T., Inoue, S., Horie, Y., Iizuka, M., and Masamune, O. 1998a. *Listeria monocytogenes* in Crohn's disease. Scand. J. Gastroenterol. 33: 430-434.

Chiba, M., Fukushima, T., Koganei, K., Nakamura, N., and Masamune, O. 1988b. *Listeria monocytogenes* in the colon in a case of fulminant ulcerative colitis. Scand. J. Gastroenterol. 33: 778-782.

Chiodini, R.J., Van Kruiningen, H.J., Thayer, W.R., and Coutu, J.A. 1986. Spheroplastic phase of mycobacteria isolated from patients with Crohn's disease. J. Clin. Microbiol. 24: 357-363.

Chiodini, R.J., Van Kruiningen, H.J., Thayer, W.R., Merkal, R.S., and Coutu, J.A. 1984. Possible role of mycobacteria in inflammatory bowel disease. I. An unclassified *Mycobacterium* species isolated from patients with Crohn's disease. Dig. Dis. Sci. 29: 1073-1079.

Collins, M.T., Lisby, G., Moser, C., Chicks, D., Christensen, S., Reichelderfer, M., Hoiby, N., Harms, B.A., Thomsen, O.O., Skibsted, U., and Binder, V. 2000. Results of multiple diagnostic tests for *Mycobacterium avium* subsp. *paratuberculosis* in patients with inflammatory bowel disease and in controls. J. Clin. Microbiol. 38: 4373-4381.

Colombel, J.F., Lemann, M., Cassagnou, M., Bouhnik, Y., Duclos, B., Dupas, J.L., Notteghem, B., and Mary, J.Y. 1999. A controlled trial comparing ciprofloxacin with mesalazine for the treatment of active Crohn's disease. Groupe d'Etudes Therapeutiques des Affections Inflammatoires Digestives (GETAID). Am. J. Gastroenterol. 94: 674-678.

Cottone, M., Pietrosi, G., Martorana, G., Casa, A., Pecoraio, G., Oliva, L., Orlando, A., Rosselli, M., Rizzo, A., and Pagliaro, L. 2001. Prevalence of cytomegalovirus infection in severe refractory ulcerative and Crohn's colitis. Am. J. Gastroenterol. 96: 773-775.

Crohn, B.B., Ginzburg, L., and Oppenheimer, G.D. 1932. Regional ileitis: a pathologic and clinical entity. J.A.M.A. 99: 1323-1329.

Dalziel, T.K. 1913. Chronic intestinal enteritis. Br. Med. J. 2: 1068-73.

Darfeuille-Michaud, A., Neut, C., Barnich, N., Lederman, E., Di Martino, P., Desreumaux, P., Gambiez, L., Joly, B., Cortot, A., and Colombel, J.F. 1998. Presence of adherent *Escherichia coli* strains in ileal mucosa of patients with Crohn's disease. Gastroenterology 115: 1405-1413.

Delneste, Y., Donnet-Hughes, A., and Schiffrin, E.J. 1998. Functional Foods: Mechanisms of action on immunocompetent cells. Nutr. Rev. 56: S93-S98.

DeSimone, C., Vesley, R., Bianchi Salvadori, B., Jirillo, E. 1993. The role of probiotics in modulation of the immune system in man and in animals. Int. J. Immunother. 9: 23-28.

D'Haens, G.R., Geboes, K., Peeters, M., Baert, F., Penninckx, F., and Rutgeerts, P. 1998. Early lesions of recurrent Crohn's disease caused by infusion of intestinal contents in excluded ileum. Gastroenterology 114: 262-267.

Dieleman, L.A., Goerres, M.S., Arends, A., Sprengers, D., Varney, R.L., Harmsen, H.J., and Sartor R.B. 2001a. *Lactobacillus* GG prevents recurrence of colitis in HLA-B27 transgenic rats after antibiotic treatment. Gastroenterology 118: A4312.

Dieleman, L.A., Hoentjen, F., Ehre, C., Mann, B.A., Sprengers, D., Harmsen, H.J., and Sartor, R.B. 2001b. Antibiotics with a selective aerobic and anaerobic spectrum have different therapeutic activities in various regions of the colon in IL-10 knock-out mice. Gastroenterology 120: A687.

Drouault, S., Corthier, G., Ehrlich, S.D., and Renault, P. 1999. Survival, physiology, and lysis of *Lactococcus lactis* in the digestive tract. Appl. Environ. Microbiol. 65: 4881-4886.

Duchmann, R., Kaiser, I., Hermann, E., Mayet, W., Ewe, K., and Meyer zum Büschenfelde, K. 1995. Tolerance exists towards resident intestinal flora but is broken in active inflammatory bowel disease (IBD). Clin. Exp. Immunol. 102: 448-455.

Duchmann, R., Märker-Hermann, E., and Meyer zum Büschenfelde, K. 1996. Bacteria-specific T-cell clones are selective in their reactivity towards different enterobacteria or *H. pylori* and increased in inflammatory bowel disease. Scand. J. Immunol. 44:71-79.

Duchmann, R., Schmitt, E., Knolle, P., Meyer zum Büschenfelde, K., and Neurath, M. 1995. Tolerance towards resident intestinal flora in mice is abrogated in experimental colitis and restored by treatment with interleukin-10 or antibodies to interleukin-12. Eur. J. Immunol. 26: 934-938.

Duffy, L.F., Daum, F., Fisher, S.E., Selman, J., Vishnubhakat, S.M., Aiges, H.W., Markowitz, J.F., and Silverberg, M. 1985. Peripheral neuropathy in Crohn's disease patients treated with metronidazole. Gastroenterology 88: 681-684.

Dunne, C., Murphy, L., Flynn, S., O'Mahony, L., O'Halloran, S., Feeney, M., Morrissey, D., Thornton, G., Fitzgerald, G., Daly, C., Kiely, B., Quigley, E.M., O'Sullivan, G.C., Shanahan, F., and Collins, J.K. 1999. Probiotics: from myth to reality. Demonstration of functionality in animal models of disease and in human clinical trials. Antonie van Leeuwenhoek 76: 279-292.

Dunne, C. 2001. Adaptation of bacteria to the intestinal niche: probiotics and gut disorder. Inflamm. Bowel. Dis. 7: 136-145.

Ekbom, A., Daszak, P., Kraaz, W., and Wakefield, A.J. 1996. Crohn's disease after *in-utero* measles virus exposure. Lancet 348: 515-517.

Ekbom, A., Wakefield, A.J., Zack, M., and Adami, H.O. 1994. Perinatal measles infection and subsequent Crohn's disease. Lancet 344: 508-510.

Elliott, S.N., Buret, A., McKnight, W., Miller, M.S.J., and Wallace, J.L. 1998. Bacteria rapidly colonize and modulate healing of gastric ulcers in rats. Am. J. Physiol. 275: G425-G432.

Elo, S., Saxelin, M., and Salminen, S. 1991. Attachement of *Lactobacillus casei* strain GG to human colon carcinoma cell line Caco-2: comparison with other dairy strains. Lett. Appl. Microbiol. 13: 154-156.

Elson, C.O., Sartor, R.B., Tennyson, G.S., and Riddell, R.H. 1995. Experimental models of inflammatory bowel disease. Gastroenterology 109: 1344-1367.

Fabia, R., Ar'Rajab, A., Johansson, M.L., Andersson, R., Willen, R., Jeppsson, B., Molin, G., and Bengmark, S. 1993. Impairment of bacterial flora in human ulcerative colitis and experimental colitis in the rat. Digestion 54: 248-255.

Feeney, M., Ciegg, A., Winwood, P., and Snook, J. 1997. A case-control study of measles vaccination and inflammatory bowel disease. The East Dorset Gastroenterology Group. Lancet 350: 764-766.

Friedman, G., and George, J. 2001. Treatment of refractory 'pouchitis' with prebiotic and probiotic therapy. Gastroenterology 118: A4167.

Giaffer, M.H., Holdsworth, C.D., and Duerden, B.I. 1991. The assessment of fecal flora in patients with inflammatory bowel disease by a simplified bacteriological technique. J. Med. Microbiol. 35: 238-243.

Giaffer, M.H., Holdsworth, C.D., and Duerden, B.I. 1992. Virulence properties of *Escherichia coli* strains isolated from patients with inflammatory bowel disease. Gut 33: 646-650.

Gionchetti, P., Rizzello, F., Venturi, A., Brigidi, P., Matteuzzi, D., Mazzocchi, G., Pogglioli, G., Miglioli, M., and Campieri M. 2000. Oral bacteriotherapy as maintenance treatment in patients with chronic pouchitis: a double-blind, placebo-controlled trial. Gastroenterology 119: 305-309.

Gionchetti, P., Rizzello, F., Venturi, A., Ugolini, F., Rossi, M., Brigidi, P., Johansson, R., Ferrieri, A., Poggioli, G., and Campieri, M. 1999. Antibiotic combination therapy in patients with chronic, treatment-resistant pouchitis. Aliment. Pharmacol. Ther. 13: 713-718.

Goldin, B.R., Gorbach, S.L., Saxelin, M., Barakat, S., Gualtieri, L., and Salminen, S. 1992. Survival of *Lactobacillus* species (strain GG) in human gastrointestinal tract. Dig. Dis. Sci. 37: 121-128.

Goodgame, R.W., Kimball, K., Akram, S., Graham, D.Y., and Ou, C.N. 1999. Randomized controlled trial of clarithromycin and ethambutol in the treatment of Crohn's disease. Gastroenterology 116, A725.

Gorbach, S.L., Chang, T.W., and Goldin, B.R. 1997. Successful treatment of relapsing *Clostridium difficile* colitis with *Lactobacillus* GG. Lancet 26: 1519.

Gorbach, S.L. 2000. Probiotics and gastrointestinal health. Am. J. Gastroenterol. 95: S2-S4.

Gotteland, M., Cruchet, S., and Verbeke, S. 2001. Effect of *Lactobacillus* ingestion on the gastrointestinal mucosal barrier alterations induced by indometacin in humans. Aliment. Pharmacol. Therap. 15: 11-17.

Graham, D.Y., Al-Assi, M.T., and Robinson, M. 1995. Prolonged remission in Crohn's disease following therapy for *Mycobacterium paratuberculosis* infection. Gastroenterology 108: A826.

Graham, D.Y., Markesich, D.C., and Yoshimura, H.H. 1987. Mycobacteria and inflammatory bowel disease. Results of culture. Gastroenterology 92: 436-442.

Greenbloom, S.L., Steinhart, A.H., Greenberg, G.R. 1998. Combination of ciprofloxacin and metronidazole for active Crohn's disease. Can. J. Gastroenterol. 12: 53-56.

Gupta, P., Andrew, H., Kirschner, B.S., and Guandalini, S. Is *Lactobacillus* GG helpful in children with Crohn's disease? Results of a preliminary, open-label study. J. Pediatr. Gastroenterol. Nutr. 31: 453-457.

Guslandi, M. 2000. *Saccharomyces boulardii* in the maintenance of Crohn's disease. Can. J. Gastroenterol. 14: A32.

Haller, D., Bode, C., Hammes, W.P., Pfeifer, A.M., Schiffrin, E.J., and Blum, S. 2000a. Non-pathogenic bacteria elicit a differential cytokine response by intestinal epithelial cell/leucocyte co-cultures. Gut 47: 79-87.

Haller, D., Blum, S., Bode, C., Hammes, W.P., and Schiffrin, E.J. 2000b. Activation of human peripheral blood mononuclear cells by nonpathogenic bacteria *in vitro:* evidence of NK cells as primary targets. Infect. Immun. 68: 752-759.

Hartley, M.G., Hudson, M.J., Swarbrick, E.T., Hill, M.J., Gent, A.E., Hellier, M.D., and Grace, R.H. 1992. The rectal mucosa-associated microflora in patients with ulcerative colitis. J. Med. Microbiol. 36: 96-103.

Helwig, U., Rizzello, F., Waterworth, C., Zucchoni, E., and Venturi, A. 1999. Effect of probiotic therapy on pro- and antiinflammatory cytokines in pouchitis. Gut 45: A01.04

Hermiston, M.L., and Gordon, J.I. 1995. Inflammatory bowel disease and adenomas in mice expressing a dominant negative N-cadherin. Science 270: 1203-1207.

Hilsden, R.J., Meddings, J.B., and Sutherland, L.R. 1996. Intestinal permeability changes in response to acetylsalicylic acid in relatives of patients with Crohn's disease. Gastroenterology 110: 1395-1403.

Hilton, E., Kolakowski, P., Smith, M., and Singer, C. 1997. Efficacy of *Lactobacillus* GG as a diarrheal preventative in travelers. J. Travel. Med. 4: 41-43.

Hobson, C.H., Butt, T.J., Ferry, D.M., Hunter, J., Chadwick, V.S., and Broom, M.F. 1988. Enterohepatic circulation of bacterial chemotactic peptide in rats with experimental colitis. Gastroenterology 94: 1006-1013.

Hockertz, S. 1997. Augmentation of host defence against bacterial and fungal infections of mice pretreated with the non-pathogenic *Escherichia coli* strain Nissle 1917. Arzneimittelforschung/Drug Research 47: 793-796.

Hollander, D. 1999. Intestinal permeability, leaky gut, and intestinal disorders. Curr. Gastroenterol. Rep. 1: 410-416.

Holma, R., Salmenperä, P., Lohi, J., Vapaatalo, H., and Korpela, R. 2001. Effects of *Lactobacillus rhamnosus* GG and *Lactobacillus reuteri* R2LC on acetic acid-induced colitis in rats. Scand. J. Gastroenterol. 6: 630-635.

Hudson, M., Piasecki, C., Sankey, E.A., Sim, R., Wakefield, A.J., More, L.J., Sawyerr, A.M., Dhillon, A.P., and Pounder, R.E. 1992. A ferret model of acute multifocal gastrointestinal infarction. Gastroenterology 102: 1591-1596.

Hugot, J.P., Chamaillard, M., Zouali, H., Lesage, S., Cezard, J.P., Belaiche, J., Almer, S., Tysk, C., O'Morain, C.A., Gassull, M., Binder, V., Finkel, Y., Cortot, A., Modigliani, R., Laurent-Puig, P., Gower-Rousseau, C., Macry, J., Colomber, J.F., Sahbatou, M., and Thomas, G. 2001. Association of NOD2 leucine-rich repeat variants with susceptibility to Crohn's disease. Nature 411: 599-603.

Hurst, R.D., Molinari, M., Chung, T.P., Rubin, M., and Michelassi, F. 1996. Prospective study of the incidence, timing and treatment of pouchitis in 104 consecutive patients after restorative proctocolectomy. Arch. Surg. 131: 497-500.

Iizuka, M., Nakagomi, O., Chiba, M., Ueda, S., and Masamune, O. 1995. Absence of measles virus in Crohn's disease. Lancet 345: 199.

Isolauri, E., Majamaa, H., Arvola, T., Rantala, I., Virtanen, E., and Arvilommi, H. 1993. *Lactobacillus casei* strain GG reverses increased permeability induced by cow milk in suckling rats. Gastroenterology 105: 1643-1650.

Isolauri, E. 2000. The use of probiotics in paediatrics. Hosp. Med. 61: 6-7.

Juntunen, M., Kirjavainen, P.V., Ouwehand, A.C., Salminen, S.J., and Isolauri, E. 2001. Adherence of probiotic bacteria to human intestinal mucus in healthy infants and during rotavirus infection. Clin. Diagn. Lab. Immunol. 8:293-296.

Kaila, M., Isolauri, E., Soppi, E., Virtanen, E., Laine, S., and Arvilommi, H. 1992. Enhancement of the circulating antibody secreting cell response in human diarrhea by a human *Lactobacillus* strain. Pediatr. Res. 32: 141-144.

Kallinowski, F., Wassmer, A., Hofmann, M.A., Harmsen, D., Heesemann, J., Karch, H., Herfarth, C., and Buhr, H.J. 1998. Prevalence of enteropathogenic bacteria in surgically treated chronic inflammatory bowel disease. Hepatogastroenterology 45: 1552-1558.

Kanauchi, O., Andoh, A., Araki, Y., Mitsuyama, K., Toyonaga, A., Sata, S., Hibi, T., Iwanaga, T., and Bamba, T. 2000. The mechanism of germinated barley foodstuff in attenuating intestinal inflammation in colitis. Gastroenterology 118: A73.

Kanauchi, O., Nakamura, T., Agata, K., Mitsuyama, K., Iwanaga, T. 1998. Effects of germinated barley foodstuff on dextran sulfate sodium-induced colitis in rats. J. Gastroenterol. 33: 179-188.

Kanazawa, K., Haga, Y., Funakoshi, O., Nakajima, H., Munakata, A., and Yoshida, Y. 1999. Absence of *Mycobacterium paratuberculosis* DNA in intestinal tissues from Crohn's disease by nested polymerase chain reaction. J. Gastroenterol. 34: 200-206.

Kangro, H.O., Chong, S.K., Hardiman, A., Heath, R.B., and Walker-Smith, J.A. 1990. A prospective study of viral and mycoplasma infections in chronic inflammatory bowel disease. Gastroenterology 98: 549-553.

Kent, T.H., Cardelli, R.M., and Stammler, F.W. 1969. Small intestinal ulcers and intestinal flora in rats given indomethacin. Am. J. Pathol. 54: 237-249.

Kim, S.C., Tonkonogy, S.L., Balish, E., and Sartor, R.B. 2001. IL-10 deficient mice monoassociated with non-pathogenic *Enterococcus faecalis* develop chronic colitis. Gastroenterology 120: A441.

Kishi, D., Takahashi, I., Kai, Y., Tamagawa, H., Iijima, H., Obunai, S., Nezu, R., Ito, T., Matsuda, H., and Kiyono, H. 2000. Alteration of V beta usage and cytokine production of CD4+ TCR beta beta homodimer T cells by elimination of *Bacteroides vulgatus* prevents colitis in TCR alpha-chain-deficient mice. J. Immunol. 165: 5891-5899.

Kobayashi, K., Blaser, M.J., Brown, W.R. 1989. Immunohistochemical examination for mycobacteria in intestinal tissues from patients with Crohn's disease. Gastroenterology 96: 1009-1015.

Kraehenbuhl, J.-P., and Neutra, M.R. 2000. Epithelial M cells: Differentiation and function. Annu. Rev. Cell. Dev. Biol. 16: 301-332.

Kreuzpaintner, G., Das, P.K., Stronkhorst, A., Slob, A.W., and Strohmeyer, G. 1995. Effect of intestinal resection on serum antibodies to the mycobacterial 45/48 kilodalton doublet antigen in Crohn's disease. Gut 37: 361-366.

Kruis, W., Fric, P., and Stolte, M. 2001. Maintenance of remission in ulcerative colitis is equally effective with *Escherichia coli* Nissle 1917 and with standard mesalamine. Gastroenterology 120: A680

Kruis, W., Schutz, E., Fric, P., Fixa, B., Judmaier, G., Stolte, M. 1997. Double-blind comparison of an oral *Escherichia coli* preparation and mesalazine in maintaining remission of ulcerative colitis. Aliment. Pharmacol. Therap. 11: 853-858.

Kuhn, R., Lohler, J., Rennick, D., Rajewsky, K., and Müller, W. 1993. Interleukin-10-deficient mice develop chronic enterocolitis. Cell 75: 263-274.

Liao, W., Cui, X.S., Jin, X.Y., Floren, C.H. 1994. Lactulose - a potential drug for the treatment of IBD. Med. Hypotheses. 43: 234-238.

Lichtman, S.N., Sartor, R.B., Keku, J., and Schwab, J.H. 1990. Hepatic inflammation in rats with experimental small intestinal bacterial overgrowth. Gastroenterology 98: 414-423.

Lichtman, S.N., Keku, J., Schwab, J.H., and Sartor, R.B. 1991. Evidence for peptidoglycan absorption in rats with experimental small bowel bacterial overgrowth. Infect. Immun. 59: 555-562.

Liu, Y., Van Kruiningen, H.J., West, A.B., Cartun, R.W., Cortot, A., and Colombel, J.F. 1995. Immunocytochemical evidence of *Listeria*, *Escherichia coli*, and *Streptococcus* antigens in Crohn's disease. Gastroenterology 108: 1396-1404.

Mack, D.R., Michail, S., Wei, S., McDougall, L., and Hollingsworth, M.A. 1999. Probiotics inhibit enteropathogenic *E. coli* adherence *in vitro* by inducing intestinal mucin gene expression. Am. J. Physiol. 276: G941-G950.

Macpherson, A., Khoo, U.Y., Forgacs, I., Philpott-Howard, J., and Bjarnason, I. 1996. Mucosal antibodies in inflammatory bowel disease are directed against intestinal bacteria. Gut 38: 365-375.

Madden, M.V., McIntyre, A.S., and Nicholls, R.J. 1994. Double-blind crossover trial of metronidazole versus placebo in chronic unremitting pouchitis. Dig. Dis. Sci. 39: 1193-1196.

Madsen, K.L., Cornish, A., Soper, P., McKaigney, C., Jijon, H., Yachimec, C., Doyle, J, Jewell, L., and De Simone, C. 2001. Probiotic bacteria enhance murine and human intestinal epithelial barrier function. Gastroenterology 121: 580-591.

Madsen, K.L., Doyle, J.S., Jewell, L.D., Taverini, M.M., and Fedorak, R.N. 1999. *Lactobacillus* species prevents colitis in Interleukin-10 gene-deficient mice. Gastroenterology 116: 1107-1114.

Mähler, M., Bristol, I.J., Leiter, E.H., Workman, A.E., Birkenmeier, E.H., Elson, C.O., and Sundberg, J.P. 1998. Differential susceptibility of inbred mouse strains to dextran sulfate sodium-induced colitis. Am. J. Physiol. 274: G544-G551.

Malchow, H.A. 1997. Crohn's disease and Escherichia coli. A new approach in therapy to maintain remission of colonic Crohn's disease? J. Clin. Gastroenterol. 25: 653-658.

Malin, M., Suomalainen, H., Saxelin, M., and Isolauri, E. 1996. Promotion of IgA immune response in patients with Crohn's disease by oral bacteriotherapy with *Lactobacillus* GG. Ann. Nutrition. Metabol. 40: 137-145.

Mantzaris, G.J., Petraki, K., Archavlis, E., Amberiadis, P., Kourtessas, D., Christidou, A., and Triantafyllou G. 2001. A prospective randomized controlled trial of intravenous ciprofloxacin as an adjunct to corticosteroids in acute, severe ulcerative colitis. Scand. J. Gastroenterol. 36: 971-974.

Mao, Y., Nobaek, S., Kasravi, B., Adawi, D., Stenram, U., Molin, G., and Jeppsson, B. 1996. The effects of *Lactobacillus* strains and oat fiber on methotrexate-induced enterocolitis in rats. Gastroenterology 111: 334-344.

Marteau, P., de Vrese, M., Cellier, C.J., Schrezenmeier, J. 2001. Protection from gastrointestinal diseases with the use of probiotics. Am. J. Clin. Nutr. 73: 430S-436S.

Marteau, P., Pochart, P., Bouhnik, Y., Zidi, S., Goderel, I., and Rambaud, J.C. 1992. Survival of *Lactobacillus acidophilus* and *Bifidobacterium* sp. in the small intestine following ingestion in fermented milk. A rational basis for the use of probiotics in man. Gastroenterol. Clin. Biol. 16: 25-28.

McNab, R., Tannock, G.W., Jenkinson, H.F. 1995. Characterization of CshA, a high molecular mass adhesin of *Streptococcus gordonii*. Dev. Biol. Stand. 85: 371-375.

Metchnikoff, E. 1907. Optimistic studies. In: The prolongation of life. E. Chalmers, ed. Heinemann, London. p. 161-183.

Meurman, J.H., Antila, H., Korhonen, A., and Salminen, S. 1995. Effect of *Lactobacillus rhamnosus* strain GG (ATCC 53103) on the growth of *Streptococcus sobrinus* in vitro. Eur. J. Oral. Sci. 103: 253-258.

Meyers, S., Mayer, L., Bottone, E., Desmond, E., and Janowitz, H.D. 1981. Occurrence of *Clostridium difficile* toxin during the course of inflammatory bowel disease. Gastroenterology 80: 697-70.

Miettinen, M., Lehtonen, A., Julkunen, I., and Matikainen, S. 2000. *Lactobacilli* and *Streptococci* activate NF-kappa B and STAT signaling pathways in human macrophages. J. Immunol. 164: 3733-3740.

Miettinen, M., Matikainen, S., Vuopio-Varkila, J., Pirhonen, J., Varkila, K., Kurimoto, M., and Julkunen, I. 1998. *Lactobacilli* and *streptococci* induce interleukin-12 (IL-12), IL-18, and gamma interferon production in human peripheral blood mononuclear cells. Infect. Immun. 66: 6058-6062.

Miettinen, M., Vuopio-Varkila, J., and Varkila, K. 1996. Production of human tumor necrosis factor alpha, interleukin-6, and interleukin-10 is induced by lactic acid bacteria. Infect. Immun. 64: 5403-5405.

Mitsuyama, K., Saiki, T., Kanauchi, O., Iwanaga, T., Tomiyasu, N., Nishiyama, T., Tateishi, H., Shirachi, A., Ide, M., Suzuki, A., Noguchi, K., Ikeda, H., Toyonada, A., and Sata, M. 1998. Treatment of ulcerative colitis with germinated barley foodstuff feeding: a pilot study. Aliment. Pharmacol. Ther. 12: 1225-1230.

Mizoguchi, A., Mizoguchi, E., Chiba, C., and Bhan, A.K. 1996. Role of appendix in the development of inflammatory bowel disease in TCR-alpha mutant mice. J. Exp. Med. 184: 707-715.

Montgomery, S.M., Morris, D.L., Pounder, R.E., Wakefield, A.J. 1998. Are concurrent measles and mumps infections in childhood a risk for inflammatory bowel disease? Gastroenterology 114: G4268.

Moss, M.T., Sanderson, J.D., Tizard, M.L., Hermon-Taylor, J., el Zaatari, F.A., Markesich, D.C., and Graham, D.Y. 1992. Polymerase chain reaction detection of *Mycobacterium paratuberculosis* and *Mycobacterium avium* subsp *silvaticum* in long term cultures from Crohn's disease and control tissues. Gut 33: 1209-1213.

Naser, S.A., Schwartz, D., and Shafran, I. 2000. Isolation of *Mycobacterium avium* subsp *paratuberculosis* from breast milk of Crohn's disease patients. Am. J. Gastroenterol. 95: 1094-1095.

Neish, A.S., Gewirtz, A.T., Zeng, H., Young, A.F., Hobert, M.E., Karmali, V., Rao, A.S., and Madara, A.L. 2000. Prokaryotic regulation of epithelial responses by inhibition of IκB-α ubiquitination. Science 289: 1560-1563.

Nielsen, L.L., Nielsen, N.M., Melbye, M., Sodermann, M., Jacobsen, M., and Aaby, P. 1998. Exposure to measles *in utero* and Crohn's disease: Danish register study. Br. Med. J. 316: 196-197.

Nissle, A. 1966. On the clarification of the relations of disease cause, symptoms and rational therapy. Med. Welt 23: 1290-1294.

Obermeier, F., Deml, L., Herfarth, H.H., Rath, H.C., Dunger, N., and Falk, W. 2001. CpG-Motifs of bacterial DNA are potent modulators of Inflammation in DSS induced colitis in mice. Gastroenterology 120: A440.

Oksanen, P., Salminen, S., Saxelin, M., Hämäläinen, P., Ihantola-Vormisto, A., Muurasniemi-Isoviita, L., Nikkari, S., Oksanen, T., Pörsti, I., Salminen, E., Siitonen, S., Stuckey, H, Toppila, A., Vapaatalo, H. 1990. Prevention of traveller's diarrhea by *Lactobacillus* GG. Ann. Int. Med. 22: 53-56.

Ölschläger, T., Altenhöfer, A., and Hacker, J. 2001. Inhibition of *Salmonella typhimurium* invasion into intestinal cells by the probiotic *E. coli* strain Nissle. Gastroenterology 120: A1682

Onderdonk, A,B., Franklin, M.L., and Cisneros, R.L. 1981. Production of experimental ulcerative colitis in gnotobiotic guinea pigs with simplified microflora. Infect. Immun. 32: 225-231.

Pein, K., and Hotz, J. 1993. Therapeutic effects of *Saccharomyces boulardii* on mild residual symptoms in a stable phase of Crohn's disease with special respect to chronic diarrhea - a pilot study. Z. Gastroenterol. 31: 129-134.

Perdigon, G., Alvarez, S., and Pesce de Ruiz, H.A. 1991. Immunoadjuvant activity of oral *Lactobacillus casei*: influence of dose on the secretory immune response and protective capacity in intestinal infections. J. Dairy. Res. 58: 485-496.

Perdigon, G., Alvarez, S., Rachid, M., Aguero, G., and Gobbato, N. 1995. Immune system stimulation by probiotics. J. Dairy. Sci. 78: 1597-1606.

Pitcher, M.C., Beatty, E.R., and Cummings, J.H. 2000. The contribution of sulphate reducing bacteria and 5-aminosalicylic acid to faecal sulphide in patients with ulcerative colitis. Gut 46: 64-72.

Pochart, P., Marteau, P., Bouhnik, Y., Goderel, I., Bourlioux, P., and Rambaud, J.C. 1992. Survival of bifidobacteria ingested via fermented milk during their passage through the human intestine: an *in vivo* study using intestinal perfusion. Am. J. Clin. Nutr. 55: 78-80.

Powrie, F., and Mason, D. 1990. OX-22high CD4+ T cells induce wasting disease with multiple organ pathology: prevention by the OX-22low subset. J. Exp. Med. 172: 1701-1708.

Prantera, C., Berto, E., Scribano, M.L., and Falasco, G. 1998. Use of antibiotics in the treatment of active Crohn's disease: experience with metronidazole and ciprofloxacin. Ital. J. Gastroenterol. Hepatol. 30:602-606.

Prantera, C., Kohn, A., Mangiarotti, R., Andreoli, A., and Luzi, C. 1994. Antimycobacterial therapy in Crohn's disease: results of a controlled, double-blind trial with a multiple antibiotic regimen. Am. J. Gastroenterol. 89: 513-518.

Prantera, C., Zannoni, F., Scrivano, M.L., Berto, E., Andreoli, A., Kohn, A., Luzi, C. 1996. An antibiotic regimen for the treatment of active Crohn's disease: A randomized, controlled clinical trial of metronidazole plus ciprofloxacin. Am. J. Gastroenterol. 91: 328-332.

Rath, H.C., Bender, D.E., Holt, L.C., Grenther, T., Taurog, J.D., Hammer, R.E., and Sartor, R.B. 1995. Metronidazole attenuates colitis in HLA-B27 / hβ2 microglobulin transgenic rats: a pathogenic role for anaerobic bacteria. Clin. Immunol. Immunopathol. 76: S45.

Rath, H.C., Herfarth, H.H., Ikeda, J.S., Grenther, W.B., Hamm, T.E., Jr., Balish, E., Wilson, K.H., and Sartor, R.B. 1996. Normal luminal bacteria, especially Bacteroides species, mediate chronic colitis, gastritis, and arthritis in HLA-B27/human beta2 microglobulin transgenic rats. J. Clin. Invest. 98: 945-953.

Rath, H.C., Ikeda, J.S., Linde, H.J., Schölmerich, J., Wilson, K.H., and Sartor, R.B. 1999. Varying cecal bacterial loads influences colitis and gastritis in HLA- B27 transgenic rats. Gastroenterology 116: 310-319.

Rath, H.C., Schultz, M., Freitag, R., Dieleman, L.A., Li, F., Linde, H.J., Schölmerich, J., and Sartor, R.B. 2001. Different subsets of enteric bacteria induce and perpetuate experimental colitis in rats and mice. Infect. Immun. 69: 2277-2285.

Rath, H.C., Wilson, K.H., and Sartor, R.B. 1999. Differential induction of colitis and gastritis in HLA-B27 transgenic rats selectively colonized with *Bacteroides vulgatus* or *Escherichia coli*. Infect. Immun. 67: 2969-2974.

Rembacken, B.J., Snelling, A.M., Hawkey, P.M., Chalmers, D.M., and Axon, A.T. 1999. Non-pathogenic *Escherichia coli* versus mesalazine for the treatment of ulcerative colitis: a randomised trial. Lancet 354: 635-639.

Reuter, B.K., Davies, N.M., and Wallace, J.L. 1997. Nonsteroidal anti-inflammatory drug enteropathy in rats: role of permeability, bacteria, and enterohepatic circulation. Gastroenterology 112: 109-107.

Robertson, D.J., Sandler, R.S. 2001. Measles virus and Crohn's disease: a critical appraisal of the current literature. Inflamm. Bowel. Dis. 7: 51-57.

Roediger, W.E., Duncan, A., Kapaniris, O., and Millard, S. 1993. Sulphide impairment of substrate oxidation in rat colonocytes: a biochemical basis for ulcerative colitis? Clin. Sci. 85: 623-627.

Roediger, W.E. 1980. The colonic epithelium in ulcerative colitis: an energy-deficiency disease? Lancet 2: 712-715.

Ruseler-van Embden, J.G., and Both-Patoir, H.C. 1983. Anaerobic gram-negative faecal flora in patients with Crohn's disease and healthy subjects. Antonie van Leeuwenhoek 49: 125-132.

Rutgeerts, P., Goboes, K., Peeters, M., Hiele, M., Penninckx, F., Aerts, R., Kerremans, R., and Vantrappen, G. 1991. Effect of faecal stream diversion on recurrence of Crohn's disease in the neoterminal ileum. Lancet 338: 771-774.

Rutgeerts, P., Hiele, M., Geboes, K., Peeters, M., Penninckx, F., Aerts, R., and Kerremans, R. 1995. Controlled trial of metronidazole treatment for prevention of Crohn's recurrence after ileal resection. Gastroenterology 108: 1617-1621.

Rutgeerts, P.J., D'Haens, G., Baert, F., Sels, F., Hiele, M., Aerden, I., Lemmens, P, Penninckx, F., D'Hoore, A., and Geboes, K. 1999. Nitromidazol antibiotics are efficacious for prophylaxis of postoperative recurrence of Crohn's disease: a placebo-controlled trial. Gastroenterology 116: G3506.

Sadlack, B., Merz, H., Schorle, H., Schimpl, A., Feller, A.C., and Horak, I. 1993. Ulcerative colitis-like disease in mice with a disrupted interleukin-2 gene. Cell 75: 253-261.

Sanderson, J.D., Moss, M.T., Tizard, M.L., Hermon-Taylor, J. 1992. *Mycobacterium paratuberculosis* DNA in Crohn's disease tissue. Gut 33: 890-896.

Sartor, R.B. 1994. Cytokines in intestinal inflammation: pathophysiological and clinical considerations. Gastroenterology 106: 533-539.

Sartor, R.B., dela Cadena, R.A., Green, K.D., Stadnicki, A., Davis, S.W., Schwab, J.H., Adam, A.A., Raymond, P., and Colman, R.W. 1996a. Selective kallikrein-kinin system activation in inbred rats differentially susceptible to granulomatous enterocolitis. Gastroenterology 110: 1467-1481.

Sartor, R.B. 1996b. Cytokine regulation of experimental intestinal inflammation in genetically engineered and T-lymphocyte reconstituted rodents. Aliment. Pharmacol. Therap. 10: 36-42.

Sartor, R.B. 1997. Review article: Role of the enteric microflora in the pathogenesis of intestinal inflammation and arthritis. Aliment. Pharmacol. Therap. 11: 17-22.

Sartor, R.B. 1999. Microbial factors in the pathogenesis of Crohn's disease, ulcerative colitis and experimental intestinal inflammation. In: Inflammatory Bowel Disease. J.B. Kirsner, ed. Williams & Wilkins, Baltimore. p. 153-178.

Satoh, H., Gurth, P.H., Grossman, N.I. 1983. Role of bacteria in gastric ulceration produced by indomethacin in the rat: cytoprotective action of antibiotics. Gastroenterology 84: 483-489.

Saxelin, M. 1997. *Lactobacillus* GG - A human probiotic strain with thorough clinical documentation. Food. Rev. Int. 13: 293-313.

Scheppach, W., Sommer, H., Kirchner, T., Paganelli, G.M., Bartam, P., and Christl, S. 1992. Effect of butyrate enemas on the colonic mucosa in distal ulcerative colitis. Gastroenterology 103: 51-56.

Schiffrin, E.J., Rochat, F., Link-Amster, H., Aeschlimann, J.M., and Donnet-Hughes, A. 1995. Immunomodulation of human blood cells following the ingestion of lactic acid bacteria. J. Dairy. Sci. 78: 491-497.

Schölmerich, J. 1998. Studien zur Rezidivprophylaxe des Morbus Crohn. Chirurg 69: 908-914.

Schultz, M., Veltkamp, C., Dieleman, L.A., Wyrick, R.B., Tonkongy, S.L., and Sartor, R.B. 2002. *Lactobacillus plantarum* 299v in the treatment and prevention of spontaneous colitis in IL-10-deficient mice. Inflamm. Bowel. Disease. 8: 71-80.

Schultz, M., Linde, H.J., Falk, W., and Schölmerich, J. 2000a. Oral administration of *Lactobacillus* GG (*L.* GG) induces an antiinflammatory, TH-2 mediated systemic immune response towards intestinal organisms. Gastroenterology 118: A781.

Schultz, M., and Sartor, R.B. 2000b. Probiotics and inflammatory bowel diseases. Am. J. Gastroenterol. 95:S19-S21.

Schultz, M., Tonkonogy, S.L., Sellon, R.K., Veltkamp, C., Godfrey, V.L., Kwon, J., Grenther, W.B., Balish, E., Horak, I., and Sartor, R.B. 1999. IL-2-deficient mice raised under germfree conditions develop delayed mild focal intestinal inflammation. Am. J. Physiol. 276: G1461-G1472.

Schultz, M., and Sartor, R.B. 1997. The contribution of bacterial flora to chronic intestinal inflammation. In: Inflammatory Bowel Disease. Caprilli, R., ed. Schattauer Verlagsgesellschaft, Stuttgart. p. 17-24

Segain, J.P., Raingeard, l.B., Bourreille, A., Leray, V., Gervois, N., Rosales, C., Ferrier, L., Bonnet, C., Blottiere, H.M., and Galmiche, J.P. 2000. Butyrate inhibits inflammatory responses through NFkappaB inhibition: implications for Crohn's disease. Gut 47: 397-403.

Sellon, R.K., Tonkonogy, S.L., Schultz, M., Dieleman, L.A., Grenther, W.B., Balish, E., Rennick, D., and Sartor, R.B. 1998. Resident enteric bacteria are necessary for development of spontaneous colitis and immune system activation in interleukin-10-deficient mice. Infect. Immun. 66: 5224-5231.

Shaffer, J.L., Hughes, S., Linaker, B.D., Baker, R.D., and Turnberg, L.A. 1984. Controlled trial of rifampicin and ethambutol in Crohn's disease. Gut 25: 203-205.

Shanahan, F. 2001. Inflammatory bowel disease: Immunodiagnostics, immunotherapeutics, and ecotherapeutics. Gastroenterology 120: 622-635.

Shanahan, F. 2000a. Probiotics and inflammatory bowel disease: is there a scientific rationale? Inflamm. Bowel. Dis. 6: 107-115.

Shanahan, F. 2000b. Therapeutic manipulation of gut flora. Science 289: 1311-1312.

Silva, M., Jacobus, N.V., Deneke, C., and Gorbach, S.L. 1987. Antimicrobial substance from a human *Lactobacillus* strain. Antimicrobial. Agen. Chemother. 31: 1231-1233.

Simon, G.L., and Gorbach, S.L. 1995. Normal Alimentary Tract Microflora. In: Infections of the Gastrointestinal Tract. M.J. Blaser, P.D. Smith, J.I. Ravdin, H.B. Greenberg, R.L. Guerrant, eds. Raven Press, New York. p. 53-70.

Stainsby, K.J., Lowes, J.R., Allan, R.N., and Ibbotson, J.P. 1993. Antibodies to *Mycobacterium paratuberculosis* and nine species of environmental mycobacteria in Crohn's disease and control subjects. Gut 34: 371-374.

Steidler, L., Hans, W., Schotte, L., Neirynck, S., Obermeier, F., Falk, W., Fiers, W., and Remaud, E. 2000. Treatment of murine colitis by *Lactococcus lactis* secreting interleukin-10. Science 289: 1352-1355.

Steinhart, A.H., Hiruki, T., Brzezinski, A., and Baker, J.P. 1996. Treatment of left-sided ulcerative colitis with butyrate enemas: a controlled trial. Aliment. Pharmacol. Ther. 10: 729-736.

Sternberg, E.M., Young, W.S., Bernardini, R., Calogero, A.E., Chrousos, G.P., Gold, P.W., and Wilder, R.L. 1989. A central nervous system defect in biosynthesis of corticotropin- releasing hormone is associated with susceptibility to streptococcal cell wall-induced arthritis in Lewis rats. Proc. Natl. Acad. Sci. USA 86: 4771-4775.

Stimpson, S.A., Esser, R.E., Carter, P.B., Sartor, R.B., Cromartie, W.J., and Schwab, J.H. 1987. Lipopolysaccharide induces recurrence of arthritis in rat joints previously injured by peptidoglycan-polysaccharide. J. Exp. Med. 165: 1688-1702.

Sundberg, J.P., Elson, C.O., Bedigian, H., and Birkenmeier, E.H. 1994. Spontaneous, heritable colitis in a new substrain of C3H/HeJ mice. Gastroenterology 107: 1726-1735.

Surawicz, C.M., McFarland, L.V., Greenberg, R.N., Rubin, M., Fekety, R., Mulligan, M.E., Gracia, R.J., Brandmaker, S., Bowen, K., Borjal, D., and Elmer, G.W. 2000. The search for a better treatment for recurrent *Clostridium difficile* disease: use of high-dose Vancomycin with *Saccharomyces boulardii*. CID 31: 1012-1017.

Surawicz, C.M., Elmer, G.W., Speelman, P., McFarland, L.V., Chinn, J., and van Belle, G. 1989. Prevention of antibiotica-associated diarrhea by *Saccharomyces boulardii*: a prospective study. Gastroenterology 96: 981-988.

Sutherland, L.R., Martin, F., Bailey, R.J., Fedorak, R.N., Dallaire, C., Rossmann, R., Saibil, F., and Lariviere, L. 1997. A randomized, placebo-controlled, double-blind trial of mesalamine in the maintenance of remission of Crohn's disease. The Canadian Mesalamine for Remission of Crohn's Disease Study Group. Gastroenterology 112: 1069-1077.

Sutherland, L., Singleton, J., Sessions, J., Hanauer, S., Krawitt, E., Rankin, G., Summers, R., Mekhjian, H., Greenberger, N., Kelly, M., Levine, J., Thompson, A., Albert, E., and Prokipchuk, E. 1991. Double blind, placebo controlled trial of metronidazole in Crohn's disease. Gut 32: 1071-1075.

Sutton, C.L., Kim, J., Yamane, A., Dalwadi, H., Wie, B., Landers, C., Targan, S.R., and Brown, J. 2000 Identification of a novel bacterial sequence associated with Crohn's disease. Gastroenterology 119: 23-31.

Swift, G.L., Srivastava, E.D., Stone, R., Pullan, R.D., Newcombe, R.G., Rhodes, J., Wilkinson, F., Rhodes, P., Roberts, G., and Lawrie, B.W. 1994. Controlled trial of anti-tuberculous chemotherapy for two years in Crohn's disease. Gut 35: 363-368.

Talbot, I.C., Kamm, M.A., and Leaker, B.R. 1992. Pathogenesis of Crohn's disease. Lancet 340: 315-316.

Tannock, G.W., Munro, K., Harmsen, H.J., Welling, G.W., Smart, J., and Gopal, P.K. 2000. Analysis of the fecal microflora of human subjects consuming a probiotic product containing *Lactobacillus rhamnosus* DR20. Appl. Environ. Microbiol. 66: 2578-2588.

Teahon, K., Smethurst, P., Levi, A.J., Menzies, I.S., and Bjarnason, I. 1992. Intestinal permeability in patients with Crohn's disease and their first degree relatives. Gut 33: 320-323.

Tejada-Simon, M.V., Ustunol, Z., and Pesta, J.J. 1999. Effects of lactic acid bacteria ingestion on basal cytokine mRNA and immunoglobulin levels in the mouse. J. Food. Prot. 52:287-291.

Thayer, W.R., Jr., Coutu, J.A., Chiodini, R.J., Van Kruiningen, H.J., and Merkal, R.S. 1984. Possible role of mycobacteria in inflammatory bowel disease. II. Mycobacterial antibodies in Crohn's disease. Dig. Dis. Sci. 29: 1080-1085.

Thomas, G.A., Swift, G.L., Green, J.T., Newcombe, R.G., Braniff-Mathews, C., Rhodes, J., Wilkinson, S., Strohmeyer, G., and Kreuzpainter, G. 1998. Controlled trial of antituberculous chemotherapy in Crohn's disease: a five year follow up study. Gut 42: 497-500.

Thompson, N.P., Montgomery, S.M., Pounder, R.E., and Wakefield, A.J. 1995a. Is measles vaccination a risk factor for inflammatory bowel disease? Lancet 345: 1071-1074.

Thompson, N.P., Pounder, R.E., and Wakefield, A.J. 1995b. Perinatal and childhood risk factors for inflammatory bowel disease: a case-control study. Eur. J. Gastroenterol. Hepatol. 7: 385-390.

Triantafillidis, J.K., Nicolakis, D., Emmanoullidis, A., Antoniou, A., Papatheodorou, K., and Cheracakis, P. 1996. Ornidazole in the treatment of active Crohn's disease: short-term results. Ital. J. Gastroenterol. 28: 10-14.

Turunen, U.M., Farkkila, M.A., Hataka, K., Seppaia, K., Sivonen, A., Ogren, M., Vuoristo, M., Valtonen, V., and Miettinen, T.A. 1998. Long-term treatment of ulcerative colitis with ciprofloxacin: A prospective, double-blind, placebo-controlled study. Gastroenterology 115: 1072-1078.

Ulisse, S., Giochetti, P., D'Alò, S.D., Russo, F.P., Pesce, I., Ricci, G., Rizzello, F., Helwig, U., Cifone, M.G., Campieri, M., and DeSimone, C. 2001. Expression of cytokines, inducible nitric oxide synthase, and matrix metalloproteinase in pouchitis: effects of probiotic treatment. Am. J. Gastroenterol. 96: 2691-2699.

Ursing, B., Alm, T., Barany, F., Bergelin, I., Ganrot-Norlin, K., Hoevels, J., Huitfeldt, B., Jarnerot, B., Krause, U., Krook, A., Lindstrom, B., Nordle, O., and Rosen, A. 1982. A comparative study of metronidazole and sulfasalazine for active Crohn's disease: the cooperative Crohn's disease study in Sweden. II. Result. Gastroenterology 83: 550-562.

van de Merwe, J.P., Schroder, A.M., Wensinck, F., and Hazenberg, M.P. 1988. The obligate anaerobic faecal flora of patients with Crohn's disease and their first-degree relatives. Scand. J. Gastroenterol. 23: 1125-1131.

Vanderhoof, J.A. 2000. Probiotics and intestinal inflammatory disorders in infants and children. J. Pediatr. Gastroenterol. Nutr. 30: S34-S38.

Veltkamp, C., Tonkonogy, S.L., De Jong, Y.P., Albright, C., Grenther, W.B., Balish, E., Trhorst, C., and Sartor, R.B. 2001. Continuous stimulation by normal luminal bacteria is essential for the development and perpetuation of colitis in Tg(epsilon26) mice. Gastroenterology 120: 900-913.

Venturi, A., Gionchetti, P., Rizzello, F., Johansson, R., Zucconi, E., Brigidi, P., Matteuzzi, D., Campieri, M. 1999. Impact on the composition of the faecal flora by a new probiotic preparation: preliminary data on maintenance treatment of patients with ulcerative colitis. Aliment. Pharmacol. Therap. 13: 1103-1108.

von Ritter, C., Sekizuka, E., Grisham, M.B., and Granger, D.N. 1988. The chemotactic peptide N-formyl methionyl-leucyl-phenylalanine increases mucosal permeability in the distal ileum of the rat. Gastroenterology 95: 651-656.

Wakefield, A.J., Fox, J.D., Sawyerr, A.M., Taylor, J.E., Sweenie, C.H., Smith, M., Emery, V.C., Hudson, M., Tedder, V.S., and Pounder, R.E. 1992. Detection of herpesvirus DNA in the large intestine of patients with ulcerative colitis and Crohn's disease using the nested polymerase chain reaction. J. Med. Virol. 38: 183-190.

Wakefield, A.J., Pittilo, R.M., Sim, R., Cosby, S.L., Stephenson, J.R., Dhillon, A.P., and Pounder, R.E. 1993. Evidence of persistent measles virus infection in Crohn's disease. J. Med. Virol. 39: 345-353.

Weber, P., Koch, M., Heizmann, W.R., Scheurlen, M., Jenss, H., and Hartmann, F. 1992. Microbic superinfection in relapse of inflammatory bowel disease. J. Clin. Gastroenterol. 14: 302-308.

Wei, B., Huang, T., Dalwadi, H., and Braun, J. 2001. Identification of *Pseudomonas fluorescens* as the microorganism expressing the Crohn's disease associated I2 gene. Gastroenterology 120: A82.

Wells, C.L., van de Westerlo, E.M., Jechorek, R.P., Feltis, B.A., Wilkins, T.D., and Erlandsen, S.L. 1996. *Bacteroides fragilis* enterotoxin modulates epithelial permeability and bacterial internalization by HT-29 enterocytes. Gastroenterology 110: 1429-1437.

Wold, A.E. 2001. Immune Effects of Probiotics. Scand. J. Nutr. 45: 76-85.

Wold, A.E., and Adlerberth, I. 2000. Pathologic effects on commensialism. In: Persistent Bacterial Infections. J.P. Nataro, M.J. Blaser, S. Cunningham-Rundles, eds. ASM Publishing Co. p. 115-144.

From: *Probiotics and Prebiotics: Where Are We Going?*
Edited by: Gerald W. Tannock

Chapter 8

Gut Microflora and Atopic Disease

Clare S. Murray and Ashley Woodcock

Abstract

In recent years many countries have experienced a rise in allergic disease, which cannot be genetic in origin. Over the same period many aspects of modern life have changed and theories have been put forward to explain this trend. In 1989, Strachan introduced his "Hygiene Hypothesis", in which he proposed that allergic diseases could be prevented by infection in early childhood. However, despite numerous studies a specific "infective protective factor" has not been identified. Recently, attention has turned towards the intestinal microflora and the possibility that colonisation with specific microbes may be more important than sporadic infections. The immune system is Th-2 skewed in newborn babies, and the intestinal microflora may act as a counter-regulator, driving towards Th-1 differentiation. Colonisation with microbes begins immediately after birth and soon outnumber the human host cells. Thus, the microflora is the earliest and by far the largest stimulus to the immune system, and outweighs that of any occasional infection.

There is evidence suggestive of an association between intestinal microflora and allergic disease and also suggestive that probiotics may improve or even prevent disease. However, these studies are on small numbers of children and long-term follow up is awaited. Longitudinal studies are necessary to establish whether the intestinal microflora plays an active role in the aetiology of allergic disease and whether manipulation can lead to a decrease in prevalence.

Introduction

Over recent decades many countries in the developed world and also the developing world have experienced a rise in asthma prevalence and probably atopic sensitisation, which cannot be genetic in origin. Over the same period many aspects of modern life have changed and several theories have been put forward to explain this increasing trend. There is circumstantial evidence for a rise in exposure to allergens in many regions (e.g. increased number of family pets, reduced indoor ventilation, increased soft furnishings) but this is probably acting on a more susceptible host. In 1989, Strachan introduced his Hygiene Hypothesis, in which he proposed that allergic diseases could be prevented by infection in early childhood. Subsequently a plausible immunological mechanism based on balance between differential T cell cytokine expression (Th 1 differentiation for infection versus Th 2 for allergy) has promoted this concept. Despite numerous studies a specific infective protective factor has not been identified.

Recently, attention has turned towards the intestinal microflora and the possibility that colonisation with specific microbes may be more important than sporadic infections. In the prevention of allergic disease, the intestinal microflora appears to promote potentially anti-allergic processes. The immune system is Th 2-skewed in newborn babies, and the intestinal microflora may act as a major postnatal counter-regulator, driving towards Th1 differentiation during infancy. Colonisation with microbes begins immediately after birth and they soon outnumber the human host cells by a factor of ten. Thus, the gut microflora is the earliest and by far the largest normal stimulus to the human immune system, and is likely to outweigh that of any occasional infection. In this chapter we give an overview of atopic disease and how it has increased over recent years. We describe the Hygiene

Hypothesis, and the current evidence for and against the concept that the intestinal microflora may be the putative protective factor against the development of allergy.

Atopy

The term 'atopy' was first used by clinicians at the beginning of the 20th century to describe a group of disorders, including asthma, eczema and hayfever, that appeared to run in families, to have characteristic wealing skin reactions to environmental allergens and to have circulating antibody in their serum, that could be transferred to the skin of non-sensitised individuals. The term is now best used to describe those individuals who readily develop antibodies of immunoglobulin E class (IgE) against common materials present in the environment.

Atopic Eczema

Atopic eczema (or atopic dermatitis) is a common problem in early childhood. Characteristic features include red, hot, oedematous skin, with weeping and oozing in the acute stages, excoriation due to intense itching and frequently secondary infection. In the chronic stages scaling, fissuring, thickening and lichenification can be seen. It is often associated in very early life with food allergies. Serum levels of IgE are usually elevated.

Allergic Rhinoconjunctivitis

Allergic rhinoconjuctivitis is characterised by sneezing, runny nose, itchy or blocked nose and itchy, irritated eyes. In addition itching of the palate and ears can occur, although nasal and conjunctival mucosa are the main focus of the pathophysiological changes. Seasonal allergic rhinoconjunctivitis is usually associated with grass pollen allergy, otherwise known as hay fever. Symptoms can also occur all year round usually secondary to house dust mite allergy.

Asthma

Asthma is defined as a generalised but reversible obstruction of the airflow in the intrathoracic airways, with bronchial hyper-responsiveness and airway inflammation. It is thought to arise as a result of complex interactions between multiple genes and the environment. Clinically it is characterised by episodic cough, breathlessness and wheezing. Many asthmatic patients are atopic, with positive skin tests to airborne allergens.

Asthma/Wheezing Phenotypes in Childhood

Significant advances have been made in recent years in understanding the different ways in which asthma is expressed in childhood, primarily by large prospective longitudinal studies. Although there is some degree of overlap, childhood asthma can be divided into 2 distinct phenotypes, transient early wheezers and persistent wheezers.

- Transient early wheezers, only wheeze early in childhood generally with viral upper respiratory tract infections. They are characterised by having diminished lung function at birth, which improves slowly with age. Often they are children whose parents smoke. The outlook is good and most children have stopped wheezing by age 3.

- Persistent wheezers may start wheezing early or later in life, but wheezing then persists into later childhood. They often have a family history of asthma or atopic disease. They usually have normal lung function soon after birth, which deteriorates with age, and is already measurably and possibly permanently reduced by the age of 6. They also usually have raised serum IgE detectable from infancy. It is these persistent wheezers who are the children that develop classical childhood allergic asthma.

Recent studies suggest that persistent wheezers have different immune responses to viral infections than transient wheezers (Martinez *et al.,* 1998). Persistent wheezers were found to have raised serum IgE levels during the acute phase of their first lower respiratory illness compared with the convalescent phase, a response that was not found in transient wheezers. In addition, transient early wheezers had a significantly reduced peripheral eosinophil count during the acute phase of the lower respiratory illness, a response not observed in the persistent wheezers. This supports the idea that wheezing episodes in transient early wheezers are not caused by an abnormal immune response to a virus but more likely attributable to a mechanical or functional alteration of the lung parenchyma and/or the airways, which makes wheezing more likely during viral respiratory tract infections in children whose airways are already narrow to start with. It is important when examining intervention studies for *allergic* asthma, that confounding data from non-allergic early infant wheezers is discounted.

Prevalence of Asthma

Asthma is the commonest chronic disease of childhood in the UK. The prevalence of wheeze and asthma has risen over recent decades (Ninan *et al.,* 1992). An Australian study found the 12-month prevalence of wheeze had increased by about 17% over the 10-year period 1982-1992 in children aged 8-10 years (Peat *et al.,* 1994). In addition they found that airway hyper-responsiveness, measured by histamine inhalation, also increased between 1.4 and 2-fold. A similar study in London found the prevalence of asthma among 7-8 year olds to have increased by 16% in the period 1978-1991, but also found a substantial reduction in the level of disability and acute severe attacks, perhaps indicating an improvement in treatment received (Anderson *et al.,* 1994). These and other studies indicate the prevalence of asthma in these countries appears to be increasing by about 1% per annum. However, a study in Rome, Italy, suggests that this rapid increase in prevalence that was seen in the 1970s and 1980s may be slowing. They report the prevalence of asthma in school children aged 6-14 years to have increased significantly between 1974 and 1992, but to have remained stable between 1992 and 1998 (Ronchetti *et al.,* 2001).

As discussed earlier, not all wheeze in childhood progresses to asthma. The studies already discussed have highlighted the increase in wheeze and asthma in older children, but until very recently it was unclear as to whether wheeze in pre-school children, which may be transient, has also increased. Kuehni *et al.* studied the prevalence of respiratory symptoms in children aged 1-5 years, in Leicestershire in the UK, between 1990 and 1998 (Kuehni *et al.,* 2001). They found a significant increase in wheeze, treatment for wheeze, diagnosis of asthma, and admission to hospital over the eight years of the study. More interestingly, they found that not only did wheezing with multiple triggers (e.g. allergens, exercise) increase but also viral associated wheeze increased. That is to say that the prevalence of persistent wheezing (allergic asthma) and transient wheezing (with viral upper respiratory tract infections) both increased. The fact that all pre-school wheezing disorders increased indicate that this cannot be accounted for solely by an increase in atopy. A change in pulmonary responsiveness to environmental triggers must have occurred independently of any change in atopy.

In recent years the International Study of Asthma and Allergies in Childhood (ISAAC) has given us a wealth of information regarding the prevalence of asthma and allergic disease around the world (Warner 1999). Differences in prevalence of symptoms between countries were found to be up to 60-fold. For asthma symptoms the highest 12-month prevalences were from centres in the UK, Australia, New Zealand and Ireland, followed by most centres in North, Central and South America. The lowest prevalences were in Eastern European countries, the Indian subcontinent and the Far East. Data from the UK indicated a 12-month prevalence of wheezing of 33.3% and a self-reported diagnosis of asthma in 20.9% of 12-14 year olds (Kaur *et al.*, 1998).

Many hypotheses have been mounted to try to explain the rising prevalence of asthma and the worldwide variations. The strong association with a Westernised society and affluence, has brought diet and hygiene into question. The effects of increasing indoor allergens because of energy saving measures, which reduce air exchanges, also provide an attractive hypothesis. Clearly genetic influences may also explain a part of the worldwide variations in prevalence.

Hygiene Hypothesis

The hygiene hypothesis was proposed by Strachan in 1989, who had demonstrated an inverse relationship between birth order in families and prevalence of hayfever (Strachan 1989). It was a novel explanation for the main epidemiological features of hayfever and the apparent rise in prevalence of atopy. In summary it stated "These observations... could be explained if allergic diseases could be prevented by infection in early childhood, transmitted by unhygienic contact with older siblings, or acquired prenatally..". Over the past century declining family size, improved household amenities and higher standards of personal cleanliness have reduced opportunities for cross infection in young families. This may have resulted in more widespread clinical expression of atopic disease.

During the 1990s a plausible immunological mechanism arose for the hygiene hypothesis. The distinction between Th1 and Th2 lymphocyte populations was discovered in laboratory animals, with the recognition that immunity to bacteria and viruses induces a Th1 pattern of cytokine release (IFN-γ, IL-2), potentially suppressing the Th2 responses (IL-4, IL-5, IL-10, IL-13) involved in IgE mediated allergy.

Since the hygiene hypothesis was proposed numerous studies have investigated the effect of infections on the development of atopic disease. It is important to note that the relationship found by Strachan between birth order and hayfever prevalence does not apply so consistently or conspicuously to asthma symptoms. Although some large studies of asthma prevalence do show a strong inverse relationship with birth order (Seidman *et al.,* 1991), most only show a weak association (Wickens *et al.,* 1999). This implies that early infection may be protective against allergy, but that other factors such as allergen exposure have a much larger impact on the end-organ response that causes asthma (Pearce *et al.,* 1999). Any relationship may be further confounded by other non-allergic wheezing syndromes (transient early wheeze) which are positively related to viral infections.

Inconsistencies in the Hygiene Hypothesis

Most cases of asthma in children in poor American inner cities are strongly associated with allergic sensitisation to allergens, in particular cockroaches (Rosenstreich *et al.,* 1997). Sensitisation in the inner cities is not less common than in affluent urban areas, which is in contrast to the hygiene hypothesis. However, these inner city children are probably not exposed to the wide array of bacteria that children who perhaps live in farming communities, who have access to natural soil, eat unprocessed foods or live in areas endemic for oro-faecal infections.

Those who live in the developing world, not invariably have serum IgE to inhalant allergens. However, these atopic responses are rarely associated with allergic disease (Yemaneberhan *et al.,* 1997). Severe helminthic infections are commonplace in such areas and potentiate Th-2 responses, and perhaps compete for IgE binding sites on effector cells, and thus prevent inflammatory consequences of exposure to allergens. A study in Venezuela found that treating these helminthic infections decreased total serum IgE, but paradoxically led to a rise in skin reactions to inhalant allergens (Lynch *et al.,* 1993). Given these inconsistencies it appears more complicated than just 'hygiene favours allergy' and ' dirt prevents allergy'.

Specific Infections and Allergy

Respiratory Syncytial Virus

One virus, namely respiratory syncytial virus (RSV), has been heavily implicated in the *initiation* of asthma. It has long been recognised that infants who have RSV bronchiolitis infection, are at increased risk of subsequent episodes of wheezing in infancy and childhood. As the majority of children will have been infected with RSV by the age of 2 (Openshaw 1995), it has been believed that it is those infants with severe disease that are at increased risk of continuing to wheeze and developing asthma (Sly *et al.*, 1989; Murray *et al.*, 1992). Some studies have suggested that non-specific bronchial hyper-responsiveness (to inhaled histamine) can persist for 10 years or more after acute bronchiolitis (Pullan *et al.*, 1982). Several hypotheses have been put forward to explain this possible phenomenon.

- RSV infections could promote a Th-2 immune response that promotes allergic inflammation in susceptible individuals (Jackson *et al.*, 1996; Alwan *et al.*, 1993).

- RSV infection affects lung development and produces airway remodelling.

- Children that acquire a severe RSV infection have a pre-existing immune defect that leads to more severe infection and perhaps promotion of allergic sensitisation.

More recently however, longitudinal studies have begun to resolve this question. Infants with RSV bronchiolitis do not seem to be at increased risk of asthma later in life (Wennergren *et al.*, 1997; Stein *et al.*, 1999). Children followed from birth to 13 years of age were no longer at increased risk of wheezing by the time they reached 13 years old. The current view is that RSV infection itself does *not* trigger persistent asthma. RSV is however associated with severe wheezing in predisposed infants (i.e. those with diminished airway function at birth and smoking mothers).

Measles

Whether measles infection is protective against atopy is controversial. A study in Guinea-Bissau in Africa, followed long term survivors of a measles epidemic, and found that measles infection early in childhood was associated with reduced risk of allergen sensitisation (Shaheen *et al.*, 1996). However this effect has not been found in further studies in Western societies (Farooqi *et al.*, 1998; Lewis *et al.*, 1998; Paunio *et al.*, 2000). It is possible that in the Guinea-Bissau cohort, those children with impaired Th1 immunity, and therefore at increased risk of allergic disease, were also at increased risk of dying during the measles epidemic. Thus the prevalence of atopy was artificially lowered in the infected survivors.

Mycobacteria

It has also been suggested that mycobacterial infection may have protective effects against atopy. Shirakawa *et al.* examined 867 Japanese schoolchildren who had complete tuberculin skin test result records at birth, 6 and 12 years (Shirakawa *et al.*, 1997). Children who exhibited strong tuberculin reactions at 6 and 12 years, indicating natural exposure to tuberculosis had fewer symptoms of asthma, rhinitis and eczema. Several European studies have since been unable to find an association (Alm *et al.*, 1997; Strannegard *et al.*, 1998; Omenaas *et al.*, 2000). A Finnish case control study of 456 children and 706 adults with active tuberculosis and age matched controls found only a slightly lower prevalence of allergic disease among tuberculosis cases, but a slightly higher prevalence of asthma (von Hertzen *et al.*, 1999). Studies are underway to determine whether there is a protective effect of mycobacteria, given by vaccination against allergy. Intervention studies in at risk babies could safely investigate the value of neonatal BCG, which has been in wide use for decades. However, there are important ethical and safety considerations for other mycobacterial vaccines (e. g. *Mycobacterium vaccae*) in healthy neonates who are at risk of (but not affected by) allergic disease.

Hepatitis A

Perhaps the most consistent evidence of an inverse relationship between infection and allergic sensitisation has emerged from studies of Hepatitis A. Hepatitis A is associated with poor hygiene and low socio-economic status. Matricardi *et al.* studied the prevalence of atopy in Italian military recruits, and found the presence of antibodies to Hepatitis A (indicating previous infection) were associated with a reduced risk of asthma and atopy (Matricardi *et al.*, 1997). In a second study in the same population the association was found to be quite specific to Hepatitis A (measles, mumps, rubella, varicella, CMV, herpes simplex and *Helicobacter pylori* were not associated)

(Matricardi *et al.,* 2000). A significant inverse relationship was, however, also seen with *Toxoplasma gondii*. Two other studies have also addressed this subject. A population survey of San Marino found a 40% relative reduction in atopy amongst those seropositive for Hepatitis A (Matricardi *et al.,* 1999). A nested case control study in Aberdeen also found a relative reduction in atopy in those seropositive for Hepatitis A or *Toxoplasma gondii*, though this did not reach statistical significance (Bodner *et al.,* 2000). This important and consistent data points to food borne or faeco-orally transmitted infections, rather than respiratory infections, being more likely determinants of the risk of allergic disease.

The Gastrointestinal Tract and Atopy

Abnormal Gut Mucosa and Allergic Disease

The gastrointestinal mucosa is abnormal in a variety of allergic diseases. In both eczema and food allergy there is evidence of an inflammatory response in the gastrointestinal tract, with faecal TNF-α and Eosinophil cationic protein (ECP) being increased (Majamaa *et al.,* 1996). In addition the permeability of the intestinal mucosa has been shown to be increased in allergic disease, both *in vitro* and *in vivo*. Breast-fed infants with multiple food allergies, improved on maternal exclusion diets, have been shown to have a 50% increase in intestinal permeability during re-challenge (de Boissieu *et al.,* 1997). *In vitro* studies in eczema have shown a 10-fold increase in permeability to horseradish peroxidase (Majamaa *et al.,* 1996). Also a study in adult asthmatics demonstrated significantly increased intestinal permeability (with Cr EDTA) in asthmatics compared with non-asthma controls (Benard *et al.,* 1996). Increased permeability could allow the systemic absorption of antigens, bypassing antigen presenting cells and thus producing systemic hyper-responsiveness (Strobel 1995). Why is the permeability of the GI tract abnormal? Could it be related to an abnormal gut microflora?

Development of Normal Gut Microflora

The intestinal microflora in healthy humans is confined to the distal ileum and colon. It is comprised of numerous species of bacteria, predominantly anaerobes. It is estimated to contain 400 different species. The population level of each species is believed to be regulated by the competition for nutrients and space. Unfortunately the study of gut microflora has been almost entirely culture based (i.e. only bacteria which grow in established culture methods have been studied). In addition studies on the development of the normal microflora have often been on frozen stools and therefore of questionable significance. There is an urgent need for PCR based studies on normal gut microflora development in human children in a variety of different geographic and economic circumstances, and in health and disease.

It is known that humans are born with sterile colons, but acquisition of the microflora begins once the fetal membranes have ruptured. However, the establishment of the microflora ecosystem is slow and evolves over several years. In humans, several factors influence, not only the rate of colonisation, but also the composition of the gut microflora, such as mode of delivery, gestational age at delivery, feeding regime and antibiotic use (Yoshioka *et al.,* 1983; Bennet *et al.,* 1987; Hall *et al.,* 1990; Adlerberth *et al.,* 1991). Newborn babies acquire their gut microflora from maternal intestinal microflora and from their environment.

Enterobacteria (especially *E.coli*), streptococci and staphylococci are often the first organisms to colonise the neonate. As these facultative bacteria expand, they consume oxygen and lower the oxidation-reduction potential enabling anaerobic bacteria (particularly bifidobacteria, clostridia and bacteroides) to proliferate. As the anaerobic bacteria expand, the facultative bacteria decline in number, but during the first few months of life relatively high numbers of both bacteria exist together.

Gut Microflora and the Immune System

The gut mucosa has two major roles, allowing the passage of nutrients and at the same time protecting against infectious agents. There are more lymphoid cells in the gut than in the spleen, peripheral lymph nodes and blood put together. This may be a reflection of the fact that the gut mucosa separates some 10^{13} human cells from approximately 10^{14} bacteria that make up the gut microflora. The gut lymphoid system differs from that of other tissues in that it predominately produces IgA antibodies. Exposure to the normal gut microflora and foods are potent stimuli to the development of the intestinal lymphoid tissues after birth. Germfree animals have been shown to have considerably fewer IgA producing plasma cells in the gut than other animals and colonisation with the gut microflora leads to development of a normal immune system in these animals.

Oral Tolerance to Ingested Antigens

The gastrointestinal tract has a primary role in programming the immune response to allergens. Substances introduced per-orally are usually ignored by the immune system, a phenomenon known as 'oral tolerance'. Digested proteins are bound at the luminal surface of epithelial cells and after absorption are processed to small peptides with low immunogenicity. Macrophages and dendritic cells present these small peptide fragments to stimulate cytokine release from T cells. Studies involving germfree animals have led to the conclusion that a normal healthy gut microflora is necessary to induce oral tolerance. Thus animals bred in a sterile environment are unable to develop normal tolerance (Moreau *et al.,* 1988; Sudo *et al.,* 1997). It has been claimed that this depends on a lack of endotoxic lipopolysaccharide (LPS) which seems necessary for the development of tolerance (Wannemuehler *et al.,* 1982; Kim *et al.,* 1995). LPS is a constituent of the external cell membrane of Gram-negative bacteria. In patients with cow's milk allergy, intact rather than intestinally processed cow's milk proteins have been shown to stimulate peripheral blood mononuclear cells to release proinflammatory cytokines. In addition it has been shown that cow's milk protein degraded by lactobacilli, rather than pepsin or trypsin, generate tolerogenic proteins from the native protein (Sutas *et al.,* 1996). Also lactobacilli have been shown to reduce the allogenicity of peas by 90%, by fermentation (Barkholt *et al.,* 1998). This substantiates the hypothesis that specific strains of intestinal microflora aid the host protection against allergic sensitisation.

A link between respiratory diseases due to inhalant allergens and events occurring in the gut is not so tenuous at it first may seem. Firstly, a study in adults administered aerosolised radiolabelled pollen grains, found that most of the pollen grains once impacted on the oro-pharyngeal mucosa, were then ingested (Wilson *et al.,* 1973). In infants, with much narrower airways, quantities that are ingested may be much greater. Secondly, eosinophilic inflammation has been observed in the gut mucosa of asthmatic and atopic subjects, thus implying that the entire mucosal immune system is involved in these individuals (Wallaert *et al.,* 1995).

Anthroposophic Lifestyle, Antibiotics and Atopic Disease

Indirect evidence is available which is suggestive of altered gut microflora being implicated in the pathogenesis of atopic disease.

Studies of children from anthroposophic families have found the prevalence of atopic disease to be much lower than in those from other families. People who follow an anthroposophic lifestyle use antibiotics restrictively, have few vaccinations and eat a diet of locally produced foods, preserved by spontaneous fermentation, thus high in live lactobacilli. *Lactobacillus plantarum,* most common in spontaneously fermented vegetables, has been shown to colonise the human gut and affect indigenous strains (Johansson *et al.,* 1993). Alm *et al.* (1999) carried out a case control study comparing children attending Steiner schools (anthroposophic schools) in Sweden and children in neighbouring normal schools. Skin prick and blood tests showed the children from the Steiner schools to have a much lower prevalence of atopy than controls (OR 0.67, 95%CI 0.43-0.91). In addition, there was an inverse relationship between the number of characteristic features of an anthroposophic lifestyle and the risk of atopy (*p* for trend 0.01).

Several studies have investigated antibiotic usage and the development of atopic disease. Farooqi carried out a retrospective study of 1934 subjects who were born into an Oxfordshire general practice (1975-1984) and had remained registered at the practice throughout their lives (Farooqi *et al.,* 1998). Treatment with antibiotics in the first two years of life was associated with an increased risk of subsequent atopic disease (OR 2.07, 95% CI 1.64 to 2.60, *p*<0.0001). Wickens *et al.* investigated children in Steiner schools in New Zealand aged 5-10 years by sending out questionnaires to parents regarding antibiotic usage and atopic disease (Wickens *et al.,* 1999). The adjusted odds ratio for asthma was 4.05 (95%CI 1.55 to 10.55) if antibiotics were used in the first year of life and 1.64 (95%CI 0.60 to 4.46) if antibiotics had been used only after the first year, compared with children who had never had antibiotics. Also the number of courses of antibiotics during the first year of life was significantly associated with an increased odds ratio for asthma: 2.27 (95%CI 1.14 to 4.51) for 1 or 2 courses and 4.02 (95%CI 1.57 to 10.31) for 3 or more courses compared with those children who had had no antibiotics in the first year.

Biological mechanisms of the association between antibiotic usage and the development of atopic disease still remain speculative, but could be due to the reduction of intensity and duration of infections. However, as many infections in early life will be due to viral infections, on which antibiotics will have no effect, it could be that antibiotics directly influence the immune system by altering gut microflora. The effect of antibiotics seems greatest in the first year of life and increases with the number of courses. This supports the idea of an effect of antibiotics on the normal development of gut microflora in infancy.

Altered Gut Microflora and Allergy

The prevalence of atopy in eastern Europe is substantially lower than in the West (Jogi *et al.,* 1998). Bjorksten's group in Sweden/Estonia was the first to propose that persistent pressure on the immune system by intestinal bacteria may prevent atopic sensitisation to allergens. They examined the intestinal microflora of 1-year old infants in Estonia (low prevalence of atopy) and Sweden (high prevalence of atopy) and found high counts of lactobacilli and eubacteria in the Estonian babies and increased numbers of clostridia in the Swedish infants (Sepp *et al.,* 1997). The microflora pattern in the Estonian infants was similar to the microflora prevailing in infants in western Europe in the 1960's and they postulated a shift in bacteria among infants in western industrialised countries. Subsequently they examined the gut microflora of allergic and non-allergic 2-year old children in Sweden and Estonia (Bjorksten *et al.,* 1999). Allergic children were less often colonised with lactobacilli and more likely to harbour aerobes, (coliforms in the Estonia, *Staphylococcus aureus* in Sweden). Bottcher *et al.* (2000) have studied faecal short chain fatty acids in allergic and non-allergic infants (indirect measure of gut microflora). Allergic infants were found to have higher levels of iso-caproic acid, associated with *Clostridium difficile,* and lower levels of propionic, iso-butyric, butyric, iso-valeric and valeric acids.

More recently, preliminary evidence suggests that differences in microflora can precede the development of atopic disease (Kalliomaki *et al.,* 2001). Intestinal microflora was examined in 76 infants at 3 weeks and 3 months of age. Children were then skin prick tested at 12 months of age to determine whether or not they were atopic. Atopic children were found to have a trend for more clostridia and fewer bifidobacteria than non-atopic children. The authors suggest that the intestinal bacteria play a crucial role in the maturation of human immunity to a non-atopic state. Custovic et al used a different approach with a semi-quantitative ELISA for measurement of *Clostridium difficile*-specific IgG in serum to investigate whether there were differences between atopic wheezy infants and non-atopic non-wheezy controls. In a prospective cohort study of children at one year, cases (atopic, history of recurrent wheeze) were matched (sex, date of birth, exposure to house dust mite, cat and dog allergen) with a control group of infants (non-atopic, no history of wheeze). Cases had significantly higher *Clostridium difficile*-specific IgG absorbance levels (GM 0.298, 95% CI 0.249-0.358) compared to controls (GM 0.235, 95% CI 0.201-0.274; mean difference 1.27, 95% CI 1.07-1.50; p=0.01) (A. Custovic, unpublished). These results are further evidence for a difference in the composition of intestinal microflora between allergic and non-allergic infants at 1 year of age, with allergic children having higher *Clostridium difficile* IgG antibody levels.

Probiotics and Atopic Disease

With evidence accumulating for abnormalities of gut microflora in relation to allergic disease, studies have started to focus on improving intestinal microbial balance in an effort either to improve established atopic disease, or to prevent it occurring. Probiotics are defined as live microbial feed supplements, which beneficially affect the host by improving its intestinal microbial balance. Humans already consume lactobacilli and bifidobacteria in live yoghurts. Are they safe in infancy, and are they effective?

Much of the initial work has come from Finland, in three different clinical situations. First, Majamaa *et al*. (1997) used probiotic supplements (*Lactobacillus GG* 5×10^8 colony forming units/g formula) in infants (age 2.5-15.7 months) with atopic eczema and cow's milk allergy, and found that atopic eczema scores improved in those infants given probiotics compared with those infants who had not received them. These infants also had elevated faecal markers of intestinal inflammation, (TNF-α and α1-antitrypsin,) which returned to normal when taking probiotics.

Subsequently, in a second study, the same group (Isolauri *et al.*) randomised 27 exclusively breastfed infants with atopic dermatitis (mean age 4.6 months), to extensively hydrolysed milk formula alone or with *Lactobacillus GG* or *Bifidobacterium lactis* Bb-12 supplementation. There was a significant improvement in skin condition in both the supplemented groups (Isolauri *et al., 2000*).

In the third study, Kalliomäki *et al*. (2201) carried out a randomised placebo-controlled trial of probiotics (*Lactobacillus GG* 1×10^{10} colony-forming units daily) in the primary prevention of atopic disease. Probiotics were given prenatally for 2-4 weeks to mothers with a first degree relative with atopic disease, and then post-natally for 6 months to their high-risk infant. Children were then followed prospectively to the age of 2 years. The frequency of atopic eczema in the probiotic group was half of that in the placebo group, relative risk 0.51 (95%CI 0.32 to 0.84). However, there was no accompanying decrease in either total or specific IgE or skin prick test reactions between the groups, suggesting that there was no change in atopy.

These preliminary studies offer a potentially exciting new treatment for allergic diseases. However, these are small studies, in a population of exclusively breast-fed children with mild allergic disease. There are major remaining questions:

- What is the defect in the intestinal flora? (deficiency of e.g. bifidobacteria, versus colonisation by e.g. clostridia?)
- Will these data transfer to other parts of the world (e.g. used with cows-milk based feeds and/or early weaning)?
- Will probiotics be effective in treating more severe disease?
- Will they prevent allergic asthma?
- What is the mechanism of action?
- Will this data be transferable to other probiotics?
- What is the safe dose/timing? Is it best to treat the mother antenatally?
- Are they safe? (is excess Th1 skewing hazardous?)

We badly need large controlled studies to explore these concepts. However we must proceed cautiously, with initial safety studies in infants with established allergic disease, before moving to primary prevention studies in at-risk children.

An Alternative 'Hygiene Hypothesis'

In evolutionary terms, it is unlikely that protection from atopic disease is provided by one bacterial or viral species alone. However, possibly a high turnover of bacteria at the mucosal level, could cause a potent and continuous immune stimulation, which may be sufficient to prevent the development of atopy. It is of interest to note that in a study of intestinal microflora in Pakistan (low prevalence of atopy) in the first 6 months of life, on average 8.5 different *E. coli* strains were found per infant (Adlerberth *et al.,* 1998). This is in marked contrast to a study carried out in Sweden (high prevalence of atopy) where infants were found to carry the same strain of *E. coli* for months (Kuhn *et al.,* 1986).

The hypothesis that a high turnover of bacterial species and strains in the gut could cause sufficient immune stimulation to protect against atopy was proposed by Wold in 1998. It has biological plausibility and would also explain the effects of family size and birth order on atopy. That is, cross infections with new bacterial strains are facilitated in large families. It may be that gut associated lymphoid tissue (GALT) may not be the only target for such an atopy protective effect, but other mucosa-associated lymphoid tissue may also be important (nasal associated lymphoid tissue, Waldeyer's ring). If a high bacteria turnover is essential to prevent atopy, then different bacteria must share common immune-modulating effects. Examples of molecules which stimulate Th-1 immunity do exist, such as endotoxic LPS and DNA containing immune-stimulating sequences (ISS).

Summary

Allergic disease has increased to epidemic levels over the last few decades, particularly in developed countries. At the same time there has been a huge change in initial gut colonisation during the last century, and there are big differences between privileged infants from the developed world (high prevalence of allergy) and newborns in underprivileged countries (Simhon *et al.,* 1982). The dominance of lactobacilli and bifidobacteria in the early microflora in the developing world has been increasingly replaced by a variety of hospital-acquired organisms in the developed world. This has led to a novel discordance between maternal and infant microflora, exacerbated by hospital deliveries, caesarean sections and special-care baby unit admissions (Gronlund *et al.,* 1999). The first bacteria to colonise a previously sterile gut may establish a permanent niche, thus putting subsequent arrivers at a competitive disadvantage (Hooper *et al.,* 2001).

There is evidence suggestive of an association between intestinal microflora and allergic disease and also studies suggesting that changing the intestinal microflora with the use of probiotics may improve disease or even prevent it occurring. However, these studies as yet are on small numbers of children and long-term follow up is awaited. Long-term longitudinal studies are necessary to establish:

- The normal development of gut microflora in human infants with modern techniques,
- whether the gut microflora plays an active role in the aetiology of allergic disease
- whether manipulation of it can safely lead to a decrease in the prevalence of allergic disease.

References

Adlerberth, I., Carlsson, B., de Man, P., Jalil, F., Khan, S.R., Larsson, P., Mellander, L., Svanborg, C., Wold, A.E., and Hanson, L.A. 1991. Intestinal colonization with *Enterobacteriaceae* in Pakistani and Swedish hospital-delivered infants. Acta Paediatr. Scand. 80(6-7): 602-610.

Adlerberth, I., Jalil, F., Carlsson, B., Mellander, L., Hanson, L.A., Larsson, P., Khalil, K., and Wold, A.E. 1998. High turnover rate of *Escherichia coli* strains in the intestinal flora of infants in Pakistan. Epidemiol. Infect. 121(3): 587-98.

Alm, J.S., Lilja, G., Pershagen, G., and Scheynius, A. 1997. Early BCG vaccination and development of atopy. Lancet 350(9075): 400-403.

Alm, J.S., Swartz, J., Lilja, G., Scheynius, A., and Pershagen, G. 1999. Atopy in children of families with an anthroposophic lifestyle. Lancet 353(9163): 1485-1488.

Alwan, W.H., Record, F.M., and Openshaw, P.J. 1993. Phenotypic and functional characterization of T cell lines specific for individual respiratory syncytial virus proteins. J. Immunol. 150(12): 5211-5218.

Anderson, H. R., Butland, B.K., and Strachan, D.P. 1994. Trends in prevalence and severity of childhood asthma. B.M.J. 308(6944): 1600-1604.

Barkholt, V., Jorgensen, P.B., Sorensen, D., Bahrenscheer, J., Haikara, A., Lemola, E., Laitila, A., and Frokiaer, H. 1998. Protein modification by fermentation: effect of fermentation on the potential allergenicity of pea. Allergy 53(46): 106-108.

Benard, A., Desreumeaux, P., Huglo, D., Hoorelbeke, A., Tonnel, A.B. and Wallaert, B. 1996. Increased intestinal permeability in bronchial asthma. J. Allergy Clin. Immunol. 97(6): 1173-1178.

Bennet, R. and Nord, C.E. 1987. Development of the faecal anaerobic microflora after caesarean section and treatment with antibiotics in newborn infants. Infection 15(5): 332-6.

Bjorksten, B., Naaber, P., Sepp, E., and Mikelsaar, M. 1999. The intestinal microflora in allergic Estonian and Swedish 2-year-old children. Clin. Exp. Allergy 29(3): 342-346.

Bodner, C., Anderson, W.J., Reid, T.S., and Godden, D.J. 2000. Childhood exposure to infection and risk of adult onset wheeze and atopy. Thorax 55(5): 383-387.

Bottcher, M.F., Nordin, E.K., Sandin, A., Midtvedt, T., and Bjorksten, B. 2000. Microflora-associated characteristics in faeces from allergic and nonallergic infants. Clin Exp Allergy 30(11): 1590-1596.

de Boissieu, D., Matarazzo, P., Rocchiccioli, F., and Dupont, C. 1997. Multiple food allergy: a possible diagnosis in breastfed infants. Acta Paediatr. 86(10): 1042-1046.

Farooqi, I.S. and Hopkin, J.M. 1998. Early childhood infection and atopic disorder. Thorax 53(11): 927-932.

Gronlund, M.M., Lehtonen, O.P., Eerola, E., and Kero, P. 1999. Fecal microflora in healthy infants born by different methods of delivery: permanent changes in intestinal flora after cesarean delivery. J. Pediatr. Gastroenterol. Nutr. 28(1): 19-25.

Hall, M.A., Cole, C.B., Smith, S.L., Fuller, R., and Rolles, C.J. 1990. Factors influencing the presence of faecal lactobacilli in early infancy. Arch. Dis. Child. 65(2): 185-188.

Hooper, L.V. and Gordon, J.I. 2001. Commensal host-bacterial relationships in the gut. Science 292(5519): 1115-1118.

Isolauri, E., Arvola, T., Sutas, Y., Moilanen, E., and Salminen, S. 2000. Probiotics in the management of atopic eczema. Clin. Exp. Allergy 30(11): 1604-1610.

Jackson, M. and Scott, R. 1996. Different patterns of cytokine induction in cultures of respiratory syncytial (RS) virus-specific human TH cell lines following stimulation with RS virus and RS virus proteins. J. Med. Virol. 49(3): 161-169.

Jogi, R., Janson, C., Bjornsson, E., Boman, G., and Bjorksten, B. 1998. Atopy and allergic disorders among adults in Tartu, Estonia compared with Uppsala, Sweden. Clin. Exp. Allergy 28(9): 1072-1080.

Johansson, M.L., Molin, G., Jeppsson, B., Nobaek, S., Ahrne, S., and Bengmark, S. 1993. Administration of different *Lactobacillus* strains in fermented oatmeal soup: *in vivo* colonization of human intestinal mucosa and effect on the indigenous flora. Appl. Environ. Microbiol. 59(1): 15-20.

Kalliomaki, M., Kirjavainen, P., Eerola, E., Kero, P., Salminen, S., and Isolauri, E. 2001. Distinct patterns of neonatal gut microflora in infants in whom atopy was and was not developing. J. Allergy Clin Immunol. 107(1): 129-134.

Kalliomaki, M., Salminen, S., Arvilommi, H., Kero, P., Koskinen, P., and Isolauri, E. 2001. Probiotics in primary prevention of atopic disease: a randomised placebo-controlled trial. Lancet 357(9262): 1076-1079.

Kaur, B., Anderson, H.R., Austin, J., Burr, M., Harkins, L.S., Strachan, D.P., and Warner, J.O. 1998. Prevalence of asthma symptoms, diagnosis, and treatment in 12-14 year old children across Great Britain (international study of asthma and allergies in childhood, ISAAC UK). B.M.J. 316(7125): 118-124.

Kim, J.H., and Ohsawa, M. 1995. Oral tolerance to ovalbumin in mice as a model for detecting modulators of the immunologic tolerance to a specific antigen. Biol. Pharm. Bull. 18(6): 854-858.

Kuehni, C.E., Davis, A., Brooke, A.M., and Silverman, M. 2001. Are all wheezing disorders in very young (preschool) children increasing in prevalence? Lancet 357(9271): 1821-1825.

Kuhn, I., Tullus, K., and Mollby, R. 1986. Colonization and persistence of *Escherichia coli* phenotypes in the intestines of children aged 0 to 18 months. Infection 14(1): 7-12.

Lewis, S.A. and Britton, J.R. 1998. Measles infection, measles vaccination and the effect of birth order in the aetiology of hay fever. Clin. Exp. Allergy 28(12): 1493-1500.

Lynch, N.R., Hagel, I., Perez, M., Di Prisco, M.C., Lopez, R., and Alvarez, N. 1993. Effect of anthelmintic treatment on the allergic reactivity of children in a tropical slum. J. Allergy Clin. Immunol. 92(3): 404-411.

Majamaa, H. and Isolauri, E. 1996. Evaluation of the gut mucosal barrier: evidence for increased antigen transfer in children with atopic eczema. J. Allergy Clin. Immunol. 97(4): 985-990.

Majamaa, H. and Isolauri, E. 1997. Probiotics: a novel approach in the management of food allergy. J. Allergy Clin. Immunol. 99(2): 179-185.

Majamaa, H., Miettinen, A., Laine, S., and Isolauri, E. 1996. Intestinal inflammation in children with atopic eczema: faecal eosinophil cationic protein and tumour necrosis factor-alpha as non- invasive indicators of food allergy. Clin. Exp. Allergy 26(2): 181-187.

Martinez, F.D., Stern, D.A., Wright, A.L., Taussig, L.M., and Halonen, M. 1998. Differential immune responses to acute lower respiratory illness in early life and subsequent development of persistent wheezing and asthma. J. Allergy Clin. Immunol. 102(6 Pt 1): 915-920.

Matricardi, P.M., Rosmini, F., Ferrigno, L., Nisini, R., Rapicetta, M., Chionne, P., Stroffolini, T., Pasquini, P., and D'Amelio, R. 1997. Cross sectional retrospective study of prevalence of atopy among Italian military students with antibodies against hepatitis A virus. B.M.J. 314(7086): 999-1003.

Matricardi, P.M., Rosmini, F., Rapicetta, M., Gasbarrini, G., and Stroffolini, T. 1999. Atopy, hygiene, and anthroposophic lifestyle. San Marino Study Group. Lancet 354(9176): 430.

Matricardi, P.M., Rosmini, F., Riondino, S., Fortini, M., Ferrigno, L., Rapicetta, M., and Bonini, S. 2000. Exposure to foodborne and orofecal microbes versus airborne viruses in relation to atopy and allergic asthma: epidemiological study. B.M.J. 320(7232): 412-417.

Moreau, M.C., and Corthier, G. 1988. Effect of the gastrointestinal microflora on induction and maintenance of oral tolerance to ovalbumin in C3H/ HeJ. mice. Infect. Immun. 56(10): 2766-2768.

Murray, M., Webb, M.S., O'Callaghan, C., Swarbrick, A.S., and Milner, A.D. 1992. Respiratory status and allergy after bronchiolitis. Arch. Dis. Child. 67(4): 482-487.

Ninan, T.K., and Russell, G. 1992. Respiratory symptoms and atopy in Aberdeen schoolchildren: evidence from two surveys 25 years apart. B.M.J. 304(6831): 873-875.

Omenaas, E., Jentoft, H.F., Vollmer, W.M., Buist, A.S., and Gulsvik, A. 2000. Absence of relationship between tuberculin reactivity and atopy in BCG vaccinated young adults. Thorax 55(6): 454-458.

Openshaw, P.J. 1995. Immunity and immunopathology to respiratory syncytial virus. The mouse model. Am. J. Respir. Crit. Care Med. 152(4 Pt 2): S59-62.

Paunio, M., Heinonen, O.P., Virtanen, M., Leinikki, P., Patja, A., and Peltola, H. 2000. Measles history and atopic diseases: a population-based cross-sectional study. J.A.M.A. 283(3): 343-346.

Pearce, N., Pekkanen, J., and Beasley, R. 1999. How much asthma is really attributable to atopy? Thorax 54: 268-272.

Peat, J.K., van den Berg, R.H., Green, W.F., Mellis, C.M., Leeder, S.R., and Woolcock, A.J. 1994. Changing prevalence of asthma in Australian children. B.M.J. 308(6944): 1591-1596.

Pullan, C.R., and Hey, E.N. 1982. Wheezing, asthma, and pulmonary dysfunction 10 years after infection with respiratory syncytial virus in infancy. B.M.J. (Clin Res Ed) 284(6330): 1665-1669.

Ronchetti, R., Villa, M.P., Barreto, M., Rota, R., Pagani, J., Martella, S., Falasca, C., Paggi, B., Guglielmi, F., and Ciofetta, G. 2001. Is the increase in childhood asthma coming to an end? Findings from three surveys of schoolchildren in Rome, Italy. Eur Respir J. 17(5): 881-886.

Rosenstreich, D.L., Eggleston, P., Kattan, M., Baker, D., Slavin, R.G., Gergen, P., Mitchell, H., McNiff-Mortimer, K., Lynn, H., Ownby, D., and Malveaux, F. 1997. The role of cockroach allergy and exposure to cockroach allergen in causing morbidity among inner-city children with asthma. N. Engl. J. Med. 336(19): 1356-1363.

Seidman, D.S., Laor, A., Gale, R., Stevenson, D.K., and Danon, Y.L. 1991. Is low birth weight a risk factor for asthma during adolescence? Arch. Dis. Child. 66(5): 584-587.

Sepp, E., Julge, K., Vasar, M., Naaber, P., Bjorksten, B., and Mikelsaar, M. 1997. Intestinal microflora of Estonian and Swedish infants. Acta Paediatr. 86(9): 956-61.

Shaheen, S.O., Aaby, P., Hall, A.J., Barker, D.J., Heyes, C.B., Shiell, A.W., and Goudiaby, A. 1996. Measles and atopy in Guinea-Bissau. Lancet 347(9018): 1792-1796.

Shirakawa, T., Enomoto, T., Shimazu, S., and Hopkin, J.M. 1997. The inverse association between tuberculin responses and atopic disorder. Science 275(5296): 77-79.

Simhon, A., Douglas, J.R., Drasar, B.S., and Soothill, J.F. 1982. Effect of feeding on infants' faecal flora. Arch. Dis. Child. 57(1): 54-58.

Sly, P.D., and Hibbert, M.E. 1989. Childhood asthma following hospitalization with acute viral bronchiolitis in infancy. Pediatr. Pulmonol. 7(3): 153-158.

Stein, R.T., Sherrill, D., Morgan, W.J., Holberg, C.J., Halonen, M., Taussig, L.M., Wright, A.L., and Martinez, F.D. 1999. Respiratory syncytial virus in early life and risk of wheeze and allergy by age 13 years. Lancet 354(9178): 541-545.

Strachan, D.P. 1989. Hay fever, hygiene, and household size. B.M.J. 299(6710): 1259-1260.

Strannegard, I.L., Larsson, L.O., Wennergren, G., and Strannegard, O. 1998. Prevalence of allergy in children in relation to prior BCG vaccination and infection with atypical mycobacteria. Allergy 53(3): 249-254.

Strobel, S. 1995. Oral tolerance: 'of mice and men'. Acta Paediatr. Jpn. 37(2): 133-140.

Sudo, N., Sawamura, S., Tanaka, K., Aiba, Y., Kubo, C., and Koga, Y. 1997. The requirement of intestinal bacterial flora for the development of an IgE production system fully susceptible to oral tolerance induction. J. Immunol. 159(4): 1739-1745.

Sutas, Y., Soppi, E., Korhonen, H., Syvaoja, E.L., Saxelin, M., Rokka, T., and Isolauri, E. 1996. Suppression of lymphocyte proliferation *in vitro* by bovine caseins hydrolyzed with *Lactobacillus casei* GG-derived enzymes. J. Allergy Clin. Immunol. 98(1): 216-224.

von Hertzen, L., Klaukka, T., Mattila, H., and Haahtela, T. 1999. Mycobacterium tuberculosis infection and the subsequent development of asthma and allergic conditions. J. Allergy Clin. Immunol. 104(6): 1211-1214.

Wallaert, B., Desreumaux, P., Copin, M.C., Tillie, I., Benard, A., Colombel, J.F., Gosselin, B., Tonnel, A.B., and Janin, A. 1995. Immunoreactivity for interleukin 3 and 5 and granulocyte/macrophage colony-stimulating factor of intestinal mucosa in bronchial asthma. J. Exp. Med. 182(6): 1897-1904.

Wannemuehler, M.J., Kiyono, H., Babb, J.L., Michalek, S.M., and McGhee, J.R. 1982. Lipopolysaccharide (LPS) regulation of the immune response: LPS converts germfree mice to sensitivity to oral tolerance induction. J. Immunol. 129(3): 959-965.

Warner, J.O. 1999. Worldwide variations in the prevalence of atopic symptoms: what does it all mean? Thorax 54 Suppl 2: S46-51.

Wennergren, G., Amark, M., Amark, K., Oskarsdottir, S., Sten, G., and Redfors, S. 1997. Wheezing bronchitis reinvestigated at the age of 10 years. Acta Paediatr. 86(4): 351-355.

Wickens, K., Crane, J., Pearce , N., and Beasley, R. 1999. The magnitude of the effect of smaller family sizes on the increase in the prevalence of asthma and hay fever in the United Kingdom and New Zealand. J. Allergy Clin. Immunol. 104(3 Pt 1): 554-558.

Wickens, K., Pearce, N., Crane, J., and Beasley, R. 1999. Antibiotic use in early childhood and the development of asthma. Clin. Exp. Allergy 29(6): 766-771.

Wilson, A.F., Novey, H.S., Berke, R.A., and Surprenant, E.L. 1973. Deposition of inhaled pollen and pollen extract in human airways. N. Engl. J. Med. 288(20): 1056-1058.

Yemaneberhan, H., Bekele, Z., Venn, A., Lewis, S., Parry, E. and Britton, J. 1997. Prevalence of wheeze and asthma and relation to atopy in urban and rural Ethiopia. Lancet 350(9071): 85-90.

Yoshioka, H., Iseki, K., and Fujita, K. 1983. Development and differences of intestinal flora in the neonatal period in breast-fed and bottle-fed infants. Pediatrics 72(3): 317-321.

From: *Probiotics and Prebiotics: Where Are We Going?*
Edited by: Gerald W. Tannock

Chapter 9

Genomic Perspectives on Probiotics and the Gastrointestinal Microflora

Olivia E. McAuliffe and
Todd R. Klaenhammer

Abstract

The explosion of genomic technologies in recent years has revolutionized every aspect of biology in an unprecedented manner. From a long and successful history of "reductionist" science, it has now become possible to understand how component parts interact collectively to create an organism. The growing collection of genomic sequence information, the high-throughput analysis of expression profiles using DNA microarrays, and the ability to deal with this information using advanced bioinformatics offer many possibilites to advance our knowledge of the microbial world. Developments such as these will enable a better understanding of the gastrointestinal (GI) tract as a complex and delicately balanced ecosystem.

Genetic characterization of probiotic cultures is essential to unequivocally define their contributions to human health. Functional genomic approaches may help improve the functionality of these strains from an industrial and health-promoting perspective, and help to scientifically substantiate some of the health claims made for probiotic strains. This paper describes some of the recent developments in the rapidly growing area of genomics, and how these advances may be exploited to identify the molecular foundations of the relationships between probiotic organisms and their hosts, and how they contribute to our health and well-being.

Introduction

The gastrointestinal (GI) tract is the body's foremost tissue boundary, interacting with nutrients, exogenous compounds and gut microflora, and its condition is influenced by the complex interplay between these environmental factors and genetic elements. Hosts and complex microbial communities coexist, interact, and compete in conditions that are often far from optimal. The GI tract is the most densely colonized region of the human body, with approximately 10^{11} bacterial cells per gram of contents of the large intestine, comprising several hundred bacterial species (Savage, 1977; Tannock, 1995). These can be divided into three classes: the "autochthonous" microflora - populations of microbes that are present at high population levels throughout the life of the individual, the "allochthonous" microflora - populations of microbes frequently, but not always, present in the GI tract, and of variable population size; and the "opportunistic" microbes - pathogens acquired and capable of persisting in the gut (Tannock, 1999). The composition of the intestinal microflora varies during infancy, but becomes a very stable and characteristic population in the healthy adult (McCartney *et al.*, 1996). While the exact mechanisms by which these bacteria shape our physiology are unknown, comparisons between germ-free and conventional animals have shown that many biochemical, physiological and immunological functions are influenced by the presence of the diverse and metabolically active bacterial community residing in the gut (Tannock, 1995; Muyzer and Smalla, 1998).

The adaptations of symbionts and commensals to life in nutritionally advantageous host niches provide a rationale for using these organisms as therapeutic agents (Hooper and Gordon, 2001). Attempts have been made in the last 2-3 decades to improve the health status of individuals by modulating the indigenous intestinal microflora through the use of live microbial dietary adjuncts, or probiotics. Probiotics have received considerable commercial and scientific attention in this time, as a subdivision of the functional foods. Fuller (1989) defined probiotics as "a live microbial feed supplement which beneficially affects the host animal by improving its intestinal microbial balance". This definition was broadened by Havenaar and Huis in't Veld (1992) to a "mono- or mixed- culture of live microorganisms which benefits man or animals by improving the properties of the indigenous microflora". A number of excellent reviews are available on probiotics (Fuller, 1992; O'Sullivan *et al.*, 1992; Lee and Salminen, 1995; Saavedra, 1995; Salminen *et al.*, 1996; Fuller and Gibson, 1997; Vaughan *et al.*, 1999; Walter *et al.*, 2000). The major consumption of probiotics by humans is in the form of dairy-based foods containing intestinal species of lactobacilli and bifidobacteria (Tannock, 1999). Probiotic functionality depends on the ability of a strain to confer health advantages on a host by consumption of viable cells (Sanders and Klaenhammer, 2001). Potential benefits attributed to probiotic strains include their maintenance of the normal microflora, increased resistance to infectious diseases (Fernandes *et al.*, 1987) through interference, exclusion or antagonism, reduction in blood pressure (Sawada *et al.*, 1990), serum cholesterol concentration (Gilliland and Walker, 1990), allergy (Isolauri *et al.*, 1993), and carcinogen and co-carcinogen production (Goldin *et al.*, 1980; Kato *et al.*, 1981), and modulation of the immune system (Kimura *et al.*, 1997; Delneste *et al.*, 1998). With such actions, probiotic strains are being considered not only as agents with preventative functionality, but also as active agents in curative and therapeutic settings (Surawicz *et al.*, 1989; Saavedra *et al.*, 1994).

However, while some consumers may purchase probiotic-containing products for these and other health benefits, many of the effects attributed to probiotics remain scientifically unsubstantiated (O'Sullivan *et al.*, 1992; de Roos and Katan, 2000) for a number of reasons. These include our lack of understanding of the complexity of the GI environment, confusion over the identity, viability, and activity of probiotic strains leading to the misidentification of cultures in laboratory and clinical investigations, and the high cost of clinical trials which often results in the "one strain versus one placebo" experiments in attempts to prove the efficacy of probiotic strains (Klaenhammer and Kullen, 1999). It is rare that health claims can be made, particularly when specific strains or strain combinations are evaluated in dietary probiotics (Sanders 1993; 1999). The situation has improved somewhat in recent years with the availability of molecular tools to properly identify probiotic species and strains. Improving this situation further requires a clear taxonomic description of the microorganisms delivered via probiotic products and a better understanding of those physiological characteristics considered paramount to their *in vivo* survival and activity in the GI tract. In this context, genetic characterization of probiotic cultures is essential to unequivocally define their contributions to the intestinal microflora and ultimately identify the genotypes that control any unique and beneficial properties. The current genomic revolution offers an unprecedented opportunity to identify the molecular foundations of the relationships between these organisms and their hosts, to understand how they contribute to our health and well-being, and most importantly, to determine how they can be exploited to improve their beneficial attributes.

Genome Technology

The science of genomics encompasses the systematic study of the structure, content and evolution of complete genomes, a discipline born in the early 1980's following a proposal to sequence the human genome. Sequencing of the 12 Mb genome of the yeast, *Saccharomyces cerevisiae*, was one of the first genome projects to get underway (Mewes *et al.*, 1997). However, the genome era is generally regarded to have begun on July 28, 1995 when The Institute for Genomic Research (TIGR; Rockville, MD, USA) published the completed nucleotide sequence of the first microbial genome, that of *Haemophilus influenzae*, comprising 1.83 Mb of sequence (Fleischmann *et al.,* 1995). This genome sequence was generated using "shotgun" sequencing in which the whole genome was randomly fragmented, the fragments sequenced, and reassembled into one coherent genome length sequence. Subsequently, genome sequences representing the three major domains of life – bacteria, archaea, and eukaryota - were elucidated. In 1998, the genome of the first multicellular organism, the 97 Mb sequence of the roundworm, *Caenorhabditis elegans* (The *C. elegans* Sequencing Consortium, 1998), was published, followed by the mustard weed, *Arabidopsis thaliana* (The *Arabidopsis* Genome Initiative, 2000), and the fruit fly, *Drosophila melanogaster* (Adams *et al.*, 2000). On June 26 2000, the compilation of the working draft, or first assembly, of the human genome by the International Human Genome Sequencing Consortium (IHGSC), and Celera Genomics (Rockville, MD) was announced, and published on February 16, 2001 (Venter *et al.*, 2001; Lander *et al.*, 2001). The project originally was planned to last 15 years, but effective resource and technological advances have accelerated the expected completion date to 2003. The total number of genes is estimated at 30,000 to 35,000 - much lower than previous estimates of 80,000 to 140,000 that had been based on extrapolations from gene-rich areas, rather than a more representative composite of gene-rich and gene-poor areas.

In view of their relatively small genome sizes and amenability to genetic manipulation, bacteria make excellent model systems for genomic analysis. The technical advancements that permitted the completion of the draft sequence of the human genome have now made it possible to determine a complete microbial genome sequence in several hours. Stimulated by the increasing pace and decreasing costs of sequencing, projects have now been initiated to sequence the complete genomes of several hundred microbes. Each new organism sequenced reveals new features about genome organization, gene regulation, gene content and the biochemical potential entrenched in the microbial world. Thousands of new genes, previously unknown to biology, are now continuously being released into public databases. It is expected that within the next decade, genomes from every significant species will have been sequenced. As of this writing (November 2001), 58 completed genome sequences appear on the TIGR website (www.tigr.org), while 22 partially sequenced genomes are available on the US Department of Energy's Joint Genome Institute (JGI) website (www.jgi.doe.gov). To date, microbial genome projects have focused on human and animal pathogens, such as *Mycoplasma pneumoniae* (Himmelreich *et al.*, 1996), *Treponema pallidum* (Fraser *et al.*, 1998), *Helicobacter pylori* (Tomb *et al.*, 1999), *Campylobacter jejuni* (Parkhill *et al.*, 2000) and *Listeria monocytogenes* (Glaser *et al.*, 2001), due to the interest in identifying pathogenicity determinants and discovery of new gene targets for rational drug design.

A number of current genome projects are focusing on microorganisms with important roles in food, including starter cultures used in the production of fermented foods and food-grade ingredients, and organisms with probiotic potential. In the development of probiotic foods intended for human consumption, strains of the industrially relevant lactic acid bacteria (LAB) have been widely used, as they are considered to be desirable members of the intestinal microflora. In addition, LAB have been consumed naturally via milk-based environments for centuries and this long history of safe use for human consumption has afforded these organisms "generally regarded as safe" or GRAS status. A list of the organisms whose genome sequences are being elucidated, or are nearing completion, is presented in Table 1. To date, only one complete sequence of a LAB has been published, that of *Lactococcus lactis* IL1403 (Bolotin *et al.,* 2001). This species has been intensely studied and has become an excellent model for research on metabolism, physiology, genetics and molecular biology of the LAB. In 2002, genome sequences of several industrially relevant LAB, including a number of probiotic cultures will be completed (Table 1). The organisms selected

Table 1. Current LAB genome sequencing projects

Species	Strain	Institution[a]	Projected Completion Date
Lactococcus lactis	IL1403	NRA, Genoscope FR	2001 – completed
Lactococcus lactis	MG1363	RUG, IFR, UCC	2001
Lactococcus cremoris		JGI/DOE	2002
Lactobacillus acidophilus	ATCC 700396	NCSU, CalPoly	2001
Lactobacillus plantarum	WCFS1	WCFS, NL	2001
Lactobacillus delbreuckii ssp. *bulgaricus*	ATCC 11842	INRA, Genoscope FR	2001
Lactobacillus sakei	23K	INRA, FR	
Lactobacillus casei	BL23	INRA, Caen Univ. FR	2002?
Lactobacillus helveticus		UW - Madison, USA	2002
Lactobacillus rhamnosus	HN001	NZDRI, ViaLactia BioSciences	2001
Lactobacillus johnsonii	NCC533	Nestle CH	2000
Lactobacillus gasseri		JGI/DOE	2002
Lactobacillus brevis		JGI/DOE	2002
Lactobacillus casei		JGI/DOE	2002
Lactobacillus delbreuckii		JGI/DOE	2002
Bifidobacterium longum		Nestle, CH	2001
Bifidobacterium breve	NCIMB8807	UCC, IRL	2002
Bifidobacterium longum		JGI/DOE	2002

			2001-2002
Streptococcus thermophilus	LMG18311	UCL, BL	
Streptococcus thermophilus		INRA, Integrated Genomics	2001
Streptococcus thermophilus		JGI/DOE	2002
Oenococcus oeni		JGI/DOE	2002
Leuconostoc mesenteroides		JGI/DOE	2002
Brevibacterium linens		JGI/DOE	2002
Pediococcus pentosaceus		JGI/DOE	2002

[a]Institutes are as follows:

INRA, Institut National de la Recherche Agronomique, FR
RUG, Rijksuniversiteit Groningen, NL
UCC, University College Cork, IRL
NZDRI, New Zealand Dairy Research Institute, NZ
UCL, Université Catholique de Louvain, BL
NCSU, North Carolina State University, USA
WCFS, Wageningen Centre for Food Science, NL
IFR, Institute for Food Research, Norwich, UK
CalPoly, California Polytechnic University, USA
JGI/DOE, Joint Genome Institute/Department of Energy

represent the phylogenetic, morphological and metabolic diversity of the LAB. Cluster sequencing of microorganisms from a specific phylogenetic branch will promote comparisons between related species and reveal avenues for the evolution of LAB into their varied environmental niches. It is expected that many basic and application-directed programs will benefit from the sequence information generated by these initiatives.

The term "microbiome" has been coined to describe the collective genome of our indigenous microbial population. Any comprehensive genetic view of the human species as a life-form should include the genes in our microbiome, which consists of commensals, oppurtunistic pathogens and probiotic organisms (Hooper and Gordon, 2001). In the near future, the completely annotated genome sequences of many indigenous and probiotic species are expected to appear in the public domain. These include a number of probiotic lactobacilli, such as *Lactobacillus plantarum*, *Lactobacillus johnsonii*, and the well-characterized *Lactobacillus acidophilus* NCFM (Sanders and Klaenhammer, 2001), as well as *Bifidobacterium* sp. and *Streptococcus thermophilus*. The availability of these sequences will allow discovery of genes and gene networks in these organisms and establish a mechanistic understanding of their beneficial activities in the GI tract. In addition, genome sequencing and analysis will promote new applications for probiotics and potentially lead to the rational selection of improved cultures. Although genetic work on probiotic cultures lags far behind that of other members of the LAB, it is a rapidly developing field. The most critical steps will correlate genotypes to phenotypes that are essential to probiotic functionality. The availability of genome sequences of the beneficial members of the microbiome will greatly accelerate this task. Whole genome sequencing will provide a platform for comparative and functional genomic investigations that will shed light on the beneficial roles and interactions of probiotic LAB within the context of the human microbiome.

Whole Genome Sequencing

High-throughput DNA sequence analysis has been made possible by the streamlining and automation of three key milestones which have been paramount in the advances made in genome sequencing – the development of the basic chain termination method (Sanger *et al.*, 1977) and the chemical degradation method (Maxam and Gilbert, 1980) for the elucidation of DNA sequence; the invention of the polymerase chain reaction (Mullis and Faloona, 1987); and the technological advancement to automated fluorescent capillary

DNA sequencers (Smith *et al.*, 1985, 1986; Hood *et al.*, 1987; Hunkapiller *et al.*, 1991). Until recently, the most common format for DNA sequencing was a horizontal or vertical slab polyacrylamide gel for electrophoretic separation. Years of research have led to the introduction of capillary sequencers that significantly increase the throughput and decrease the time and cost required to sequence. When the Human Genome project was initiated, the cost of sequencing a single DNA base was almost $10; today sequencing costs have fallen about 100-fold to $0.10-0.20 a base. The Applied BioSystems PRISM® 3700 (Foster City, CA; www.appliedbiosystems.com) and the Molecular Dynamics MegaBACE 1000 (Sunnyvale, CA; www.mdyn.com) are two of the available capillary sequencers designed for use in production-scale DNA analysis, which have enabled the completion of the human genome sequence. With these high-throughput systems, sequencing output has risen from months and years to sequence a single gene, to a rate exceeding 1000 nucleotides per second, 24 hours a day, 7 days a week (Meldrum, 2001).

Genetic and physical maps are the basis on which most genome sequencing projects are built. Generation of maps is an important first step to whole genome sequencing, particularly when the "clone-by-clone" or hierarchical sequencing strategy is employed, which requires the generation of an ordered library of the genome. The location of genes in a genome can be specified according to physical distance, or by the relative position defined by recombination frequencies. Genetic maps describe the relative order of genetic markers in a genome, generally repeat sequences or restriction enzyme polymorphisms, with the distance between markers expressed in units of recombination. A physical, or chromosomal, map is that generated by the assembly of contiguous sequences, or contigs - a set of sequences ordered into one linear stretch on the basis of overlaps at each end. Assembly of contigs to produce a physical map requires either alignment of randomly isolated clones, or chromosome walking - the sequential isolation of adjacent clones to identify overlapping clones containing the same sequence. The generation of a genetic map prior to sequencing is an invaluable aid to the assembly of the physical map of a genome. Information of this kind is critical for comparing the genomes of related species and correlating genetic and phenotypic data.

Until recently, the technology necessary for large-scale genome sequencing projects was not available and early attempts were limited to organisms with relatively small genomes, such as the bacteriophage λ. A major restraint to these early projects was the requirement for an ordered library of the genome before sequencing could begin, as in the "clone-by-clone" approach. This

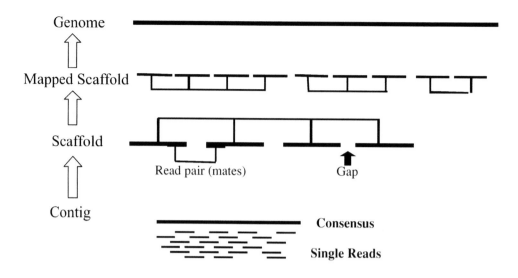

Genome

Mapped Scaffold

Scaffold

Read pair (mates)   Gap

Contig

Consensus

Single Reads

Figure 1. Strategy for assembly of whole genome sequences.

time-consuming effort consists of assembling an ordered set of overlapping fragments or clones that together represent the entire genome of the organism. Genome mapping methods, as described above, are used to determine the physical placement of each clone in the genome. By means of a procedure known as "shotgun-ing", each clone is subsequently broken into smaller random fragments that are sequenced. These fragments are reassembled by overlapping regions to give the sequence of the original clone and repeated for each clone in the library until the sequence is completed. The new, rapid approach of random shotgun sequencing has become the method of choice for genome sequencing of organisms of various sizes and base composition, due to the high-throughput nature of the emerging technologies and computational capabilities. In recent times, this method has proved to be as useful and more efficient than the more traditional method described above and has been the dominant method for generating genome sequences, including sequencing the human genome at Celera Genomics (Venter *et al.*, 2001). This method involves the construction of small and large insert libraries with DNA from the organism of interest followed by sequencing of random clones from the libraries to a pre-determined level of coverage of the genome. The coverage, or depth, of sequence refers to the average number of times a single nucleotide is represented by a high quality (accuracy of at least 99%) base in random raw sequence. In general, a single, random DNA fragment library is prepared by mechanical or sonic shearing of the genomic DNA and insertion into some suitable vector system. The ends of a large number of randomly selected fragments are sequenced from both insert ends using a set of universal primers until every part of the genome has been sequenced several times on average (Broder and Venter, 2000). Accurately paired clone-end sequences are a key tool for reassembling genomes more completely than single-stranded sequencing approaches allow, at comparable levels of sequence coverage (Figure 1). For example, sequencing a 2 MB microbial genome (2 million base pairs) at 8X coverage, with an average of 500 bp per sequencing read, would require the generation of 16,000 clones which would be sequenced in both the forward and reverse directions, resulting in 32,000 sequencing reactions.

While the random shotgun method of sequencing is undoubtedly faster and eliminates the need for initial mapping, it does rely on vast computational power to re-assemble the huge number of fragments generated by sequencing. For this task, the PHRED/PHRAP/CONSED software tools, developed at the University of Washington, are in use in the worldwide genomics community. The PHRED program reads DNA sequencer chromatogram data, calls bases, assigns probability scores to the accuracy of each base call as the chromatogram is read using a base-calling algorithm, and writes the base calls and quality values to output files in either FASTA or SCF format (Ewing and Green, 1998; Ewing *et al.*, 1999). These files can then be used by the PHRAP sequence assembly program to increase the accuracy of the assembled sequence. Contig assembly is one of the most difficult and critical functions in DNA sequence analysis. Problems arise in assembly due to the presence of repeat sequences, gaps and extrachromosomal elements. Finally, CONSED is a graphical tool for editing PHRAP assemblies (Gordon *et al.*, 1998) that allows the user to call up and interactively edit individual reads and/or contigs assembled by PHRAP, guided by error probabilities. CONSED and PHRAP are recommended for use together. Editing of individual bases is possible by calling up the chromatogram file and visually comparing two or more traces. CONSED can also be used to pick primers and templates for whichever location is specified by the user and in addition, automates the process of choosing reads for the closing of the genome that will help to bridge the physical gaps that remain between the assembled contigs (the AUTOFINISH program; http://www.phrap.org).

A number of new sequencing strategies are currently being developed. A promising single molecule sequencing strategy is nanopore sequencing, in which a channel formed by a protein molecule such as α-hemolysin is embedded in an electrically polarized membrane (Deamer and Akeson, 2000). As a single molecule of DNA is pulled through the nanopore by electrophoresis, the nucleotides transiently block ions from moving through resulting in a drop in current. However, at present, passage of a single molecule through a pore is too rapid to allow base pair resolution. Another example is pyrosequencing, a novel non-electrophoretic method that takes advantage of four enzymes cooperating in a single tube to determine the nucleotide composition of a DNA fragment in real time (Meldrum, 2000). Detection is based on the amount of visible light produced by coupling the pyrophosphate that is released during nucleotide incorporation with the enzymes sulfurylase and luciferase. Unincorporated nucleotides are degraded in the reaction mixture by the enzyme apyrase. Fully automated systems are currently being developed for pyrosequencing (http://www.pyrosequencing.com). Mass spectrometric techniques such as MALDI-TOF (matrix-assisted laser desorption-ionization time-of-flight) may one day permit rapid analysis for the size separation step in DNA sequencing. Such methods would significantly increase the speed of the separation, detection, and data acquisition processes for sequence analysis over conventional gel electrophoresis methods (Fitzgerald and Smith, 1995).

Bioinformatics and Genome Annotation

Annotation is the process of interpreting raw sequence data into useful biological information. This process can be divided into a number of phases; the gene-finding phase, the function assignment phase followed by feature identification, such as promoter, terminator and operator regions, transfer RNAs, ribosomal RNAs and repetitive sequences. Annotation also involves identifying novel genomic characteristics such as nucleotide biases, origins of replication, putative regions of gene transfer, insertion elements and plasmids. Basically, the first analysis phase involves the prediction of all putative open reading frames (ORF's) in the genome, from start codon to stop codon. Gene discovery in prokaryotes differs from that in eukaryotes as prokaryotic genomes tend to be gene-rich, typically containing 90% coding sequence and lacking introns. Usually, ORFs longer than a certain threshold are considered potential genes. Genes that are shorter than the threshold and genes on the opposite strand of longer ORF's often lead to ambiguities; however, these can be resolved by analyzing the compositional differences between coding and non-coding regions. Bioinformatics involves the development of computational methods to supplement experimental approaches in correlating genotype to phenotype and has led to the development of a new generation of gene prediction programs for improved, accurate gene identification including GeneMark (Borodovsky and McIninch, 1993), GeneScan (Ramakrishna and Srinivasan, 1999) and Glimmer

(Salzberg *et al.*, 1998) which incorporate statistical approaches into Hidden Markov models (HMM) and Interpolated Markov models (IMM). Such programs are effective in identifying coding regions in microbial sequences. GLIMMER 2.0 is one such gene finder that finds 97-98% of all genes in prokaryotic genomes without any human intervention, and is highly proficient in resolving overlapping genes (Delcher *et al.*, 1999). The system is designed to be quickly and easily trained using a well-characterized subset of genes from the genome sequence of interest, generating an organism-specific gene finder. As functional genomic strategies, such as microarrays and knockout experiments, are based heavily on the results of genome annotation, the accuracy of gene prediction is critical.

When gene structures have been identified, the second phase of genome annotation is to assign biochemical and physiological functions to each gene. Four types of information are important to elucidate the function of a gene: the loss or gain of a phenotype, the structure and biochemical properties of the protein product, the distribution of the gene products within the organism and the location in the genome. In prokaryotes, clusters of genes that are close enough to be co-transcribed as operons, or remnants of operons, provide information to infer functionally coupled genes (Gaasterland and Oprea, 2001). Overbeek *et al.* (1999) have demonstrated significant correlation between close genomic proximity and genes that participate in the same metabolic pathway for 30 prokaryotic genomes. Genomic proximity can be used to identify possible functional relatedness for genes of unknown function that are putatively co-transcribed with other well-characterized genes. However, the dominant method of functional assignment remains the analysis of the gene-encoded predicted proteins, based on their sequence similarity to previously identified proteins in databases such as GenBank, maintained by the National Center for Biotechnology Information (NCBI, www.ncbi.nlm.nih.gov). Many of the putative genes identified to date have yet to be assigned a function; in fact, 30-60% of ORF's within all of the recently elucidated microbial genomes belong to putative proteins, with an "unknown" function. Protein sequences tend to be more highly conserved than their genomic counterparts. Therefore, an initial functional screen can be performed through a protein-level pairwise sequence comparison of annotated genes with proteins of known function from other organisms in GenBank or other databases, using BLAST or FASTA searches. This approach, however, tends to assign function even when critical amino acids are missing from the annotated protein. Although, in general, conservation of sequence implies conservation of structure, and conserved structure implies conserved function, key changes in amino acids may alter the function of a protein significantly. To address this issue, databases have been constructed to capture conserved protein sequence motifs associated with function. In particular, the PFAM (Protein Family Database; Bateman and Birney, 2000) and TIGRFAM (www.tigr.org/TIGRFAMs/) databases of protein functional motifs classify protein sequence domains into functional categories and then use multiple sequence examples to extract sequence patterns that uniquely describe the function. These resources are an invaluable source of information for protein annotation. Numerous other databases exist, including the Enzyme Commission (EC) hierarchical classification of enzymes as defined by the International Union of Biochemistry and Molecular Biology. In this database, well-classified enzymes are assigned a unique number describing enzymatic activity ranging from general to specific functions. Databases such as KEGG (Ogata *et al.*, 1999) uses EC designations in efforts to make links from the gene catalogs generated by genome sequencing projects to the biochemical pathway. KEGG is publicly available at www.genome.ad.jp/kegg/. It's aims include computerization of all aspects of cellular functions in terms of the pathway of interacting molecules or genes, maintenance of gene catalogs for all organisms and link each gene product to a pathway component and to enable pathway comparison, reconstruction and analysis.

A complementary genomic perspective on protein classification has been the assembly of clusters of orthologous groups (COGs; Tatusov *et al.* 1997), by identifying the best hit for each gene in complete pairwise comparisons of a set of genomes. The relationships between genes from different genomes are best represented as a system of gene families that include both orthologs and paralogs. Orthologs are genes in different species that evolved from a common ancestral gene by speciation; in contrast, paralogs are genes related by duplication within a genome (Gogarten and Olendzenski, 1999). Horizontal gene transfer can also result in orthologs and paralogs being present in the same genome. Generally, orthologs retain the same function during the course of evolution, whereas paralogs evolve new functions, even if related to the original one. Consequently, identification of orthologous and/or paralogous protein sequences is critical for reliable prediction of gene functions in newly sequenced genomes, which is easily performed with PSI-BLAST (position-specific iterated BLAST), a program that performs BLAST searches, successively building and using patterns of conserved functional residues. A complete list of orthologs is also imperative for any meaningful comparison of genomic organization. Each COG consists of individual orthologous genes from three or more phylogenetic lineages, i.e. any two proteins from different lineages that belong to the same COG are orthologs. Comparison of all the proteins from the presently sequenced genomes has led to the identification of 3166 COGs as of November 2001 (http://www.ncbi.nlm.nih.gov/COG/xindex.html).

TIGR has established a Comprehensive Microbial Resource (CMR; http://www.tigr.org/tigr-scripts/CMR2/CMRHomePage.spl) that provides online access to the complete sequences and associated resources for all microbial genomes in the public domain, collected on a single underlying database, the Omniome. In addition to the CMR resources, each genome is linked to alternate genome pages maintained by NCBI or the sequencing center that generated the data. The TIGRFAM databases can be contrasted with similar annotations such as PFAM and COGs, while information on paralogous genes and standard enzyme commission designations can also be retrieved. Alternatively, a list of a total of 210 prokaryotic genome projects is provided by the Genomes OnLine Database (GOLD; http://wit.integratedgenomics.com/GOLD), which also provides access to a wide range of analytical databases such as KEGG (Ogata *et al.*, 1999).

Protein function assignment allows some general predictions to be made about the content of genomes; however, it is important to make the distinction between such predictions and classification of genes as members of the same family, as there are instances when the closest match may be very misleading in terms of function. Gene identification and whole genome annotation is an ongoing process as protein families will continually gain new members, split into smaller families and gain refined patterns of structural motifs. Biochemical function information assigned to one family member will become a source of putative functional information about other members. Annotation allows gene function to be inferred from the sequence, and not functionally demonstrated. However, while this is a most useful tool, similarity is not function. Undoubtedly, errors in annotation exist which can result in inaccurate classification of newly identified genes. Often, annotated genes are arranged into commonly identified metabolic pathways, and certain key activities are "missing". This may suggest that the organism under study may only contain part of the particular pathway. However, given the large number of genes with unknown function, it is possible that another protein has evolved to catalyze the missing reaction, a process known as non-orthologous gene replacement. While most investigators use automated annotation methods, it still remains imperative that annotation be verified by human evaluation, in attempt to limit the errors in protein and function assignment being deposited in databases.

Comparative Genomics

While analysis of a single genome provides enormous insight into the biology of an organism, substantially more information can be learned from comparing genomes of organisms. The goal of comparative genomics is to unravel the nature of gene relationships, regulatory patterns and gene function by focusing on specific differences and similarities in gene structure and organization across multiple genomes. Sequence conservation provides a powerful way to discern the real functional constraints on genes and gene products, while sequence differences hold the key to understanding how nature generates such diversity of form and function with such an economy of genes (Lander, 1996). Comparative analysis relies on the observation that even genomes of distantly related organisms encode evolutionary related proteins with relatively high sequence similarity (Henikoff *et al.*, 1997). It is clear from the endless stream of sequence information flooding into databases that apparently disparate bacterial species have many features in common especially within gram-positive and gram-negative lineages (Perego and Hoch, 2001). Many critical biochemical pathways in the cell have remained highly similar over evolutionary time across species. Similarities at the biochemical level are mirrored at the genomic level, since the genes encoding proteins which govern the individual reactions have been highly conserved, which is reflected in the high level of sequence similarity at the protein level. It has also been observed that the order of genes, referred to as co-linearity, within genomes may also be highly conserved. Conservation of gene order between chromosome segments of two or more organisms is known as synteny. Even though recombination does occur at a relatively high frequency in nature, the conservation of gene order does appear to provide obvious evolutionary advantages. Gene order is an important, well-conserved feature in prokaryotic organisms, where several ORF's may be organized in one operon, under the control of one promoter. Translocations or inversions of gene clusters and operons may occur between two organisms, but clustering of genes can allow the assignment of functions to unknown ORF's. Comparative analysis may also involve determining the presence or absence of particular genes in different species. A method developed by Pellegrini *et al.* (1999) clusters genes according their distribution patterns across species, referred to as phylogenetic profiling. It was established that genes that work in the same pathways frequently have correlated distribution patterns. The functions of some unknown genes can be therefore predicted on the basis of their having similar distribution patterns to genes with known function. The correlated loss of genes in certain species can also provide insight into gene function.

The availability of whole genome sequences opens great new possibilities for understanding the evolutionary process. Sequence analysis has revolutionized the study of evolution, as it is now possible to draw phylogenetic trees relating organisms on the basis of similarities in their genes rather than often unreliable phenotypic data (Lander and Weinberg, 2000). As early as 1965, Zuckerkandl and Pauling demonstrated how the evolutionary history of a species could be inferred from comparisons of gene sequences. Subsequently, molecular analysis of rRNA molecules provided the first evolutionary classification of microbes (Woese and Fox, 1977; Woese, 1987). Use of comparative genomics with sequence generated from closely-related species is a new approach to gaining insights into chromosome evolution. Such evolutionary comparison among organisms identifies sequences that play important functional roles in protein structure or gene regulation and hence, remain unaltered over periods of evolutionary time. In addition, such comparison should allow the identification of genes which were crucial to the creation of new species, as these genes are likely to have undergone strong selection and more rapid sequence evolution. The accumulating information on rRNA sequences provides a growing resource for comparative identification of probiotic cultures, both established and potentially new candidates (Kullen and Klaenhammer, 1999). Comparative analysis of genome sequences of similar organisms occupying various niches, for example, *Lactobacillus plantarum* and *Lactobacillus casei* defined in both probiotic and fermentation roles, will be facilitated by the cluster sequencing of numerous members of the LAB. Comparative genomics will provide an important view of microbial adaptation and expand our understanding of molecular evolution related to metabolic diversity and ecological niche, resolving many of the taxonomic and phylogenetic debates which surround these organisms.

With the enormous advances in genome sequencing in recent times, a great deal of sequence work continues to be accomplished as researchers fill in the gaps left in the genome maps of eukaryotic and prokaryotic organisms. It is not just a matter of developing the software tools to manage sequence data or for locating genes, but developing tools for moving from the sequence data to *in silico* analysis to understanding the biology of the organism under study. Such analyses will reveal the mechanisms underlying the biological systems within these organisms. The next issue to address is that a substantial portion of the genes in every organism correlate with no known function. The immediate challenge is to find efficient ways to identify the function of these genes.

Table 2. Genes identified with potential probiotic functionality

Proposed Function	Potential Gene(s) Involved	Gene Products	Reference(s)
Acid resistance	*atpBEFHAGDC*	F_1F_0-ATPase	Kullen and Klaenhammer, 1999
Bile resistance	*bsh*	Bile salt hydrolase	Moser and Savage, 2001
Alleviation of lactose intolerance	*lacZ*	β-galactosidase	Mustapha *et al.*, 1997
Pathogen exclusion	*pln* operon	Plantaricin production	Diep *et al.*, 1996
Colonization/Adhesion	*cnb*	Collagen-binding protein	Roos *et al.*, 1996
	bspA	Basic surface protein A	Turner *et al.*, 1999
	dlt	Lipotechoic acid	Granato *et al.*, 1999
	mub	Mucus-binding adhesion	Roos, 1999
	slpA/B	S-layer protein	Boot and Pouwels, 1996
Aggregation	*aggH*	Auto-aggregation	Roos *et al.*, 1999

Impact of Genome Sequencing and Comparative Genomics on Probiotic and Other Microbial Systems

At this juncture, the genomes of a diverse array of microbes have already been sequenced. For some species, several strains have been sequenced, facilitating whole genome comparisons, examining genomic composition, gene organization, and gene families within and across the major groups of microbial organisms. Comparative genomics has been used extensively to examine the genomes of already sequenced organisms which are closely related. When comparing the genomes of two strains of *Helicobacter pylori*, a similar genomic organization and gene order was found, but 6-7% of ORF's in each strain were missing (Alm and Trust, 1999). Similar "plasticity zones" were found in comparisons of two recently completed *Chlamydia* genomes (Read *et al.*, 2000). Whole genome comparison of human isolates of *Campylobacter jejuni* revealed that intraspecies and interspecies horizontal genetic exchange for this species is common (Dorrell *et al.*, 2001). With the first genome sequences of probiotic organisms scheduled to appear in the near future, it is anticipated that considerable DNA sequence information will be available for comparative genomic analysis. With this information, it will be feasible to identify genetic similarities and differences *in silico* and by whole genome array comparisons between closely-related species. By simultaneously analyzing the genes from the various probiotic organisms under study, it will become possible to define those genes that are uniquely important for survival in a given environmental niche, whether in a food system or in the GI tract, and those which are redundant in that context. Comparing genomes will also define conserved genetic elements that enable common metabolic functions and those conferring unique traits that enable adaptation to such diverse environmental habitats. Key genes and gene networks of interest will be those that direct important functional properties of probiotic lactic acid bacteria; some of these important properties and the genes potentially responsible are listed in Table 2. These include genes involved in adherence and colonization, survival through the stomach and GI tract, antimicrobial activity and substrate metabolism.

Despite the limited knowledge of the mechanisms of adhesion, complexes such as lipotechoic acids and S-layer proteins have been shown to be involved to some extent in the adhesion of LAB to the intestinal mucosa (Boot and Pouwels, 1996; Granato *et al.*, 1999; Sleyter and Beveridge, 1999). It is well-documented that probiotic strains do not appear to colonize a host permanently - strains of bacteria incorporated into probiotic products can usually only be detected in the faeces of human subjects while they continue to be consumed (Goldin *et al.*, 1992). Genomic comparisons between probiotic organisms and commensals may identify the differences in the colonization capabilities between these organisms and the results of such analysis may lead to the identification of more suitable probiotic strains in the future. Another interesting genomic comparison will be that of adherent strains with the genome sequences of pathogens, as the ability to adhere is often regarded as an early event in the virulence of an organism. Other critical gene sets will be those which may be responsible for pathogen exclusion, e.g. antimicrobial and bacteriocin production, bile resistance, detoxification, immunostimulation and acid and stress tolerance (Table 2). Comparative analysis of genomes thus far has revealed that horizontal transfer is the rule rather than the exception. Comparisons of both commensal organisms that reside in the gut and those probiotic strains which have a long history of human consumption, may reveal if similar exchanges between these two groups have taken place over evolutionary time. In the context of probiotic organisms, these analyses will be particularly interesting, as they will not only provide information about the role of horizontal gene transfer in the general population, but allow the evaluation of whether or not this process plays a role in the adaptation of these organisms to their specific niches.

Comparisons between those species used as probiotics, such as *L. acidophilus* and *L. johnsonii* versus food lactobacilli, such as *L. delbreukii* subsp. *bulgaricus* and *L. casei*, which occupy quite diverse environments, is expected to define key gene sets that are important in probiotic roles and which can be used to guide strain selection for multiple roles as either probiotics or as fermentation cultures. Before probiotic cultures can be of benefit to the consumer, they must undergo manufacturing and processing conditions in which various environmental signals and stresses are encountered. The collective gene content, and the manner in which these genes are regulated, are responsible for the critical phenotypic traits and behavior that influence the growth, survival, activity, and benefits of probiotic cultures. Probiotic bacteria must survive in the particular food type in high numbers, survive gastric pH, survive intestinal bile acids, adhere to or interact with the intestinal surface, and compete in the intestinal environment. Because the efficacy of probiotics is related to the numbers of organisms consumed, it is critical that cell viability is maximized during processing and delivery. Sequence analysis promises to identify two major categories of gene systems; firstly, those which are required for survival and activity in vastly different and altering environments, such as in food systems versus the GI tract, and secondly, the responsive gene systems that react to the various stimuli encountered within the food carrier or GI tract (Kullen and Klaenhammer, 1999). Such molecular strategies can be used to identify strains with the desired properties, with the intention of optimizing processing procedures to maximize both survival and probiotic characteristics.

A potentially successful probiotic strain is expected to have several desirable properties in order to exert its beneficial effect. Despite progress in probiotic research in recent times, not all of the available probiotic strains on the market have adequate scientific documentation (Sanders and Huis in't Veld, 1999). Therefore, it is necessary to establish rational criteria for screening and selection of candidate microorganisms (Havenaar and Huis in't Veld, 1992). Genomic sequencing strategies will enhance our ability to select functional probiotic strains and strain combinations using genetic screening (Kullen and Klaenhammer, 1999). It is most likely that these activities will become the basis of successful strain-screening and selection procedures in the future, and that newly established probiotic strains will have to undergo significantly more scrutiny than those currently available.

Functional Genomics

The first phase of the science of genomics, driven by the human genome project, has ended with the availability of rapid and routine whole genome sequencing. The emphasis in this new endeavor thus far has been on the accumulation and annotation of new sequence data. However, the most profound advances will come from utilizing sequence information to evaluate how the genome determines function. Functional genomics aims to discover the biological function of particular genes and to ascertain how sets of genes and their products work together. The recent explosion in bacterial genome sequences now enables true studies in "functional genomics", including the analysis of the transcriptome – the complement of mRNAs transcribed from a cell's genome and their relative levels of expression in a particular cell under a defined set of conditions; the proteome – the complete complement of proteins encoded by the genome; and the metabolome - the quantitative complement of all the low molecular weight molecules present in cells in a particular physiological or developmental state (Figure 2). In many laboratories, whole genome analyses based on the ORF's identified in genome sequencing projects are underway. By using approaches that include gene chips, microarrays and proteomic analysis, it should be possible to move from a static picture of a genome, interpreted from DNA and protein sequence, to the identification of gene networks and a better understanding of the dynamic nature of the regulation of gene expression in a microbial cell.

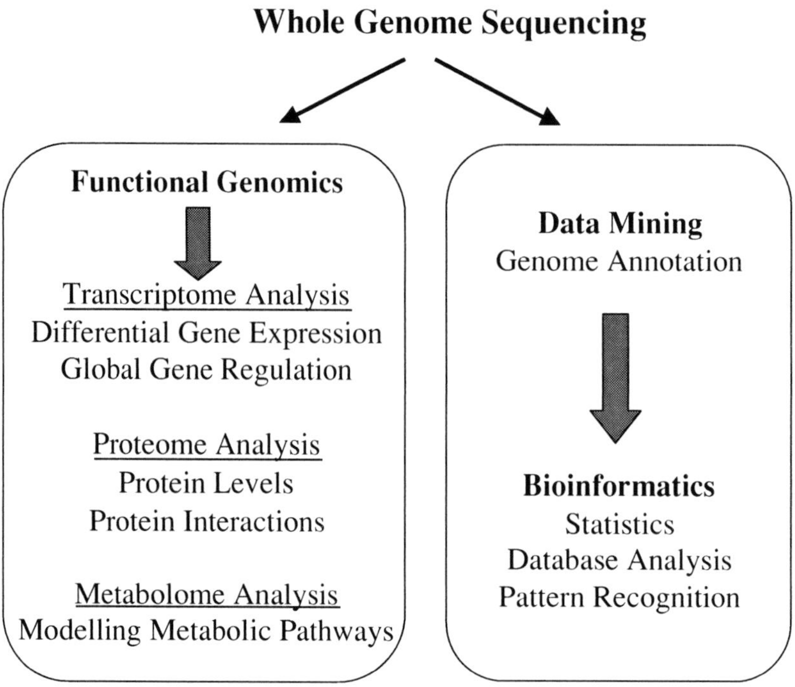

Whole Genome Sequencing

Functional Genomics

Transcriptome Analysis
Differential Gene Expression
Global Gene Regulation

Proteome Analysis
Protein Levels
Protein Interactions

Metabolome Analysis
Modelling Metabolic Pathways

Data Mining
Genome Annotation

Bioinformatics
Statistics
Database Analysis
Pattern Recognition

Figure 2. Analysis in the post-genomic era.

Microarrays

The ability to measure RNA expression profiles across entire genomes provides a level of information not previously attainable. DNA microarrays offer such a platform; the simultaneous measurement of the expression level of many genes in a single hybridization assay. Each array consists of a reproducible pattern of thousands of different DNA sequences attached to a solid support, such as glass. RNA or cDNA is fluorescently or radioactively labeled and hybridized to cDNA sequences on the array and then detected by high-resolution laser scanning. Hybridization signals for each DNA sequence on the array are determined using a process that allows gene expression profiling, comparative genomics and genotyping. The intensity of the signal observed is assumed to be proportional to the amount of transcript present in the RNA population that is being studied. This methodology is particularly powerful in quantifying differences in transcript levels between cell types or treatments. The steps involved in microarray analysis of gene expression include array construction, probe preparation, hybridization of the probe to the array, scanning and detection of the hybridized array and normalization and analysis of the data.

Array Construction

Several methods have been described for producing microarrays and are reviewed elsewhere (Eisen and Brown, 1999; Lipshutz *et al.*, 1999). Two basic technologies have been developed; delivery technologies, or spotted DNA microarrays, and synthesis technologies as seen with oligonucleotide chips. Spotted DNA arrays, originally developed by Brown and colleagues (Schena *et al.*, 1998), are made by printing PCR-amplified DNA fragments, prepared from genomic or cDNA, onto a glass surface. Each gene is represented by a single DNA fragment, greater than several hundred base pairs in length, which may span the entire length of a gene. Many companies are producing ultralow-volume (nanoliter) spotters and gridders to allow researchers to fabricate their own DNA arrays "in-house". Most instruments designed for printing these arrays rely on pin- or needle-based fluid transfer to spot the samples from a reservoir to the substrate, derivatized glass slides coated with either amino silane or poly-L-lysine. The pin diameter and shape, solution viscosity and substrate characteristics will determine spot size, shape and concentration. An alternative system is the "pin and ring" system where the ring holds a small droplet of the DNA solution as a meniscus picked up from the well of a microtiter plate, and the pin punches a smaller droplet from this reservoir onto the substrate. This system is reported to give more uniform density across each spot, but uses more DNA, and produces lower spot densities than other pin-based transfer methods. Companies such as Affymetrix (http://www.affymetrix.com), BioRobotics (http://www.BioRobotics.com) and Genetic Microsystems (http://www.geneticmicro.com) offer such instruments. However, while pin transfer is the current state of the art, advances are being made with inkjet-like printer technology to reduce the spot volume into the picoliter range. Ink jetting does not require direct surface contact, and therefore, this method is theoretically amenable to very high throughput. Because of the ease of use and affordability, microspotting is likely to become the microarray technology of choice for basic research laboratories.

The GeneChip® oligonucleotide array technology has been developed at Affymetrix Inc. (Santa Clara, California; www.affymetrix.com), employing their proprietary oligonucleotide synthesis technique. Oligonucleotide arrays display specific oligonucleotide probes at precise locations in a high-density, information-rich format. Basically, oligonucleotides of defined sequence are fabricated *in situ* on the surface of a glass wafer in a manner analogous to conventional solid-phase oligo synthesis but modified to include a light-sensitive deprotection step. By using photomasks, similar in concept to those used in the semiconductor industry, a series of photospecific bases can be added to select points on the array to create a series of oligos with different sequences. The oligonucleotides are anchored at the 3' end, thereby maximizing the availability of single-stranded nucleic acid for hybridization. Using a combination of a pair of 25-mer oligonucleotides that are designed to be a perfect match (PM) and a mismatch (MM) to the gene targets provides a balance of the highest sensitivity and specificity in the presence of a complex background, especially for low abundance transcripts. Each set of oligonucleotides is offset by one base, so that they can be arranged in order by analyzing overlaps, a process referred to as tiling. MM control oligonucleotides are identical to their PM partners except for a single base change that differs in a central position. The presence of these oligonucleotides allows cross-hybridization and local background signals to be estimated and subtracted from the PM signal to eliminate "noise", which becomes imperative when the specific signal intensity is relatively low and close to background signal intensity. Extensive experimentation by Affymetrix has shown that 15-20 diverse 25-mer oligonucleotides representing each gene target provides a very effective balance between signal intensity and related sequence discrimination, enabling expression monitoring of thousands of targets in complex samples. They also demonstrated that this strategy is most effective with shorter (e.g. 25-mer) probes since a single mismatch is sufficient to destabilize the hybridization. A longer probe cannot take advantage of a similar approach since it requires much longer stretches of mutations or deletions to achieve a similar discriminatory effect. In addition, the innate redundancy of these arrays minimizes the level of false positive results.

An emerging technology for fabrication of oligonucelotide arrays is currently being developed at NimbleGen Systems Inc. (http://www.nimblegen.com). These DNA array systems are based upon the company's proprietary maskless array synthesizer (MAS) technology; custom DNA arrays can be manufactured in less than three hours that have the consistency of traditionally manufactured arrays, allowing researchers to do massive parallel analysis of gene expression in many different genomes. The MAS system is a benchtop, fully integrated, high-density, DNA array fabrication instrument. NimbleGen builds its arrays using photo-deposition chemistry with its proprietary maskless UV light projector, employing a solid-state array of miniaturized aluminum mirrors to pattern up to 780,000 individual pixels of light. By using the maskless UV projector to control the patterning of UV light on the glass in the reaction chamber, unparalleled precision and control over DNA array fabrication chemistry and structure can be achieved. As the name suggests, this technology eliminates the need for photomasks, which are expensive and time-consuming to design and build.

Probe Preparation

A major challenge in prokaryotic expression analysis using microarrays is the specific labeling of messenger RNA (mRNA) for hybridization. Unlike eukaryotic labeling strategies that rely on the presence of the polyA tail to enrich for mRNA, a reliable method to enrich or specifically label prokaryotic mRNA has not been available to date. However, a number of research groups have found that it is possible to use total RNA for labeling. In the first published report on microbial gene expression analysis using oligonucleotide chips, de Saizieu *et al.* (1998) demonstrated that the presence of rRNA in the sample does not prevent detection of gene transcripts present down to a level of several copies per cell. This group used a two-step direct RNA labeling technique that introduces label by coupling with psoralen-biotin, producing biotinylated cRNA. This biotin cRNA is stained with a fluorophore conjugated to avidin after hybridization and detected by laser scanning.

For spotted arrays, the simplest approach is direct labeling of the target DNA with a fluorescent group. Probes are generally labeled by incorporating distinguishable fluorescently tagged nucleotides during oligo-primed reverse transcription of mRNA. Different fluorophores, such as the fluorescent cyanine dyes Cy3 and Cy5, and to a lesser extent, fluorescein and rhodamine, are used to label cDNAs from control and experimental samples of RNA. Criteria for selecting a fluorophore include a narrow excitation and emission peak, a high level of photon-emission, resulting in better sensitivity, and resistance to photobleaching. Concerns regarding the effect of the bulkiness of the fluorescent group on the efficiency of incorporation of different dyes to different templates have also led to the development of alternate labeling methods such as Genisphere® 3DNA Submicro system (http://www.genisphere.com). In this system, rather than incorporating dyes into cDNA, a dye-labeled reagent is hybridized to the random hexamers used to initiate reverse transcription. The ability to use multiple dyes in a similar experiment will allow better comparison of several mRNA samples simultaneously; at present, a number of multi-laser scanners are on the market. Such multiplexing will increase the accuracy of comparative analysis by eliminating factors such as chip-to chip variation, discrepancies in reaction conditions and other such problems that arise when comparing separate experiments.

Hybridization

Because of the variation in size and sequence of array-bound DNAs, a single stringent hybridization condition that is optimal for each spot on the array is not feasible. It is, however, possible to find temperature and salt conditions that give acceptably strong signals for the desired hybridization products and much weaker signals for mismatches. Hybridization conditions also depend on the application. Expression monitoring experiments tend to require long overnight hybridizations, with lower stringencies, higher salt concentrations and lower temperatures. This enhances annealing of low-copy number sequences. For mutation detection applications, the identification of single-base mismatches requires greater hybridization stringencies over short time periods (Marshall and Hodgson, 1998). Important parameters such as the detection level, sensitivity to expression ratios and the correlation between transcript concentration and hybridization signal can be determined by spiking control RNA transcripts in a background of a total messenger population (van Hal *et al.*, 2000). The fluorescently-labeled cDNAs are combined together prior to hybridization and incubated with the array. Relative amounts of a particular gene transcript in both samples are determined by measuring the signal intensities detected for both fluorophores and calculating signal ratios.

Scanning and Image Processing

The first step after hybridization is to capture an image of the array and assign numerical data to each element. To determine which DNA spots correlate with changes in gene expression, the slides are first scanned to produce visual images. For imaging, several types of microarray readers are on the market, including CCD cameras, non-confocal laser scanners and confocal laser scanners. The most popular, the confocal laser scanner, an example of which is the ScanArray® 3000 from GSI Lumonics (Watertown, MA, USA) has the advantage that the light collection efficiency and resolution is usually much higher than that of other systems. Also, these scanners have a small depth of focus which reduces artifacts but also requires more scanning precision. The confocal system scans the array, measuring fluorescent signal from target nucleic acid bound at the surface and generates a separate TIFF image for both Cy3 and Cy5 labeled probes. These images are analyzed to calculate the relative expression levels of each gene and to identify differentially expressed genes. A number of software packages are available that enable easy and accurate visualization and quantitation of gene expression data. QuantArray® (Packard Biochip Technologies, MA, USA), developed for Windows NT, provides automated analysis of up to five-color microarray images without the need to manually draw grids. Other packages include GenePix Pro 3.0.5 (Axon Instruments Inc.,) and ScanAlyze (Stanford University).

Data Normalization and Analysis

Following image processing, the data generated for the arrayed genes must be further analyzed before differentially expressed genes can be identified. The first step is the normalization of the relative fluorescence intensities in each of the two scanned channels, Cy3 and Cy5. This is necessary to adjust for differences in labeling and detection efficiencies for the fluorescent labels and for differences in the quantity of starting RNA from the two samples examined in the assay. Normalization strategies used are described elsewhere (Hegde *et al.*, 2000). Transcript profiling by microarray analysis produces expression ratios, which span a range from "induced" to "repressed". Thus, to analyze these data, the problem of deciding which expression ratios to regard as "significant" must be dealt with. It is possible to approach this problem by simply choosing a post-normalization cut-off value of two-fold up- and down-regulation to define differential expression. To separate genes that are truly differentially expressed from stochastic changes, three or more independent samples may be compared from the same cell type or treatment, and the variability used to evaluate the noise component of their signals. Alternatively, the reproducibility of individual genes duplicated on a given microarray can be used.

The first expression monitoring experiment of genomes indicated that new computational tools would be necessary for the analysis and visualization of the massive quantities of data generated by such parallel analysis. In one comparison of genomic expression patterns between two conditions, hundreds of genes may show measurable changes in expression. When several experiments have been done, it is possible to identify which genes are correlated in their responses to a certain cell treatment or perturbation. A variety of algorithms have therefore been designed to identify and highlight "clusters" of genes with similarities in their expression pattern, and clusters of experimental samples with similar patterns of gene expression. Programs such as CLUSTER and TREEVIEW make it possible to perform pairwise average-linkage cluster analysis. Every variable gene from a series of hybridizations is compared to every other variable gene, and the most highly correlated gene pair is identified. The data for this gene pair are replaced by an average, a correlation matrix is calculated and the process is iterated. TREEVIEW displays the output from CLUSTER in a graphical format. When a coregulated class of genes is known, supervised clustering algorithms trained to recognize known members of the class, can assign uncharacterized genes to that class.

There are a number of important differences between both spotted and oligonucleotide arrays due to the differences in assay and gene representation. Spotted arrays hybridized simultaneously with two differentially labeled samples intrinsically normalize for noise and background in a pairwise comparison. The transcriptional read-out for these paired samples is provided as expression ratios and requires that different samples of an experimental set be hybridized with the same control or reference sample (Harrington *et al.*, 2000). High-density oligonucleotide array assays allow flexibility in sample comparisons and provide estimates of the levels of gene transcripts in individual samples. One of the main advantages of using oligonucleotide arrays is that the oligomer probes are designed to uniquely represent the corresponding gene thus minimizing cross-hybridization between similar sequences. The potential for cross-hybridization between genes with significant levels of sequence similarity is high using spotted array. However, oligonucleotide arrays require gene sequence information for specifying the *de novo* synthesis of the oligomers on the array, whereas spotted arrays can be produced from both known and unknown cDNA and PCR fragments.

The most common application of microarray technology is transcriptional profiling, or the gene by gene determination of differences in transcript abundance between two mRNA preparations. An extension of this is the production of virtual expression arrays (VEA; Garner 1999), where RNA of related or even less related species can be used for differential gene expression, using less stringent hybridization conditions. The measurements obtained from microarrays can also provide essential functional gene annotation data by firstly answering the question of whether or not a gene is biologically real, and secondly, by specifying the environmental and physiological conditions in which a gene is expressed and in association with what other genes. Hence, microarray analysis can elucidate the true nature of the "expressed genome" by confirming the expression of genes of unknown function. However, DNA microarray systems are versatile tools which may also be used for more than expression profiling; other applications include screening of mutations and comparative analysis. In addition to examining gene expression, DNA microarrays have also been used to compare interstrain, intraspecific variations in bacteria at the genomic levels, referred to as genomotyping. Identifying genomic rearrangements, such as deletions, inversions and translocations, in mutant strains and closely related species, is a microarray application that examines DNA rather than RNA. This technique has been termed array-based comparative genome hybridization. This technique has been employed in the previously mentioned comparative genomic studies involving the pathogens, *Helicobacter pylori* (Alm and Trust, 1999; Salama *et al.*, 2000) and *Campylobacter jejuni* (Dorrell *et al.*, 2001). This array application is technically more challenging than transcript profiling due to the increased complexity of genomic DNA over the transcribed gene set, and because of the abundance of repetitive sequences in genomic DNA.

While the arrays described above employ either ORF-specific oligonucleotides or ORF-specific amplicons, which reduces errors in cross hybridization, these arrays can be costly to fabricate due to the number of specific PCR primers required. A variation on the standard array technology and a less costly approach is to use "shotgun arrays" derived from random genomic libraries (Hayward *et al.*, 2000). In this case, the library is typically made in a vector containing universal primer sites, such as SP6 and T7 priming sites, adjacent to the cloning site. Each individual clone from the library is then independently amplified and the amplicons are spotted in the array. Genome coverage is achieved by employing a "minimum tiling pathway" – arraying enough clones to be representative of the genome. These types of arrays are ideal for genotyping as they also contain DNA from intergenic regions, and for systematic comparisons to identify common genetic elements shared by closely-related strains. In addition, these arrays can easily be fabricated for representative strains that will not be subjected to whole genome sequencing. However, one of the disadvantages of this approach is that the sequence of each array element is unknown; each amplicon may contain more than one ORF, making it difficult to pinpoint which genes are responsible for the signal observed.

Microarray Analysis of Probiotics and the Gut Microflora

Microarray technology will have several beneficial uses to analyze the impact of the gut microflora and probiotics on their hosts. One primary area of research will be to assess the functionality of these organisms. Several claimed health benefits of probiotics are scientifically unsubstantiated and gene expression analysis could help to identify relevant biological functions by large-scale expression analysis. With the completion of the genome sequences of the organisms involved, it should be possible to design whole genome arrays to analyze global gene expression patterns and correlate the presence and expression of specific gene sets in the genome to probiotic performance, e.g. adherence, competition, or immunostimulation. The environmental conditions which indigenous and probiotic microorganisms find themselves exposed to, either in a food system or in the GI tract, can either positively or negatively influence their survival and behavior. From an industrial perspective, gene expression arrays can be used to evaluate the impact of environmental signals which are encountered throughout processing and production of probiotic products, and in the GI tract. It should then be possible to identify genes and gene networks that respond to changes in growth and environmental conditions, e.g. low pH, redox potential, heat and bile. By monitoring expression, it may be possible to develop culture-preconditioning treatments to promote desirable patterns of gene expression, and consequently, positively impact culture manufacturing and performance. Future probiotic strain selection could also be facilitated by the ability to identify strains with the appropriate genetic complement and expression profile.

As so little is known about the range of intestinal functions that are shaped by the components of the microflora, microarray analysis could also be used to examine the impact of probiotics and commensals on the biology of the human intestine. Such experiments have been performed recently, to study the impact of *Bacteroides thetaiotaomicron*, a prominent component of the normal mouse and human intestinal microflora, on their colonized host (Hooper *et al.*, 2001). Adult germ-free mice were colonized with *B. thetaiotaomicron* and ileal RNA was isolated after 10 days. Affymetrix chips, representing approximately 25,000 mouse genes, were used to compare ileal gene expression in age-matched germ-free and colonized animals. The microarray data revealed the unanticipated breadth of this commensal's impact on expression of genes involved in modulating fundamental intestinal functions, including nutrient absorption, mucosal barrier fortification, xenobiotic metabolism, angiogenesis and postnatal intestinal maturation (Hooper *et al.,* 2001). In addition, colonization by this organism was demonstrated to produce no detectable inflammatory response, as deduced by the absence of a discernible induction or repression of immune response genes represented on the microarrays. These findings provide some perspective about the essential nature of the interactions between the indigenous microflora and their hosts. Numerous studies have been published reporting the simultaneous expression of host and pathogen gene sets using microarray methods, including *Bordetella pertussis* (Belcher *et al.*, 2000),

Salmonella typhimurium (Rosenberger *et al.*, 2000), and *Pseudomonas aeruginosa* (Ichikawa *et al.*, 2000). The use of high-density arrays to survey genome wide-transcriptional responses in host cells after exposure to microbial pathogens is a powerful approach to understanding host-pathogen interactions. Similar strategies could be adopted to examine the interactions that occur between commensal and probiotic organisms, and with their hosts. For example, examining the reactions of probiotic cultures after exposure to human intestinal cell lines, such as Caco-2 or HT-29 cells, may reveal important responses that either promote or negate probiotic functionality.

Molecular tools to define the effects of probiotic interventions on the composition of the host's microflora have recently become available (Satokari *et al.*, 2001) and made possible the identification of many non-culturable microorganisms. Furthermore, the development of microarrays containing rRNA gene sequences from different members of the gut microflora is a potential starting point for determining whether identifiable changes in species composition can be associated with particular health or disease states and subsequently, for designing hypothesis-based therapeutic trials of probiotic supplements.

DNA microarray-assisted gene expression analysis offers a powerful tool to identify the genes affected by mutations which may potentially affect some important probiotic characteristics. The classical approach for elucidating the function of a gene is by phenotypic mutant analysis which can be a time-consuming and cumbersome task. Data generated from microarray experiments could potentially identify genes and gene networks directly or indirectly affected in a mutant, and give important clues in the elucidation of the function of the mutated genes. This type of experimentation removes the strong bias associated with expression analysis of only a small subset of genes.

One extension of this approach is the application of transcript imaging to assess the risks of organisms that have undergone genetic modification. Whether genetic modification is by mutation and selection, e.g. UV irradiation, or by directed genetic engineering, microarray analysis offers the opportunity to screen for possible risk factors appearing in the altered transcriptome. In fact, this analysis should also provide strong evidence in support of genetic engineering (precise changes) rather than the traditional "fry and try" approaches that result in pleotrophic mutational effects.

Proteomics

The methods described above are most suitable for contrasting gene expression levels across cell types and treatments of a chosen genome, or subset of a genome, but generally do not provide data on the absolute levels of expression. Transcript levels do not necessarily translate into protein expression or activity. Thus, microarray experiments should be accompanied by analyses at the protein level. Proteomics, the large scale analysis of the proteins that are present in a cell, is developing rapidly (Blackstock and Weir, 1999). Microarrays and proteomics are complementary technologies while focusing on different steps of the same process: the expression of genetic information into functional molecules, cells and organisms. The field of proteomics has evolved over the last two decades from its conceptual origins to the display of large numbers of proteins from a specific cell line or organism using 2-D gel electrophoresis (Anderson and Anderson, 1998; Wilkins *et al.*, 1996). A completed genome sequence establishes a basis for high throughput analysis of the the complete set of proteins expressed by the genome, or the proteome. Proteomics can address several problems that cannot be solved by DNA analysis including protein synthesis, the relative abundance of protein products, subcellular localization, and post-translational modifications. Proteomics characterizes large sets of proteins, with newly developed mass spectrometry tools, global two-hybrid techniques, and novel computational tools and methods to process, analyze and interpret the data (Fields, 2001). Mass spectrometry measures the masses of peptides, typically derived from a trypsin digestion, which are then compared to the predicted masses of peptides from *in silico* translations of sequences in genomic databases. Advances in automation, increased sensitivity and high throughput analysis in combination with greatly expanded databases, have extended the application and utility of mass spectrometry.

Two predominant methods are used to perform proteomic analysis. Proteins are first separated by isolelectric focusing and then analyzed directly to generate "virtual" 2D gels. Mass mapping or fingerprinting of peptide fragments using matrix-assisted laser desorption/ionization are performed (MALDI; Berndt *et al.*, 1999; Henzel *et al.*, 1997). The second method is tandem mass spectrometry of peptide fragments to obtain specific sequence information using electrospray ionization, as seen more recently with the development of MALDI time-of-flight mass spectrometry (Shevchenko *et al.*, 2000). This technique is more effective for analysis of higher molecular weight proteins. Two-dimensional gel electrophoresis and protein mass spectrometry represent an integrated technology by which several thousand protein species can be separated, detected and quantified in a single operation. Hundreds of the detected proteins can then be identified in a highly automated fashion by sequential analysis of the peptide mixtures generated by digestion of individual gel spots (Gygi *et al.*, 2000). A promising alternative method is the direct analysis of complex peptide mixtures generated after digestion of protein samples collected by liquid chromatography (LC-MS/MS). The sequence information for thousands of peptides can be recorded in a single LC-MS/MS analysis. The number of proteins that can be identified is limited only by the length of the analysis and the complexity of the mixtures. A key feature of mass spectrometry analysis of gel-separated proteins and peptides is the ability to reveal some structural information about a particular peptide of interest, such as *de novo* sequence information or patterns of post-translational modification.

To complement this technology, the yeast two-hybrid system has contributed significantly to the functional characterization of protein-protein interactions. Initially finding protein partners that interact with just one protein, the system has been scaled up to thousands of putative interactions involving hundreds of different proteins, as demonstrated recently for the yeast, *Saccharomyces cerevisiae* (Uetz *et al.*, 2000). This system is based on the specific-binding of a protein-DNA binding domain fusion construct to a cognate protein-DNA activation domain fusion construct to increase transcription of a targeted reporter gene (Fields and Song, 1989). Other proteomic strategies currently available include phage display (Li, 2000), chip-based technologies (Borrebaeck *et al.*, 2001), and 3-D analysis such as X-ray crystallography/NMR spectroscopy.

This field has moved rapidly to now allow identification of most cellular proteins and further investigate protein interactions on a massive scale. One of the most important areas of research needed on probiotics is to understand the protein composition of the cell surface. The first intimate interactions of a probiotic cell with the intestinal epithelium will occur at the cell surface. Proteins that mediate those interactions are likely to play an important role in attachment, retention and immunomodulation. Understanding which surface proteins are expressed, produced and carry out positive roles in probiotic cultures will be one important challenge to be faced by proteomics.

High-throughput Gene Inactivation Strategies

Knockout studies are one important approach to investigate the function of DNA sequences and the proteins they encode. The basis for this type of analysis is that inactivation of genes in living organisms can sometimes reveal the biological function of specific genes, by detecting changes that occur as a result of the mutation. The most directed method for targeted mutagenesis is to specifically disrupt the ORF by homologous recombination (Thomas and Capeechi, 1987). To date, this has been shown to work effectively in yeast and mice, in addition to a number of bacterial species. A simple alternative to homologous recombination is to induce transient loss of gene function using RNA-mediated interference (RNAi; Fjose *et al.,* 2001). For reasons which are as yet not fully understood, the presence of a small copy number of double-stranded RNA molecules in a cell can dramatically reduce the level of transcript of the associated gene, resulting in a loss-of-function phenotype. This method is suitable for application of genomic scale reverse genetics, and has been applied to whole chromosome mutagenesis in *C. elegans* (Fraser *et al.,* 2000), by designing thousand of constructs to specifically knock out expression of each identified ORF.

Genetic Tools for Probiotic Organisms

Developments in the genetics of lactic acid bacteria (LAB) have been primarily led by research on lactococci, with advances in lactobacilli and other LAB following later. Although knowledge about the genetics of lactobacilli is limited in comparison to other bacteria, innovations in molecular biology, such as plasmid identification, vector construction, transformation, and gene isolation have facilitated significant progress in recent years. However, these efforts have been dispersed over a collection of newly recognized, and collectively significant probiotic lactobacilli (e.g. *L. acidophilus, L. reuteri, L. gasseri, L. johnsonii, L. plantarum, L. casei*), resulting in only incremental steps in our understanding of the genetic programming and potential of the *Lactobacillus* species, overall (Kullen and Klaenhammer, 1999). Genetic work on bifidobacteria is in its infancy having yet to realize functional cloning vectors and reasonably efficient transformation systems. Current and future genomic sequencing projects promise to deliver a wealth of information into the field, providing the much-needed genetic information and tools. Development of genetic tools and efficient gene inactivation strategies will be essential across a broad range of probiotic species, if the functions of genes predicted by annotation are to be correlated with biological activity. Integration and insertions systems are critical genetic tools for insertional mutagenesis and directed manipulations, such as gene stabilization, fusion, amplification deletion and replacement. In lactobacilli, a number of new technologies are emerging that utilize IS-elements, attP/integrase systems, or homologous recombination strategies via suicide or temperature-sensitive replicons. These systems are reviewed elsewhere (Kullen and Klaenhammer, 1999). Gene expression and secretion systems will also be essential for further industrial development (Vaughan *et al.*, 1999). The considerable advances in the genetic accessibility and protocols for lactobacilli and bifidobacteria, in conjunction with the emerging genome sequence information will provide real opportunities for establishing the mechanisms that underly probiotic functionality.

Following the identification of specific targets within genomic regions, functional genomics approaches including gene inactivation, overexpression, and complementation can be used to investigate the function of selected genes and gene networks. These genes and gene networks can be correlated with specific phenotypic traits by construction of isogenic derivatives that can be used in comparative studies to assess human probiotic roles, fermentation behavior and culture stability. In addition to identifying gene function, knockout strategies can also be used for directing beneficial genetic modifications. One example of this approach was the genetic modification of *L. johnsonii* La1 to eliminate the production of D-lactate. *L. johnsonii* La1, a commercial probiotic culture, has been studied intensively in clinical and nutritional studies that have evaluated its survival in the GI tract and positive immunomodulation of the host (Link-Amster *et al.*, 1994; Schiffrin *et al.,* 1995). The D-lactate producing strain is used commercially to produce yogurt. Generally the presence of this compound in the yogurt poses no risk to the vast majority of consumers. However, production of D-lactate may result in D-lactic acidosis and encephalopathy in patients with short bowel syndrome and other intestinal conditions. Inactivation of the single copy of the D-lactate dehydrogenase gene of La1 by a double cross-over deletion resulted in rerouting pyruvate to mainly L-lactate (Mollet, 1999). This food-grade isogenic derivative exhibited no other phenotypic changes and behaved similarly to the parent strain. In this instance, it would be very interesting to evaluate the parent and the derivative by transcript profiling to further evaluate the safety of this recombinant deriviative. This non-D-lactate producing strain can also potentially be used for infant nutrition applications, as newborns fail to completely metabolize ingested or *in situ* produced D-lactate.

Table 3. Applications of genomic technology in probiotic research

Technology	Target Application in Probiotic Research	Outcome[a]
Whole Genome Sequencing	Comparative genomics	Superior strain selection
	Evolutionary analysis	Improved strain identification
DNA Microarrays	Transcription profiling of probiotic strains	Identification of genes expressed in food and GI environment crucial to probiotic funtionality
	Comparison of intraspecific variation	Genomotyping
	Shotgun arrays	Comparison and identification of putative probiotic strains without WGS
	Mutation analysis	Risk assessment of recombinant probiotic strains
Proteomics	Identification of cellular proteins	Understanding of mechanisms involved in host-cell interactions
Genetic Technologies	Gene knockout/deletion/ replacement	Assignment of biological function to predicted annotations
	Creation of isogenic derivatives	Side-by-side behavior comparisions in industrial and clinical settings
	Genetic modifications	Directing beneficial changes to probiotic strains

[a] WGS, whole genome sequencing

Genetically engineered probiotics and commensals are attractive vehicles that are being considered for live delivery of drugs, antimicrobial compounds, and vaccines to defined host niches, within the GI tract or other mucosal surfaces of the body. The criteria which make these organisms attractive as candidates for these types of applications include their GRAS status, their survival though the GI tract and their ability to be retained for some time after feeding. Examples include the genetic engineering of a strain of *Lactococcus lactis* to produce IL-10, a cytokine which down-regulated the pro-inflammatory response in two mouse models of IBD (Steidler *et al.*, 2000). *Streptococcus gordonii*, engineered to produce an antibody fragment with antimicrobial properties, helped to alleviate vaginal *Candida albicans* infections in rats (Beninati *et al.*, 2000). Oral inoculation of lactobacilli producing tetanus toxin fragment C induced local and systemic immune responses to the expressed antigen (Shaw *et al.*, 2000). Genomic information will greatly facilitate the further exploitation of probiotic cultures into new and exciting therapeutic applications.

Concluding Remarks

Genomic studies are now approaching "industrial" speed and scale, following advances in DNA sequencing and the increasing availability of high-throughput methods for studying genes, the proteins they encode, and the pathways in which they are involved. Accumulation of genome sequences has exploded over the past 18 months, with more than 90% of the information at GenBank having been deposited within this time. Because of this vastly increased capacity, the next phases involving bioinformatic and functional analyses of the information, is set to move more rapidly than previously anticipated. The availability of the complete parts list of organisms - catalogs of all genes and proteins - is redirecting the fundamental aims of biological research toward a more holistic and global perspective, from studying single genes to studying the role of all genes or all proteins simultaneously in a systematic fashion. Therefore, while biology in the twentieth century focused on the reductionist analysis of the individual components of complex systems, this century will witness biology evolving to understand how the component parts interact collectively to create an organism (Lander and Weinberg, 2000). As the number of completed genomes continues to grow, the benefits of combined evolutionary and genomic analysis will become more apparent. The post-genomic era holds great promise for identifying the mechanistic basis of organism development and metabolic processes, and advances in "data-mining" will dramatically improve our understanding of areas such as regulation of gene expression, protein structure determination and comparative evolution. Genomics and proteomics will contribute greatly to our understanding of the molecular mechanisms underlying the relationships between food components and ingredients, probiotic microorganisms and the human body, including the gut and immunocompetent cells, and the mechanisms underlying the interactions of the microbial community in the intestinal tract (German *et al.*, 1999). Functional genomic analysis of key genes and phenotypes in probiotic organisms is expected to identify essential probiotic traits and contribute positively to efforts investigating the benefits of probiotic cultures that are delivered in food and dairy products (Table 3). Food biotechnology, in general, will benefit from a functional genomics approach, which will create novel opportunities to ensure the safety of foods, to improve the quality of fermented products, and to substantiate health claims related to the ingestion of specific microbes. Furthermore, industrial production of ingredients will be optimized by a more complete understanding of secretion processes, stress-responses and complex regulatory mechanisms. Ultimately, it is the consumer who will benefit from these developments as science unravels the genetic capacities and capabilities of the beneficial organisms in our microbiome.

Acknowledgements

Support for probiotic research at NCSU is provided by the North Carolina Dairy Foundation, Rhodia, Inc. (Madison, WA), the Southeast Dairy Foods Research Center, and Dairy Management, Inc. The authors wish to thank Eric Altermann, Rodolphe Barrangou, Michael Callanan, and Evelyn Durmaz for their helpful discussion and critical review of the manuscript.

References

Adams, M.D., Celniker, S.E., Holt, R.A., Evans, C.A., Gocayne, J.D., Amanatides, P.G., Scherer, S.E., Li, P.W. *et al.* 2000. The genome sequence of *Drosophila melanogaster*. Science. 287: 2185-2195.

Alm, R.A., and Trust, T.J. 1999. Analysis of the genetic diversity of *Helicobacter pylori*: the tale of two genomes. J. Mol. Med. 77: 834-846.

Anderson, N.L., and Anderson, N.G. 1998. Proteome and proteomics: new technologies, new concepts, and new words. Electrophoresis. 19: 1853-1861.

Bateman, A., and Birney, E. 2000. Searching databases to find protein domain organization. Adv. Protein Chem. 54: 137-157.

Belcher, C.E., Drenkow, J., Kehoe, B., Gingeras, T.R., McNamara, N., Lemjabber, H., Basbaum, C., and Relman, D.A. 2000. The trancriptional responses of respiratory epithelial cells to *Bordetella pertussis* reveal host-defensive and pathogen counter-defensive strateiges. Proc. Natl. Acad. Sci. USA. 97: 13847-13852.

Beninati, C., Oggioni, M.R., Boccanera, M., Spinosa, M.R., Maggi, T., Conti, S., Magliani, W., De Bernardis, F., Teti, G., Cassone, A., Pozzi, G., and Polonelli, L. 2000. Therapy of mucosal candidiasis by expression of an anti-idiotype in human commensal bacteria. Nat. Biotechnol. 18: 1060-1064.

Berndt, P., Hobohm, O., and Langen, H. 1999. Reliable automatic protein identification from matrix-assisted laser desorption/ionization mass spectrometric peptide fingerprints. Electrophoresis. 20: 3521-3526.

Bolotin, A., Wincker, P., Mauger, S., Jaillon. O., Malarme, K., Weissenbach, J., Ehrlich, S.D., and Sorokin, A. 2001. The complete genome sequence of the lactic acid bacterium *Lactococcus lactis* ssp. *lactis* IL1403. Genome Res. 11: 731-753.

Boot, H.J., and Pouwels, P.H. 1996. Expression, secretion and antigenic variation of bacterial S-layer proteins. Mol. Microbiol. 21: 1117-1123.

Borodovsky, M., and McIninch, J. 1993. Recognition of genes in DNA sequence with ambiguities. Biosystems. 30: 161-171.

Borrebaeck, C.A., Ekstrom, S., Hager, A.C., Nilsson, J., Laurell, T., and Marko-Vanga, G. 2001. Protein chips based on recombinant antibody fragments: a highly sensitive approach as detected by mass spectrometry. Biotechniques. 30: 1126-1130.

Blackstock, W.P., and Weir, M.P. 1999. Proteomics: quantitative and physical mapping of cellular proteins. Trends Biotechnol. 17: 121-127.

Broder, S., and Venter, J.C. 2000. Whole genomes: the foundation of new biology and medicine. Curr. Opin. Biotechnol. 11: 581-5.

Deamer, D., and Akeson, M. 2000. Nanopores and nucleic acids: prospects for ultrarapid sequencing. Trends Biotechnol. 18: 147-151.

Delcher, A.L., Harmon, D., Kasif, S., White, O., and Salzberg, S.L. 1999. Improved microbial gene identification with GLIMMER. Nucleic Acids Res. 27: 4636-41.

Delneste, Y., Donnet-Hughes, A., and Schiffrin, E.J. 1998. Functional foods: mechanisms of action on immunocompetent cells. Nutr. Rev. 56: S93-S98.

Diep, D.B., Havarstein, L.S., and Nes, I.F. 1996. Characterization of the locus responsible for the bacteriocin production in *Lactobacillus plantarum* C11. J. Bacteriol. 178: 4472-4483.

Dorrell, N., Mangan, J.A., Laing, K.G., Linton, D., Al-ghusein, H., Barrell, B.G., Parkhill, J., Stoker, N.G., *et al.* 2001. Whole genome comparison of *Campylobacter jejuni* human isolates using a low cost microarray reveals extensive genetic diversity. Genome Res. 11: 1706-1715.

de Roos, N.M., and Katan, M.B. 2000. Effects of probiotic bacteria on diarrhea, lipid metabolism, and carcinogenesis: a review of papers published between 1988 and 1998. Am. J. Clin. Nutr. 71: 405-411.

de Saizieu, A., Certa, U., Warrington, J., Gray, C., Keck, W., and Mous, J. 1998. Bacterial transcript imaging by hybridization of total RNA to oligonucleotide arrays. Nat. Biotechnol. 16: 45-48.

Eisen, M.B., and Brown, P.O. 1999. DNA arrays for analysis of gene expression. Methods Enzymol. 303: 179-205.

Ewing, B., and Green, P. 1998. Base-calling of automated sequencer traces using phred. II. Error probabilities. Genome Res. 8: 186-194.

Ewing, B., Hillier, L., Wendl, M. C., and Green, P. 1999. Base-calling of automated sequencer traces using phred. I. Accuracy assessment. Genome Res. 8: 175-185.

Fernandes, C.F., Shahani, K.M., and Amer, M.A. 1987. Therapeutic role of dietary lactobacilli and lactobacillic fermented dairy products. FEMS Microbiol. Rev. 46: 343-356.

Fields, S. 2001. Proteomics: proteomics in genomeland. Science. 291: 1221-1224.

Fields, S., and Song, O. 1989. A novel genetic system to detect protein-protein interactions. Nature. 340: 245-246.

Fitzgerald, M.C., and Smith, L.M. 1995. Mass spectrometry of nucleic acids: the promise of matrix-assisted laser desorption-ionization (MALDI) mass spectrometry. Annu. Rev. Biophys. Biomol. Struct. 24: 116-140.

Fjose, A., Ellingsen, S., Wargelius, A., and Seo, H.C. 2001. RNA interference: mechanisms and applications. Biotecnol. Ann. Rev. 7: 31-57.

Fleischmann, R.D., Adams, M.D., White, O., Clayton, R.A., Kirkness, E.F., Kerlavage, A.R., Bult, C.J., Tomb, J.F., Dougherty, B.A., Merrick, J.M., *et al.* 1995. Whole-genome random sequencing and assembly of *Haemophilus influenzae* Rd. Science. 269: 496-512

Fraser, A.G., Kamath, R.S., Zipperlin, P., Martinez-Campos, M., Sohrmann, M., and Ahringer, J. 2000. Functional genomic analysis of *C. elegans* chromosome I by systematic RNA interference. Nature. 408:325-330.

Fraser, C.M., Norris, S.J., Weinstock, G.M., White, O., Sutton, G.G., Dodson, R., Gwinn, M., and Hickey, E.K. 1998. Complete genome sequence of *Treponema pallidum*, the syphilis spirochete. Science. 281: 375-388.

Fuller, R. 1989. Probiotics in man and animals. J Appl Bacteriol. 66: 365-378.

Fuller, R. 1992. History and development of probiotics. In: Probiotics: The Scientific Basis. R. Fuller, ed. Chapman & Hall, New York. p. 1-8.

Fuller, R., and Gibson, G. 1997. Modification of the intestinal microflora using probiotics and prebiotics. Scan. J. Gastroenterol. 222: 28-31.

Gaasterland, T., and Oprea, M. 2001. Whole-genome analysis: annotations and updates. Curr. Opin. Struct. Biol. 11: 377-381.

Garner, S. 1999. Virtual Expression Arrays (VEAs) – assist design of future arrays. Lab Chips and Microarrays for Biotechnological Applications, Conference January 13-15, 1999, Zurich, Switzerland.

German, B., Schiffrin, E.J., Reniero, R., Mollet, B., Pfeifer, A., and Neeser, J.-R. 1999. The development of functional foods: lessons from the gut. Trends Biotechnol. 17: 492-499.

Gilliland, S.E., and Walker, D.K. 1990. Factors to consider when selecting a culture of *Lactobacillus acidophilus* as a dietary adjunct to produce a hypocholesterolemic effect in humans. J. Dairy Sci. 73: 905-911.

Glaser, P., Frangeul, L., Buchrieser, C., Rusniok, C., Amend, A., Baquero, F., Berche, P., Bloecker, H. et al. 2001. Comparative genomics of *Listeria* species. Science. 294: 849-852.

Gogarten, J.P., and Olendzenski, L. 1999. Orthologs, paralogs and genome comparisons. Curr. Opin. Genet. Dev. 9: 630-636.

Goldin, B.R., Gorbach, S.L., Saxelin, M., Barakat, S., Gualtieri, L., and Salminen, S. 1992. Survival of *Lactobacillus* species (strain GG) in human gastrointestinal tract. Dig. Dis. Sci. 37: 121-128.

Goldin, B.R., Swenson, L., Dwyer, J., Sexton, M., and Gorbach, S.L. 1980. Effect of diet and *Lactobacillus acidophilus* supplements on human faecal bacterial enzymes. J. Nat. Cancer. Inst. 64: 255-261.

Gordon, D., Abajian, C., and Green, P. 1998. Consed: a graphical tool for sequence finishing. Genome Res. 8: 195-202.

Granato, D., Pervotti, F., Masserey, I., Fouvet, M., Golliard, M., Servin, A., and Brassart, D. 1999. Cell surface-associated lipotechoic acid acts as an adhesion factor for attachment of *Lactobacillus johnsonii* La1 to human enterocyte-like Caco-2 cells. Appl. Environ. Microbiol. 65: 1071-1077.

Gygi, S.P., and Aebersold, R. 2000. Mass spectrometry and proteomics. Curr. Opin. Chem. Biol. 4: 489-494.

Gygi, S.P., Corthals, G.L., Zhang, Y., Rochon, Y., and Aebersold, R. 2000. Evaluation of two-dimensional gel electrophoresis-based proteome analysis technology. Proc. Natl. Acad. Sci. USA. 97: 9390-9395.

Harrington, C.A., Rosenow, C., and Retief, J. 2000 Monitoring gene expression using cDNA microarrays. Curr. Opin. Microbiol. 3: 285-291.

Havenaar, R., and Huis in't Veld, J. 1992. Probiotics: a general view. In: The Lactic Acid Bacteria: Vol I: The Lactic Acid Bacteria in Health and Diseae. E.J.B Wood, ed. Elsevier Applied Science. p. 151-170.

Hayward, R.E., DeRisi, J.L., Alfadhi, S.A., Kaslow, D.C., Brown, P.O., and Rathod, P.K. 2000. Shotgun DNA microarrays and stage-specific gene expression in *Plasmodium falciparum* malaria. Mol. Microbiol. 35: 6-14.

Hegde, P., Qi, R., Abernathy, K., Gay, C., Dharap, S., Gaspard, R., Hughes, J.E., Snesrud, E., Lee, N., and Quakenbush, J. 2000. A concise guide to cDNA microarray analysis. Biotechniques. 29: 548-556.

Henikoff, S., Greene, E.A., Pietrokovski, S., Bork, P., Attwood, T.K., and Hood, L. 1997. Gene families: the taxonomy of protein paralogs and chimeras. Science. 278: 609-614.

Henzel, W.J., Billeci, T.M., Stults, J.T., and Wong, S.C. 1997. Identifying proteins from two-dimensional gels by molecular mass searching of peptide fragments in protein sequence databases. Proc. Natl. Acad. Sci. USA. 90: 5011-5015.

Hieter, P., and Boguski, M. 1997. Functional genomics: it's all how you read it. Science. 278: 601-602.

Himmelreich, R., Hilbert, H., Plagens, H., Pirkl, E., Li, B.C., and Herrmann, R. 1996. Complete sequence analysis of the genome of the bacterium *Mycoplasma pneumoniae*. Nucleic Acids Res. 24: 4420-4449

Hood, L.E., Hunkapiller, M.W., and Smith, L.M. 1987. Automated DNA sequencing and analysis of the human genome. Genomics. 1: 201-212.

Hooper, L.V., and Gordon, J.I. 2001. Commensal host-bacterial relationships in the gut. Science. 292: 1115-1118.

Hooper, L.V., Wong, M.H., Thelin, A., Hansson, L., Flak, P.G., and Gordon, J.I. 2001. Molecular analysis of commensal host-microbial relationships in the intestine. Science. 291: 881-884.

Huis in't Veld, J.H., Havenaar, R., and Marteau, P. 1994. Establishing a scientific basis for probiotic R&D. Trends Biotechnol. 12: 6-8

Hunkapiller, T., Kaiser, R.J., Koop, B.K., and Hood, L. 1991. Large-scale and automated DNA sequence determination. Science. 254: 59-67.

Ichikawa, J.K., Norris, A., Bangers, M.G., Geiss, G.K., van't Wout, A.B., Bumgarner, R.E., and Lory, S. 2000. Interaction of *Pseudomonas aeruginosa* with epithelial cells: identification of differentially expressed genes by expression microarray analysis of human cDNA's. Proc. Natl. Acad. Sci. USA. 97: 9659-9664.

Isolauri, E., Majamaa, H., Arvola, T., Rantala, I., Virtanen, E., and Arvilommi, H. 1993. *Lactobacillus casei* GG revereses increased intestinal permeability induced by cow milk in sucking rats. Gastroenterol. 105: 1643-1650.

Kato, I., Kobayashi, S., Yokokura, T., and Mutai, M. 1981. Antitumour activity of *Lactobacillus casei* in mice. Gann 72: 517-523.

Kimura, K., McCartney, A.L., McConnell, M.A., and Tannock, G.W. 1997. Analysis of fecal populations of bifidobacteria and lactobacilli and investigation of the immunological responses of their human hosts to the predominant strains. Appl. Environ. Microbiol. 63: 3394-3398.

Klaenhammer, T.R. and Kullen, M.J. 1999. Selection and design of probiotics. International J. Food. Microbiol. 50: 45-57.

Kullen, M.J., and Klaenhammer, T.R. 1999. Genetic modification of intestinal lactobacilli and bifidobacteria. In: Probiotics: A Critical Review. G.W. Tannock, ed. Horizon Scientific Press. Wymondham. p. 65-83.

Kullen, M.J., and Klaenhammer, T.R. 1999. Identification of the pH-inducible, proton-translocating F_1F_0-ATPase (atpBEFHAGDC) operon of *Lactobacillus acidophilus* by differential display: gene structure, cloning and characterization. Mol. Microbiol. 33: 1152-1161.

Lander, E.S. 1996. The new genomics: global views of biology. Science. 274: 536-539.

Lander E.S., and Weinberg R.A. 2000. Genomics: journey to the center of biology. Science. 287:1777-82.

Lander, E.S., Linton, L.M., Birren, B., Nusbaum, C., Zody, M.C., Baldwin, J., Devon, K., Dewar, K. *et al.* 2001. Initial sequencing and analysis of the human genome. Nature. 409: 860-921.

Lee, Y.-K., and Salminen, S. 1995. The coming age of probiotics. Trends Food Sci. Technol. 6: 241-245.

Li, M. 2000. Applications of display technology in protein analysis. Nat. Biotechnol. 18: 1251-1256.

Link-Amster, H., Rochat, F., Saudan, K.Y., Mignot, O., and Aeschlimann, J.M. 1994. Modulation of a specific humoral immune response and changes in intestinal flora mediated through fermented milk intake. FEMS Immunol. Med. Microbiol. 10: 55-63.

Lipshutz, R.J., Fodor, S.P., Gingeras, T.R., and Lockhart, D.J. 1999. High density synthetic oligonucleotide arrays. Nat. Genet. 21: 20-24.

Marshall, A., and Hodgson, J. 1998. DNA chips: an array of possibilities. Nat. Biotechnol. 16: 27-31.

Maxam, A. M., and Gilbert, W. 1980. Sequencing end-labeled DNA with base-specific chemical cleavages. Methods Enzymol. 65: 499-560.

McCartney, A.L., Wenzhi, W., and Tannock, G.W. 1996. Molecular analysis of the composition of the bifidobacterial and lactobacillus microflora of humans. Appl. Environ. Microbiol. 62: 4608-4613.

Meldrum, D, R. 2000. Automation for genomics, part two: sequencers, microarrays, and future trends. Genome Research. 10: 1288-1303.

Meldrum, D. R. 2001. Sequencing genomes and beyond. Science. 292: 515-517.

Mewes, H.W., Albermann, K., Bahr, M., Frishman, D., Gleissner, A., Hani, J., Heumann, K., Kleine, K. et al. 1997. Overview of the yeast genome. Nature. 387: 7-65.

Moser, S.A., and Savage, D.C. 2001. Bile salt hydrolase activity and resistance to toxicity of conjugated bile salts are unrealted properties in lactobacilli. Appl. Environ. Microbiol. 3476-3480.

Mullis, K.B., and Faloona, F.A. 1987. Specific synthesis of DNA *in vitro* via a polymerase-catalyzed chain reaction. Methods Enzymol. 155: 335-50.

Mustapha, A., Jiang, T., and Savaiano, D. 1997. Improvement of lactose digestion by humans following ingestion of unfermented acidophilus milk: influence of bile sensitivity, lactose transport, and acid tolerance of *Lactobacillus acidophilus*. J. Dairy Sci. 80: 1537-1545.

Muyzer, G., and Smalla, K. 1998. Application of denaturing gradient gel electrophoresis (DGGE) and temperature gradient gel electrophoresis (TGGE) in microbial ecology. Antonie Van Leeuwenhoek. 73: 127-141.

Ogata, H., Goto, S., Sato, K., Fujibuchi, W., Bono, H., and Kanehisa, M. 1999. KEGG: Kyoto Encyclopedia of Genes and Genomes. Nucleic Acids Res. 27: 29-34.

O'Sullivan, M.G., Thornton, G., O'Sullivan, G.C. and Collins, J.K. 1992. Probiotic bacteria: myth or reality. Trends Food Sci. Technol. 3: 309-314.

Overbeek, R., Fonstein, M., D'Souza, M., Pusch, G.D., and Maltsev, N. 1999. The use of gene clusters to infer functional coupling. Proc. Natl. Acad. Sci. USA. 96: 2896-2901.

Parkhill, J, Wren, B.W., Mungall, K., Ketley, J.M., Churcher, C., Basham, D., Chillingworth, T., and Davies, R.M. 2000. The genome sequence of the food-borne pathogen *Campylobacter jejuni* reveals hypervariable sequences. Nature. 403: 665-668.

Pellegrini, M., Marcotte, E.M., Thompson, M.J., Eisenberg, D., and Yeates, T.O. 1999. Assigning protein function by comparative genome analysis: protein phylogenetic profiles. Proc. Natl. Acad. Sci. USA. 96: 4285-4288.

Perego, M., and Hoch, J.A. 2001. Functional genomics of gram-positive microorganisms: review of the meeting, San Diego, California, 24-28 June, 2001. J. Bacteriol. 183: 6973-6978.

Ramakrishna, R., and Srinivasan, R. 1999. Gene identification in bacterial and organellar genomes using GeneScan. Comput Chem. 23: 165-174.

Read, T.D., Brunham, R.C., Shen, C., Gill, S.R., Heidelberg, J.F., White, O., Hickey, E.K., Peterson, J. et al. 2000. Genome sequences of *Chlamydia trachomatis* MoPn and *Chlamydia pneumoniae* AR39. Nucleic Acids Res. 28: 1397-1406.

Roos, S., Alejung, P., Robert, N., Byong, L., Wadstrom, T., Lindberg, M., and Jonsson, H. 1996. A collagen-binding protein from *Lactobacillus reuteri* is part of an ABC transporter system? FEMS Microbiol. Lett. 144: 33-38.

Roos, S. 1999. Adhesion and autoaggregation of *Lactobacillus reuteri* and description of a new *Lactobacillus* species with mucus binding properties. Ph.D. Thesis, Upsala, Sweden. Swedish University of Agricultural Sciences.

Roos, S., Lindgren, S., and Jonsson, H. 1999. Autoaggregation of *Lactobacillus reuteri* is mediated by a putative DEAD-box helicase. Mol. Microbiol. 32: 427-436.

Rosenberger, C.M., Scott, M.G., Gold, M.R., Hancock, R.E., and Finlay, B.B. 2000. *Salmonella typhimurium* infection and lipopolysaccharide stimulation induce similar changes in macrophage gene expression. J. Immunol. 164: 5894-5904.

Saavedra, J.M. 1995. Microbes to fight microbes: a not so novel approach to controlling diarrheal disease. J. Pediat. Gastroenterol. Nutr. 21: 125-129.

Saavedra, J.M., Bauman, N.A., Oung, I., Perman, J.A., and Yolken, R.H. 1994. Feeding of *Bifidobacterium bifidum* and *Streptococcus thermophilus* to infants in hospital for prevention of diarrhoea and shedding of rotavirus. Lancet. 344: 1046-1049.

Salzberg, S.L., Delcher, A.L., Kasif, S., and White, O. 1998. Microbial gene identification using interpolated Markov models. Nucleic Acids Res. 26: 544-548.

Sanders, M.E. 1993. Summary of conclusions from a consensus panel of experts on health attributes of lactic cultures: significance of fluid milk products containing cultures. J. Dairy Sci. 76: 1819-1828.

Sanders, M.E. 1999. Probiotics – scientific status summary. Food Technol. 53: 67-77.

Sanders, M.E., and Huis in't Veld, J. 1999. Bringing a probiotic-containing functional food to the market: microbiological, product, regulatory and labelling issues. Antonie van Leeuwenhoek. 76: 293-315.

Sanders, M.E., and Klaenhammer, T.R. 2001. The scientific basis of *Lactobacillus acidophilus* NCFM functionality as a probiotic. J. Dairy Sci. 84: 319-331.

Sanger, F., Nicklen, S., and Coulson, A.R. 1977. DNA sequencing with chain-terminating inhibitors. Biotechnol. 24: 104-108.

Salama, N., Guillemin, K., McDaniel, T.K., Sherlock, G., Tompkins, L., and Falkow, S. 2000. A whole genome microarray reveals genetic diversity among *Helicobacter pylori* strains. Proc. Natl. Acad. Sci. USA. 97: 14668-14673.

Salminen, S., Isolauri, E., and Salminen, E. 1996. Clinical uses of probiotics for stabilizing the gut mucosal barrier: successful strains and future challenges. Antonie van Leeuwenhoek. 70: 347-358.

Satokari, R.M., Vaughan, E.E., Akkermans, A.D., Saarela, M., and de Vos, W.M. 2001. Bifidobacterial diversity in human feces detected by genus-specific PCR and denaturing gradient gel electrophoresis. Appl. Environ. Microbiol. 67: 504-513.

Savage, D.C. (1977). Microbial ecology of the gastrointestinal tract. Ann. Rev. Microbiol. 31: 107-133.

Sawada, H., Furushiro, M., Hirai, K., Motoike, M., Watanabe, T., and Yokokuram T. 1990. Purification and characterization of an antihypersensitive compound from *Lactobacillus casei*. Agric. Biol. Chem. 54: 3211-3219.

Schena, M., Heller, R.A., Theriault, T.P., Konrad, K., Lachenmeier, E., and Davis, R.W. 1998. Microarrays: biotechnology's discovery platform for functional genomics. Trends Biotechnol. 16: 301-306.

Schiffrin, E.J., Rochat, F., Link-Amster, H., Aeschlimann, J.M., and Donnet-Hughes, A. 1995. Immunomodulation of human blood cells following the ingestion of lactic acid bacteria. J. Dairy Sci. 78: 491-497.

Shaw, D.M., Gaerthe, B., Leer, R.J., Van der Stap, G.M.M., Smittenaar, C., Heijne den Bak-Glashouwer, M.-J., Thole, J.E.R., Tielen, F.J. *et al.* 2000. Engineering the microflora to vaccinate the mucosa: serum immunoglobulin G responses and activated draining cervical lymph nodes following mucosal application of tetanus toxin fragment C-expressing lactobacilli. Immunol. 100: 510-518.

Shevchenko, A., Chemushevich, I., Wilm, M., and Mann, M. 2000. De novo peptide sequencing be nanoelectrospray tandem mass spectrometry using triple quadrupole and quadrupole/time-of-flight instruments. Methods Mol. Biol. 146: 1-16.

Sleyter, U.B., and Beveridge, T.J. 1999. Bacterial S-layers. Trends Microbiol. 7: 253-260.

Smith, L.M., Fung, S., Hunkapiller, M.W., Hunkapiller, T.J., and Hood, L. 1985. The synthesis of oligonucelotides containing an aliphatic amino group at the 5' terminus: synthesis of flourescent DNA primers for use in DNA sequence analysis. Nucleic Acids Res. 13: 2399-2412.

Smith, L.M., Sanders, J.Z., Kaiser, R.J., Hughes, P., Dodd, C., Connell, C.R., Heiner, C., Kent, S.B.H., and Hood, L.E. 1986. Fluorescence detection in automated DNA sequence analysis. Nature. 321: 674-679.

Steidler, L., Hans, W., Schotte, L., Neirynck, S., Obermeier, F., Falk, W., Fiers, W., and Remaut, E. 2000. Treatment of murine colitis by *Lactococcus lactis* secreting interleukin-10. Science. 289: 1352-1355.

Surawicz, C.M., McFarland, L.V., Elmer, G., and Chinn, J. 1989. Treatment of recurrent *Clostridium difficile* colitis with vancomycin and *Saccharomyces boulardii*. Am. J. Gastroenterol. 84: 1285-1287.

Tannock, G.W. 1995. The Normal Microflora. Chapman & Hall, London, U.K.

Tannock, G.W. 1999. Probiotics: A Critical Review. G.W. Tannock, ed. Horizon Scientific Press. Wymondham. p. 5-14.

Tatusov, R.L., Koonin, E.V., and Lipman, D.J. 1997. A genomic perspective on protein families. Science. 278: 631-7.

The *Arabidopsis* Genome Initiative. 2000. Analysis of the genome sequence of the flowering plant *Arabidopsis thaliana*. Nature. 408: 796-815.

The *C. elegans* Sequencing Consortium (1998). Genome sequence of the nematode *C. elegans*: a platform for investigating biology. Science. 282: 2012-2018.

Thomas, K., and Capecchi, M. 1987. Site-directed mutagenesis by gene-targeting in mouse embryo-derived stem cells. Cell. 51:503-512.

Tomb, J.F., White, O., Kerlavage, A.R., Clayton, R.A., Sutton, G.G., Fleischmann, R.D., Ketchum, K.A., and Klenk, H.P. 1997. The complete genome sequence of the gastric pathogen *Helicobacter pylori*. Nature. 388: 539-47.

Turner, M.S., Woodberry, T., Hafner, L.M., and Giffard, P.M. 1999. The *bspA* locus of *Lactobacillus fermentum* BR11 encodes an L-cysteine uptake system. J. Bacteriol. 181: 2192-2198.

Uetz, P., Giot, L., Cagney, G., Mansfield, T.A., Judson, R.S., Knight, J.R., Lockshon, D., Narayan, V., et al. 2000. A comprehensive analysis of protein-protein interactions in *Saccharomyces cerevisiae*. Nature. 403: 623-627.

van Hal, N.L., Vorst, O., van Houwelingen, A.M., Kok, E.J., Peijnenburg, A., Aharoni, A., van Tunen, A.J., and Keijer, J. 2000. The application of microarrays in gene expression analysis. J. Biotechnol. 78: 271-280.

Vaughan, E.E., Mollet, B., and de Vos, W.M. 1999. Functionality of probiotics and intestinal lactobacilli: light in the gastrointestinal tract. Curr. Opin. Biotechnol. 10: 505-510.

Venter, J.C., Adams, M.D., Myers, E.W., Li, P.W., Mural, R.J., Sutton, G.G., Smith, H. O., Yandell, M. et al. 2001. The sequence of the human genome. Science. 291: 1304-1351.

Walter, J. Tannock, G.W., Tilsala-Timisjarvi, A., Rodtong, S., Loach, D.M., Munro, K., and Alatossava, T. 2000. Detection and idenitification of gastrointestinal *Lactobacillus* species by using denaturing gradient gel electrophoresis and species-specific PCR primers. Appl. Environ. Microbiol. 66: 297-303.

Wilkins, M.R., Sanchez, J.C., Williams, K.L., and Hochstrasser, D.F. 1996. Current challenges and future applications for protein maps and post-translational vector maps in proteome projects. Electrophoresis. 17: 830-838.

Woese, C.R. 1987. Bacterial evolution. Microbiol. Rev. 51: 221-271.

Woese, C.R., and Fox, G.E. 1977. Phylogenetic structure of the prokaryotic domains: the primary kingdoms. Proc. Natl. Acad. Sci. USA. 74: 5088-5090.

Zuckerkandl, E., and Pauling, L. 1965. Molecules as documents of evolutionary history. J. Theor. Biol. 8: 357-366.

From: *Probiotics and Prebiotics: Where Are We Going?*
Edited by: Gerald W. Tannock

Chapter 10

Intestinal Microflora and Homeostasis of the Mucosal Immune Response:Implications for Probiotic Bacteria?

Stephanie Blum and Eduardo J. Schiffrin

Abstract

The intestinal microflora can be considered a post-natally acquired organ that is composed of a large diversity of bacteria that perform important functions for the host and can be modulated by environmental factors, such as nutrition. Specific components of the intestinal microflora, including lactobacilli and bifidobacteria, have been associated with beneficial effects on the host, such as promotion of gut maturation and integrity, antagonisms against pathogens and immune modulation. Beyond this, the microflora seems to play a significant role in the maintenance of intestinal immune homeostasis and prevention of inflammation. The contribution of the intestinal epithelial cell in the first line of defense against pathogenic bacteria and microbial antigens has been recognized. However, the interactions of intestinal epithelial cells with indigenous bacteria are less well understood. This chapter will summarize the increasing scientific attention to mechanisms of the innate immune response of the host towards different components of the microflora, and suggest a potential role for selected probiotic bacteria in the regulation of intestinal inflammation.

Introduction

The mammalian intestine is inhabited by a complex and diverse microbial community which is in intimate association with the host epithelium. The colonization of the gastrointestinal (GI) tract by complex bacterial societies is determined by different habitats and ecological conditions. Although there is still much ignorance concerning the composition and temporal dynamics of the microbial gut ecosystem, the development of nucleic acid-based methods has significantly contributed to the specific detection of bacteria independent of culturing. In fact, the analysis of 16S rDNA from human fecal samples revealed that a large proportion of intestinal bacteria have escaped description so far. Cultivation-independent PCR-TGGE (temperature gradient gel electrophoresis) or DGGE (denaturing gradient gel electrophoresis) analysis combined with measurements of ecological diversity can be applied for monitoring diet- and antibiotic-induced alterations of complex intestinal microbial ecosystems (Zoetendal *et al.,* 1998; Vaughan *et al.,* 2000; Walter *et al.,* 2000).

Although in permanent interaction with environmental microorganisms, partially food derived, the indigenous microflora has a remarkably stable composition throughout most of the life span (Kimura *et al.,* 1997; Tannock *et al.,* 2000). A drastic change in microflora composition certainly occurs immediately after birth, when the so far sterile fetus becomes exposed to the external environment. In addition to the maternal bacteria encountered during delivery and skin contact during breast-feeding, environmental microorganisms contribute to the neonatal microflora (Moreau *et al.,* 1986). It has been observed that before attaining a steady state around weaning, there is a defined sequence of dominant bacterial genera. Nevertheless, it appears that also the type of neonatal feeding may influence the composition of the intestinal microflora as differences have been observed between breast-fed and formula-fed babies (Benno *et al.,* 1986). During adulthood, perturbations of the GI microflora equilibrium are rare and mainly associated with pathological conditions such as enteral infections, antibiotherapy or immune suppression. However, in the healthy elderly a significant reduction of potentially protective bacteria, such as bifidobacteria, has been reported (Hopkins *et al.,* 2001).

The composition of the gut microflora may transiently vary as a consequence of a major bacterial inoculum in the diet (Pochart *et al.,* 1992; Bouhnik *et al.,* 1992). Furthermore, administration of oligosaccharides, non-digestible by human enzymes but fermented by bacteria in the colon, was shown to result in an enhanced growth of bifidobacteria (Bouhnik *et al.,* 1999).

The mechanisms that lead to a dynamic equilibrium of the GI tract microflora and the host are not entirely known, but seem to comprise both host and microbial factors that may differ at different levels of the intestine. A recent example demonstrates that *Bacteroides fragilis* is able to modulate its surface antigenicitiy by producing at least eight different capsular polysaccharides. Based on the combination of different surface carbohydrates, the microorganism might escape from immunosurveillance and by this means maintain its ecological niche in the intestinal tract (Krinos *et al.,* 2001).

Microflora and the Secretory Immune Response

It is generally accepted that the microflora in the human intestinal tract has a major impact on gastrointestinal and mucosal immune functions (Cebra, 1999). Bacterial colonization of the gut by indigenous bacteria was shown to alter intestinal physiology of the host by modulation of genes, implicated in nutrient absorption, mucosal defense and xenobiotic metabolism (Hooper *et al.,* 2001). The lower exposure of the neonate and infant to intestinal microbial challenge in the last decades, as indicated by epidemiological data, has been associated with a higher incidence of allergic diseases (Anderson *et al.,* 2001; Matricardi *et al.,* 2000; Isolauri, 1997). Thus, it appears that bacterial challenge of the host is an important pre-requisite for the development of homeostasis of the intestinal immune system and maintenance of oral tolerance (Weiner, 1997). There is also increasing evidence that the breakdown of tolerance to the microflora could lead to, or perpetuate, inflammatory bowel disease (Duchmann *et al.,* 1995). This may imply that immunosurveillance of the bacterial content of the intestine contributes to i) the development of immunological tolerance and control of severe inflammatory reaction, ii) the control of colonization and iii) the appropriate defense against external antigens, pathogenic bacteria or viruses. Host defense against the autochthonous microflora is still poorly understood.

It has been reported that indigenous bacteria can be recognized by the host immune system and elicit local and systemic antibody responses (Kimura *et al.,* 1997; Apperloo-Renkema *et al.,* 1993). The production of secretory immunoglobulin A [(s)IgA] is the best defined effector component of the intestinal mucosa. In cooperation with innate defense factors, such as mucus, sIgA in the intestinal lumen will accomplish 'immune exclusion' to protect the mucosal surface. This occurs in the absence of complement activation and is thus a non-inflammatory process.

To date, it is still unknown whether the secretory immune response plays a role in determining the composition of the microflora. While a proportion of cells of the resident microflora is covered by IgA antibodies, the remainder are devoid of antibody coating (Van der Waaij *et al.,* 1996). Interestingly, local or systemic antibody responses do not seem to lead to the elimination of indigenous bacteria from the intestine (Apperloo-Renkema *et al.,* 1993). There is recent experimental evidence in mice suggesting that most of the intestinal IgA against indigenous bacteria is specifically induced in response to their presence, and that its production is independent of T-cell and germinal centre participation (Macpherson *et al.,* 2000). This IgA, mainly directed against bacterial protein antigens, appears to be derived from B1 lymphocytes that develop in the subepithelial compartment and are spread throughout the lamina propria (Herzenberg *et al.,* 2000). The IgA antibodies protect the host from invasion by indigenous bacteria, but do not spontaneously appear in the serum. In case of bacterial infection, specific IgG can be produced by T cell-dependent pathways. It is hypothesized that specific T cell-independent IgA forms part of the normal mucosal response against the continuous antigenic load of indigenous bacteria and might represent an evolutionary ancient pathway of the immune system. However, these observations have not so far been confirmed in humans.

In case of failure of this first line of protection, penetrating antigens need to be removed from the lamina propria (LP) by antibodies locally produced by terminally differentiated B cells and T cells. Production and secretion of IgA in the LP has been shown to be regulated by (i) endogenous mediators, such as TGF-β, IL-5 and IL-10, mainly produced by regulatory T cells (Lebman *et al.,* 1990) and (ii) is associated with intestinal bacterial colonization (Kett *et al.,* 1995). The main IgA subclass of the human jejunum is IgA1; whereas IgA2 is predominant in the colon. This might reflect the distribution of food antigens versus bacterial antigens in the normal gut. In the case of bacterial overgrowth, the composition is changed with an increase of IgA2 in the small bowel, suggesting that LPS might play a role in antibody class switch (Kett *et al.,* 1995).

Innate Defenses of the Intestinal Mucosa

The single layer of epithelial cells lining the intestinal tract has to protect the underlying compartments from both the normal microflora and invading pathogens. Moreover, the intestinal mucosa has to cope with a large antigenic load, including dietary and bacterial antigens, without triggering constant and severe inflammation. The potential for cumulative damage might explain the rapid turn-over of intestinal epithelial cells (IEC) and the requirement for mechanisms of cytoprotection and repair to preserve barrier integrity. Trefoil peptides secreted to the apical surface of the epithelium interact synergistically with intestinal mucin glycoproteins to reinforce a physicochemical barrier (Kindon *et al.,* 1995). These peptides are also involved in reconstitution of the epithelium after injury. Many factors may modulate the production of intestinal mucins for innate defenses. A recent publication underlines the protective effects of *Lactobacillus* species by stimulation of intestinal mucin synthesis (Mack *et al.,* 1999).

Antimicrobial peptides, such as α-defensins secreted from Paneth cells, or β-defensins secreted by the epithelial cell itself, are abundantly found in host defense reactions in the gastrointestinal tract (Hecht, 1999). There is now strong evidence that in addition to constitutively secreted peptide antibiotics, others are induced upon contact with microorganisms or by pro-inflammatory cytokines. The characteristic local expression pattern of defensin might indicate that specialized surfaces express a characteristic antimicrobial peptide pattern which defines the composition of the microflora and the density of microorganisms present on that surface. Antimicrobial activity of defensins is based on pore formation, membrane depolarization and interference with bacterial metabolism. In addition, some defensins induce a secretory chloride response in IEC (Lencer *et al.,* 1997), others display chemotactic activity for T cells, serving as a link between innate and adaptive immunity (Chertov *et al.,* 1996; Lillard, Jr. *et al.,* 1999). Finally, among the peptides promoting restoration of the epithelium are transforming growth factor (TGF)-β, keratinocyte growth factor (KGF) and hepatocyte growth factor (HGF) produced by epithelial cells and subjacent mesenchymal cells or myofibroblasts (Dignass *et al.,* 1993; Goke *et al.,* 1998).

Microbial Recognition

The first recognition of microbial determinants is achieved by host cellular defense molecules, the so-called pattern recognition receptors (PRR). PRRs are germ-line encoded and recognize molecular structures shared by a variety of bacteria (Stahl *et al.,* 1998). In the gut mucosa PRRs are found on macrophages and dendritic cells, which are widely distributed beneath the epithelial surface where they guard the sites of antigen entry. In addition, newly described PRRs are expressed by the intestinal epithelial cell.

A classical PRR is the mannose receptor (MR), expressed on tissue macrophages and immature dendritic cells (DC) (Fraser *et al.,* 1998). MRs recognize the pattern of carbohydrates that decorate the surface of Gram-negative and Gram-positive bacteria, yeasts, parasites and mycobacteria (Stahl *et al.,* 1998). Ligation of the MR results either in endocytosis or phagocytosis of the ligand-receptor complex and subsequent clearance of the infectious agent. The MR appears to play a critical role in the processing of microbe derived glycolipids in conjunction with CD1b (Park *et al.,* 2000). Thus, the MR is involved in both antigen clearance in the tissues and stimulation of clonal adaptive immune responses.

Another class of PRRs, the human Toll-like receptors (TLRs), is related to the *Drosophila* Toll protein, which is required for ontogenesis and antimicrobial resistance (Medzhitov *et al.,* 2000). Generally, TLRs are type I transmembrane receptors with cytoplasmic domains that resemble the mammalian IL-1 receptor (IL-1R). Ten TLR molecules have been described so far in mammals, and it is assumed that each of the TLRs recognizes a discrete subset of molecules widely shared by microbial pathogens. Reaction of bacterial products with TLRs results in cellular signaling leading to NF-κB or AP-1 activation (Medzhitov *et al.,* 2000). TLR4 was shown to be essential for the recognition of Gram-negative bacteria, acting as a co-receptor for CD14 in the cellular response to LPS (Yoshimura *et al.,* 1999; Aderem *et al.,* 2000). TLR2 is involved in cell responsiveness to Gram-positive bacteria, including peptidoglycans (Yoshimura *et al.,* 1999), lipoteichoic acid and bacterial lipoproteins (Brightbill *et al.,* 1999). This suggests that different microbial agents might activate different Toll members, leading to the activation of different target genes. More recently the differential expression of TLR2, TLR3 and TLR4 on intestinal epithelial cell lines and activation of specific signal transduction pathways after stimulation of IEC with LPS has been reported (Cario *et al.,* 2000). The same group showed that upon stimulation of TLRs with LPS or peptidoglycan, TLRs selectively move from the apical surface through the cytoplasm towards the basolateral membrane (Cario *et al.*, 2002). Thus, TLRs positioned at the apical pole of the epithelial cell seem to monitor the sensitive balance of the luminal microbial community. These data provide further evidence, that IEC might play a key role in the recognition and transduction of signals derived from luminal bacteria.

Sensing the Danger of Infection

Bacterial recognition systems used by epithelial cells are likely to have developed to maintain mucosal surfaces in a state of homeostasis with the normal microflora. A sophisticated system is required to discriminate between indigenous and enteropathogenic bacteria. Discrimination seems to depend on bacterial feature recognition by PRRs and the cellular compartment where the bacterial presence is detected. For instance, the recognition of flagellin in the basolateral membrane of the intestinal epithelium was shown to initiate a defense response, subsequent to TLR5 binding (Philpott *et al.*, 2001).

In addition to sensing bacterial presence via PRR it has been proposed that a second signal is required to initiate an appropriate response to pathogens (Matzinger, 1998). Bacterial invasion provides a signal to initiate inflammatory responses and often results in cellular damage, leading to activation of immunological defense. These are key elements for control of infection and clearance of the infecting microbe. Non-invading pathogens are also recognized by the epithelium, if cytopathic or enterotoxic effects are induced (Beatty *et al.*, 1999). The alarm signals sent by stressed, damaged or parasitized cells comprise constitutive and inducible molecules that can initiate different kinds of immunity in different tissues and to different pathogens. Endogenous 'danger signals' comprise heat shock proteins, nucleotides, extracellular matrix breakdown products or necrotic cell derived molecules. Exogenous 'danger signals' are LPS, peptidoglycans or unmethylated CpG sequences in microbial DNA. These molecules will activate *a priori* dendritic cells, necessary for the initiation of primary and secondary immune responses (Pulendran *et al.*, 2001).

Recognition of Indigenous Bacteria by Intestinal Epithelial Cells

The intestinal epithelium is increasingly recognized as a constitutive component of the innate and adaptive response of the host towards luminal bacteria (Molmenti *et al.*, 1993). Several groups have demonstrated that IEC may exert accessory function for antigen (AG) presentation. They express MHC class II molecules and may activate CD4$^+$ T cells (Mayer, 1998; Hershberg *et al.*, 1998).

Classical MHC class-I molecules (HLA A, B, C) are expressed in co-association with β2-microglobulin and they present peptides to CD8$^+$ intraepithelial lymphocytes (IEL). Since peptides contained within the groove of the MHC class Ia molecules are predominantly derived from degradation of intracellular molecules, CD8$^+$ IEL have important roles in monitoring deleterious intracellular events that might occur during viral infection, cellular stress or neoplastic transformation. In addition to polymorphic MHC class Ia molecules, non-classical MHC class-Ib molecules, such as CD1d (Blumberg *et al.*, 1991), HLA-G (Pazmany *et al.*, 1996), HLA-E, HLA-H (Hfe) (Parkkila *et al.*, 1997), MICA (Bauer *et al.*, 1999; Bahram *et al.*, 1994) or the human homologue of the rodent neonatal Fc receptor (FcRn), involved in bidirectional transport of IgG across the epithelium, are also expressed on human epithelial cells (Israel *et al.*, 1997). Restriction in expression to certain tissues and the lack of polymorphism in MHC class-Ib molecules suggests that they bind a limited array of very distinct ligands. This is especially relevant for intestinal epithelial cells which express several MHC class-Ib molecules. CD1d, which is predominantly expressed on IEC of the upper crypt and villi, presents exogenous and/or endogenous lipid antigens to T cells. IELs which express CD1d in co-association with β2-microglobulin were shown to induce secretion of IL-4 and IFNγ by NK-T cells, suggesting an important immunoregulatory function. Whether CD1d:ligand interaction on epithelial cells induces immunoregulatory cytokines needs further investigation. Expression of MICA on IEC is observed as a consequence of different stress signals and is thus thought to be involved in recognition of danger signals by CD94/NKG2 positive immune cells, leading to cytolysis of the damaged cell (Groh *et al.*, 1998).

It has also been shown that epithelial cells express complement factors (Andoh *et al.*, 1993), complement inhibitors (Guignot *et al.*, 2000), cytokine receptors (Bocker *et al.*, 1998) and can secrete cytokines and chemokines in response to pathogenic bacteria (Jung *et al.*, 1995; Eckmann *et al.*, 1993). More recently intracellular receptors for LPS have been described which were shown to be involved in the activation of NF-κB, leading to the secretion of pro-inflammatory mediators (Bertin *et al.*, 1999; Beatty *et al.*, 1999).

Recent advances in the characterization of microbial-epithelial interactions suggest that the innate epithelial response can also recognize and discriminate between different indigenous bacteria. These observations confirm the frontline role of the intestinal epithelial cells (IEC) in the recognition of components of the intestinal microflora. However, it is obvious that bacterial signals need to be processed by a network of different mucosal immune cells, resulting in an integrated response that dictates the host reaction against a constantly changing microbial environment in the intestine.

Human *in vitro* co-cultures, produced by the co-cultivation of human intestinal epithelial cell lines, such as CaCO-2 or HT-29 cells, and human peripheral blood mononuclear cells (PBMC) using a transwell culture technique, proved to be useful models to investigate the molecular basis of microbial:epithelial interactions (Haller *et al.*, 2000). Our group recently demonstrated that IEC can recognize non-pathogenic bacteria in the presence of PBMC. Furthermore, a characteristic response to a given non-pathogenic indigenous strain could be observed distinguishing between two major cytokine/chemokine responses of IEC. Non-pathogenic Gram-negative *Escherichia coli* and certain Gram-positive lactobacilli triggered a NF-κB mediated inflammatory response resulting in the production of TNFα, IL-1 β, IL-8 and MCP-1 (type 1). This initial epithelial pro-inflammatory reaction was only transient and self-limiting, as the PBMC in the basolateral compartment were able to switch off the initial inflammatory response of the epithelium. A detailed analysis of the role of different leukocyte subpopulations in the bacterial:mucosal cross-talk revealed that i) activation of epithelial cells by indigenous bacteria was promoted by T and B cells and that ii) macrophages acquired an immunosuppressive phenotype and were able to control inflammatory cytokine expression in IEC by the predominant secretion of IL-10, an inhibitory cytokine (Nathens *et al.*, 1995). Stimulation of CaCO-2/PBMC co-cultures with a non-invasive enteropathogenic *E. coli* (EPEC O111:H6) highlighted the differences in epithelial response to indigenous and pathogenic bacteria: whereas the pro-inflammatory cytokine induction was transient with the indigenous strains, inflammation was not inhibited in the case of EPEC treatment (Haller *et al.*, 2002).

TGFβ dependent direct epithelial homeostatic loop   **IL-10 dependent indirect homeostatic loop**

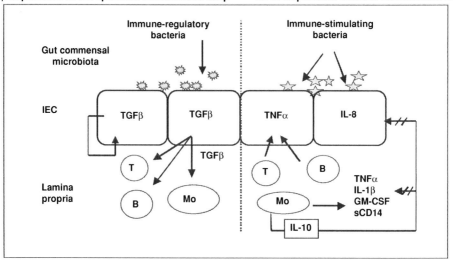

Figure 1. Proposed model for homeostasis regulation after activation of IEC by non-pathogenic bacteria. **Left side**: Direct TGFβ-dependent epithelial homeostatic effect induced by immune-regulatory non-pathogenic bacteria. TGFβ protects the barrier integrity. In addition, TGFβ promotes regulatory T cells (Th₃). **Right side**: Indirect IL-10-dependent homeostatic loop induced by immune-stimulatory bacteria. The transient pro-inflammatory response is blunted by lamina propria (LP) cells, particularly LP macrophages, which predominantly secrete IL-10.

A second class of lactobacilli, including *Lactobacillus johnsonni* La1 and *Lactobacillus gasseri*, induced the immunoregulatory cytokine TGFβ in IEC in the absence of any pro-inflammatory event (type 2). Of note, both indigenous bacterial stains were of human intestinal origin. TGFβ is produced by both immune and non-immune cells and exhibits a broad range of functions, the most important being the modulation of immune responses. In the intestinal immune system, TGFβ displays an important role in the maintenance of intestinal barrier integrity and induction of oral tolerance. Both types of epithelial responses against non-pathogenic bacteria seem to indicate the importance of either a self-limiting (type 1) or non-inflammatory (type 2) cellular immune response in the context of the antigen rich intestinal environment. It appears that certain components of the gut microflora may contribute to maintain a low level of 'physiological' intestinal inflammation, whereas others directly promote the production of immunoregulatory cytokines (Figure 1). These examples of an integrated epithelial/lamina propria response strongly suggest a role for the indigenous microflora in homeostatic responses. Furthermore, these results provide direct evidence of the beneficial effects of specific probiotic strains on intestinal immune homeostasis.

Failure of Immunoregulation and Chronic Intestinal Inflammation

There is evidence that chronic intestinal inflammation, such as in Crohn's disease, is caused by excessive immune response to mucosal antigens and elements of the normal bacterial microflora, inappropriately controlled by the normal counter regulatory mechanisms. This includes the induction of T cell anergy or clonal deletion and the expansion of antigen specific regulatory T cells (CD4$^+$CD25$^+$) in the lamina propria that produce suppressor cytokines, such as TGFβ (Th3 cells) or IL-10 (Tr1 cells) (Groux *et al.*, 1997; Singh *et al.*, 2001; Strober *et al.*, 2001b).

The development of intestinal inflammation in knockout and transgenic rodents has confirmed how genetic and environmental factors are both responsible for mucosal inflammation. Spontaneous inflammation of the GI tract has been demonstrated in transgenic rats expressing the human HLA-B27 transgene, in mice deficient for IL-2, IL-2Rα, IL-10, TCRα, TGF-β1 (Blumberg *et al.*, 1999) or CD4$^+$CD45RB^{hi} reconstituted SCID mice (Leach *et al.*, 1996). The various models of mucosal inflammation are characterized by either the overproduction of key Th1 effector cytokines (IL-12, IFNγ or TNFα) or the underproduction of regulatory cytokines (IL-10, TGFβ) (Strober *et al.*, 2001a). However, regardless of the immunological basis of experimental inflammation, the latter is dependent on the presence of the normal bacterial microflora, as inflammation does not develop under germ-free conditions (Bhan *et al.*, 1999). This suggests that inflammatory bowel diseases (IBD) are likely due to an abnormal response to the normal intestinal content rather than to intestinal pathogens (Duchmann *et al.*, 1995). Experimental animal studies demonstrated the importance of both IL-10 and TGFβ in immunoregulation at mucosal sites, more recently suggesting that TGFβ-producing cells are the primary suppressor cells, but that IL-10 is necessary for these cells to expand in a Th1 dominant environment that would otherwise inhibit the expansion of these cells (Kitani *et al.*, 2000; Groux *et al.*, 1997). In addition to regulatory T cells, TGFβ signaling in IEC plays a crucial role in maintaining mucosal immune homeostasis, as TGFβ regulates expression of MHC class II molecules and activity of metalloproteinases (Hahm *et al.*, 2000). Inhibition of TGFβ signaling due to overexpression of the Smad7 protein in target cells was shown to maintain the chronic production of pro-inflammatory cytokines (Monteleone *et al.*, 2001).

NF-κB modulation has become an obvious target for anti-inflammatory therapy. It has been recently reported that some non-pathogenic bacteria can prevent NF-κB activation through inhibition of IκB-α ubiquitination. This may suggest that luminal microflora can send positive and negative signals to mucosal epithelial cells as part of the interactions with the host (Neish *et al.*, 2000).

Indigenous bacteria, which have the capacity to induce immune-regulatory cytokines, such as TGFβ in the intestinal epithelium or IL-10 in lamina propria macrophages, may have the potential to modulate gastrointestinal inflammation. This could represent a nutritional strategy to contribute to the reconstitution of intestinal homeostasis in the case of its impairment due to genetic defects (Hugot *et al.*, 2001; Ogura *et al.*, 2001). Thus, indigenous bacterial strains, including probiotic bacteria, selected for their properties to activate an epithelial inhibitory response may be a rational nutritional intervention in IBD patients.

References

Aderem, A. and Ulevitch, R.J. 2000. Toll-like receptors in the induction of the innate immune response. Nature. 406: 782-787.

Anderson, W.J. and Watson, L. 2001. Asthma and the hygiene hypothesis. N. Engl. J. Med. 344: 1643-1644.

Andoh, A., Fujiyama, Y., Bamba, T., and Hosoda, S. 1993. Differential cytokine regulation of complement C3, C4, and factor B synthesis in human intestinal epithelial cell line, Caco-2. J. Immunol. 151: 4239-4247.

Apperloo-Renkema, H.Z., Jagt, T.G., Tonk, R.H., and van der, W.D. 1993. Healthy individuals possess circulating antibodies against their indigenous faecal microflora as well as against allogenous faecal microflora: an immunomorphometrical study. Epidemiol. Infect. 111: 273-285.

Bahram, S., Bresnahan, M., Geraghty, D.E., and Spies, T. 1994. A second lineage of mammalian major histocompatibility complex class I genes. Proc. Natl. Acad. Sci. USA. 91: 6259-6263.

Bauer, S., Groh, V., Wu, J., Steinle, A., Phillips, J.H., Lanier, L.L., and Spies, T. 1999. Activation of NK cells and T cells by NKG2D, a receptor for stress-inducible MICA. Science. 285: 727-729.

Beatty, W.L., Meresse, S., Gounon, P., Davoust, J., Mounier, J., Sansonetti, P.J., and Gorvel, J.P. 1999. Trafficking of Shigella lipopolysaccharide in polarized intestinal epithelial cells. J. Cell Biol. 145: 689-698.

Benno, Y., Suzuki, K., Suzuki, K., Narisawa, K., Bruce, W.R., and Mitsuoka, T. 1986. Comparison of the fecal microflora in rural Japanese and urban Canadians. Microbiol. Immunol. 30: 521-532.

Bertin, J., Nir, W.J., Fischer, C.M., Tayber, O.V., Errada, P.R., Grant, J.R., Keilty, J.J., Gosselin, M.L., Robison, K.E., Wong, G.H., Glucksmann, M.A., and DiStefano, P.S. 1999. Human CARD4 protein is a novel CED-4/Apaf-1 cell death family member that activates NF-kappaB. J. Biol. Chem. 274: 12955-12958.

Bhan, A.K., Mizoguchi, E., Smith, R.N., and Mizoguchi, A. 1999. Colitis in transgenic and knockout animals as models of human inflammatory bowel disease. Immunol. Rev. 169: 195-207.

Blumberg, R.S., Saubermann, L.J., and Strober, W. 1999. Animal models of mucosal inflammation and their relation to human inflammatory bowel disease. Curr. Opin. Immunol. 11: 648-656.

Blumberg, R.S., Terhorst, C., Bleicher, P., McDermott, F.V., Allan, C.H., Landau, S.B., Trier, J.S., and Balk, S.P. 1991. Expression of a nonpolymorphic MHC class I-like molecule, CD1D, by human intestinal epithelial cells. J. Immunol. 147: 2518-2524.

Bocker, U., Damiao, A., Holt, L., Han, D.S., Jobin, C., Panja, A., Mayer, L., and Sartor, R.B. 1998. Differential expression of interleukin 1 receptor antagonist isoforms in human intestinal epithelial cells. Gastroenterology. 115: 1426-1438.

Bouhnik, Y., Pochart, P., Marteau, P., Arlet, G., Goderel, I., and Rambaud, J.C. 1992. Fecal recovery in humans of viable *Bifidobacterium* sp ingested in fermented milk. Gastroenterology. 102: 875-878.

Bouhnik, Y., Vahedi, K., Achour, L., Attar, A., Salfati, J., Pochart, P., Marteau, P., Flourie, B., Bornet, F., and Rambaud, J.C. 1999. Short-chain fructo-oligosaccharide administration dose-dependently increases fecal bifidobacteria in healthy humans. J. Nutr. 129: 113-116.

Brightbill, H.D., Libraty, D.H., Krutzik, S.R., Yang, R.B., Belisle, J.T., Bleharski, J.R., Maitland, M., Norgard, M.V., Plevy, S.E., Smale, S.T., Brennan, P.J., Bloom, B.R., Godowski, P.J., and Modlin, R.L. 1999. Host defense mechanisms triggered by microbial lipoproteins through toll-like receptors. Science. 285: 732-736.

Cario, E., Brown, D., McKee, M., Lynch-Devaney, K., Gerken, G., and Podolsky, D.K. 2002. Commensal-associated molecular patterns induce selective toll-like receptor-trafficking from apical membrane to cytoplasmic compartments in polarized intestinal epithelium. Am. J. Pathol. 160: 165-173.

Cario, E. and Podolsky, D.K. 2000. Differential alteration in intestinal epithelial cell expression of toll-like receptor 3 (TLR3) and TLR4 in inflammatory bowel disease. Infect. Immun. 68: 7010-7017.

Cebra, J.J. 1999. Influences of microbiota on intestinal immune system development. Am. J. Clin. Nutr. 69: 1046S-1051S.

Chertov, O., Michiel, D.F., Xu, L., Wang, J.M., Tani, K., Murphy, W.J., Longo, D.L., Taub, D.D., and Oppenheim, J.J. 1996. Identification of defensin-1, defensin-2, and CAP37/azurocidin as T-cell chemoattractant proteins released from interleukin-8-stimulated neutrophils. J. Biol. Chem. 271: 2935-2940.

Dignass, A.U. and Podolsky, D.K. 1993. Cytokine modulation of intestinal epithelial cell restitution: central role of transforming growth factor beta. Gastroenterology. 105: 1323-1332.

Duchmann, R., Kaiser, I., Hermann, E., Meyet, W., Ewe, K., and Meyer zum B,schenfelde, K.-H. 1995. Tolerance exists towards resident intestinal flora but is broken in active inflammatory bowel disease (IBD). Clin. Exp. Immunol. 102: 448-455.

Eckmann, L., Jung, H. C., Sch,rer-Maly, C., Panja, A., Morzycka-Wrobleska, E., and Kagnoff, M. K. 1993. Differential cytokine expression by human intestinal epithelial cell lines: regulated expression of interleukin 8. Gastroenterology. 105: 1689-1697.

Fraser, I.P., Koziel, H., and Ezekowitz, R.A. 1998. The serum mannose-binding protein and the macrophage mannose receptor are pattern recognition molecules that link innate and adaptive immunity. Semin. Immunol. 10: 363-372.

Goke, M., Kanai, M., and Podolsky, D.K. 1998. Intestinal fibroblasts regulate intestinal epithelial cell proliferation via hepatocyte growth factor. Am. J. Physiol. 274: G809-G818.

Groh, V., Steinle, A., Bauer, S., and Spies, T. 1998. Recognition of stress-induced MHC molecules by intestinal epithelial gammadelta T cells. Science. 279: 1737-1740.

Groux, H., O'Garra, A., Bigler, M., Rouleau, M., Antonenko, S., de Vries, J.E., and Roncarolo, M.G. 1997. A CD4+ T-cell subset inhibits antigen-specific T-cell responses and prevents colitis. Nature. 389: 737-742.

Guignot, J., Peiffer, I., Bernet-Camard, M.F., Lublin, D.M., Carnoy, C., Moseley, S.L., and Servin, A.L. 2000. Recruitment of CD55 and CD66e brush border-associated glycosylphosphatidylinositol-anchored proteins by members of the Afa/Dr diffusely adhering family of *Escherichia coli* that infect the human polarized intestinal Caco-2/TC7 cells. Infect. Immun. 68: 3554-3563.

Hahm, K.B., Im, Y.H., Lee, C., Parks, W.T., Bang, Y.J., Green, J.E., and Kim, S.J. 2000. Loss of TGF-beta signaling contributes to autoimmune pancreatitis. J. Clin. Invest. 105: 1057-1065.

Haller, D., Bode, C., Hammes, W.P., Pfeifer, A.M., Schiffrin, E.J., and Blum, S. 2000. Non-pathogenic bacteria elicit a differential cytokine response by intestinal epithelial cell/leucocyte co-cultures. Gut. 47: 79-87.

Haller, D., Serrant, P., Perruisseau, G., Bode, C., Hammes, W. P., Schiffrin, E., and Blum, S. 2002. IL-10 producing CD14low monocytes inhibite lymphocyte-dependent activation of intestinal epithelial cells by commensal bacteria. Microbiology & Immunology. 46 (3).

Hecht, G. 1999. Innate mechanisms of epithelial host defense: spotlight on intestine. Am. J. Physiol. 277: C351-C358.

Hershberg, R.M., Cho, D.H., Youakim, A., Bradley, M.B., Lee, J.S., Framson, P.E., and Nepom, G.T. 1998. Highly polarized HLA class II antigen processing and presentation by human intestinal epithelial cells. J. Clin. Invest. 102: 792-803.

Herzenberg, L.A., Baumgarth, N., and Wilshire, J.A. 2000. B-1 cell origins and VH repertoire determination. Curr. Top. Microbiol. Immunol. 252: 3-13.

Hooper, L.V. and Gordon, J.I. 2001. Commensal host-bacterial relationships in the gut. Science. 292: 1115-1118.

Hopkins, M.J., Sharp, R., and Macfarlane, G.T. 2001. Age and disease related changes in intestinal bacterial populations assessed by cell culture, 16S rRNA abundance, and community cellular fatty acid profiles. Gut. 48: 198-205.

Hugot, J.P., Chamaillard, M., Zouali, H., Lesage, S., Cezard, J.P., Belaiche, J., Almer, S., Tysk, C., O'Morain, C.A., Gassull, M., Binder, V., Finkel, Y., Cortot, A., Modigliani, R., Laurent-Puig, P., Gower-Rousseau, C., Macry, J., Colombel, J.F., Sahbatou, M., and Thomas, G. 2001. Association of NOD2 leucine-rich repeat variants with susceptibility to Crohn's disease. Nature. 411: 599-603.

Isolauri, E. 1997. Intestinal involvement in atopic disease. J. Royal Soc. Med. 90: 15-20.

Israel, E.J., Taylor, S., Wu, Z., Mizoguchi, E., Blumberg, R.S., Bhan, A., and Simister, N.E. 1997. Expression of the neonatal Fc receptor, FcRn, on human intestinal epithelial cells. Immunology. 92: 69-74.

Jung, H. C., Eckmann, L., Yang, S. K., Panja, A., Fierer, J., Morzycka-Wrobleska, E., and Kagnoff, M. K. 1995. A distinct array of pro-inflammatory cytokines is expressed in human colon eoithelial cells in response to bacterial invasion. J. Clin. Invest. 95: 55-65.

Kett, K., Baklein, K., Bakken, A., Kral, J. G., Fausa, O., and Brandtzaeg, P. 1995. Intestinal B-cell isotype response in relation to bacterial load: evidence for immunoglobulin A subclass adaptation. Gastroenterology. 109: 819-825.

Kimura, K., McCartney, A.L., McConnell, M.A., and Tannock, G.W. 1997. Analysis of fecal populations of bifidobacteria and lactobacilli and investigation of the immunological responses of their human hosts to the predominant strains. Appl. Environ. Microbiol. 63: 3394-3398.

Kindon, H., Pothoulakis, C., Thim, L., Lynch-Devaney, K., and Podolsky, D.K. 1995. Trefoil peptide protection of intestinal epithelial barrier function: cooperative interaction with mucin glycoprotein. Gastroenterology. 109: 516-523.

Kitani, A., Fuss, I.J., Nakamura, K., Schwartz, O.M., Usui, T., and Strober, W. 2000. Treatment of experimental (Trinitrobenzene sulfonic acid) colitis by intranasal administration of transforming growth factor (TGF)-beta1 plasmid: TGF-beta1-mediated suppression of T helper cell type 1 response occurs by interleukin (IL)-10 induction and IL-12 receptor beta2 chain downregulation. J. Exp. Med. 192: 41-52.

Krinos, C.M., Coyne, M.J., Weinacht, K.G., Tzianabos, A.O., Kasper, D.L., and Comstock, L.E. 2001. Extensive surface diversity of a commensal microorganism by multiple DNA inversions. Nature. 414: 555-558.

Leach, M.W., Bean, A.G., Mauze, S., Coffman, R.L., and Powrie, F. 1996. Inflammatory bowel disease in C.B-17 scid mice reconstituted with the CD45RBhigh subset of CD4+ T cells. Am. J. Pathol. 148: 1503-1515.

Lebman, D. A., Lee, F. D., and Coffman, R. L. 1990. Mechanism for transforming growth factor beta and IL-2 enhancement for IgA expression in liposaccharide-stimulated B cell cultures. J. Immunol. 144: 952-959.

Lencer, W.I., Cheung, G., Strohmeier, G.R., Currie, M.G., Ouellette, A.J., Selsted, M.E., and Madara, J.L. 1997. Induction of epithelial chloride secretion by channel-forming cryptdins 2 and 3. Proc. Natl. Acad. Sci. USA. 94: 8585-8589.

Lillard, J.W., Jr., Boyaka, P.N., Chertov, O., Oppenheim, J.J., and McGhee, J.R. 1999. Mechanisms for induction of acquired host immunity by neutrophil peptide defensins. Proc. Natl. Acad. Sci. USA. 96: 651-656.

Mack, D.R., Michail, S., Wei, S., McDougall, L., and Hollingsworth, M.A. 1999. Probiotics inhibit enteropathogenic *E. coli* adherence *in vitro* by inducing intestinal mucin gene expression. Am. J. Physiol. 276: G941-G950.

Macpherson, A.J., Gatto, D., Sainsbury, E., Harriman, G.R., Hengartner, H., and Zinkernagel, R.M. 2000. A primitive T cell-independent mechanism of intestinal mucosal IgA responses to commensal bacteria. Science. 288: 2222-2226.

Matricardi, P.M. and Bonini, S. 2000. High microbial turnover rate preventing atopy: a solution to inconsistencies impinging on the hygiene hypothesis? Clin. Exp. Allergy. 30: 1506-1510.

Matzinger, P. 1998. An innate sense of danger. Semin. Immunol. 10: 399-415.

Mayer, L. 1998. Current concepts in mucosal immunity. I. Antigen presentation in the intestine: new rules and regulations. Am. J. Physiol. 274: G7-G9.

Medzhitov, R. and Janeway, C., Jr. 2000. The Toll receptor family and microbial recognition. Trends Microbiol. 8: 452-456.

Molmenti, E.P., Ziambaras, T., and Perlmutter, D.H. 1993. Evidence for an acute phase response in human intestinal epithelial cells. J. Biol. Chem. 268: 14116-14124.

Monteleone, G., Kumberova, A., Croft, N.M., McKenzie, C., Steer, H.W., and MacDonald, T.T. 2001. Blocking Smad7 restores TGF-beta1 signaling in chronic inflammatory bowel disease. J. Clin. Invest. 108: 601-609.

Moreau, M.C., Thomasson, M., Ducluzeau, R., and Raibaud, P. 1986. Kinetics of establishment of digestive microflora in the human newborn infant as a function of the kind of milk. Reprod. Nutr. Dev. 26: 745-753.

Nathens, A.B., Rotstein, O.D., Dackiw, A.P., and Marshall, J.C. 1995. Intestinal epithelial cells down-regulate macrophage tumor necrosis factor-alpha secretion: a mechanism for immune homeostasis in the gut-associated lymphoid tissue. Surgery. 118: 343-350.

Neish, A.S., Gewirtz, A.T., Zeng, H., Young, A.N., Hobert, M.E., Karmali, V., Rao, A.S., and Madara, J.L. 2000. Prokaryotic regulation of epithelial responses by inhibition of IkappaB-alpha ubiquitination. Science. 289: 1560-1563.

Ogura, Y., Bonen, D.K., Inohara, N., Nicolae, D.L., Chen, F.F., Ramos, R., Britton, H., Moran, T., Karaliuskas, R., Duerr, R.H., Achkar, J.P., Brant, S.R., Bayless, T.M., Kirschner, B.S., Hanauer, S.B., Nunez, G., and Cho, J.H. 2001. A frameshift mutation in NOD2 associated with susceptibility to Crohn's disease. Nature. 411: 603-606.

Park, S.H. and Bendelac, A. 2000. CD1-restricted T-cell responses and microbial infection. Nature. 406: 788-792.

Parkkila, S., Waheed, A., Britton, R.S., Feder, J.N., Tsuchihashi, Z., Schatzman, R.C., Bacon, B.R., and Sly, W.S. 1997. Immunohistochemistry of HLA-H, the protein defective in patients with hereditary hemochromatosis, reveals unique pattern of expression in gastrointestinal tract. Proc. Natl. Acad. Sci. USA. 94: 2534-2539.

Pazmany, L., Mandelboim, O., Vales-Gomez, M., Davis, D.M., Reyburn, H.T., and Strominger, J.L. 1996. Protection from natural killer cell-mediated lysis by HLA-G expression on target cells. Science. 274: 792-795.

Philpott, D.J., Girardin, S.E., and Sansonetti, P.J. 2001. Innate immune responses of epithelial cells following infection with bacterial pathogens. Curr. Opin. Immunol. 13: 410-416.

Pochart, P., Marteau, P., Bouhnik, Y., Goderel, I., Bourlioux, P., and Rambaud, J.C. 1992. Survival of bifidobacteria ingested via fermented milk during their passage through the human small intestine: an *in vivo* study using intestinal perfusion. Am. J. Clin. Nutr. 55: 78-80.

Pulendran, B., Palucka, K., and Banchereau, J. 2001. Sensing pathogens and tuning immune responses. Science. 293: 253-256.

Singh, B., Read, S., Asseman, C., Malmstrom, V., Mottet, C., Stephens, L.A., Stepankova, R., Tlaskalova, H., and Powrie, F. 2001. Control of intestinal inflammation by regulatory T cells. Immunol. Rev. 182: 190-200.

Stahl, P.D. and Ezekowitz, R.A. 1998. The mannose receptor is a pattern recognition receptor involved in host defense. Curr. Opin. Immunol. 10: 50-55.

Strober, W., Fuss, I., and Kitani, A. 2001a. Regulation of experimental mucosal inflammation. Acta Odontol. Scand. 59: 244-247.

Strober, W., Nakamura, K., and Kitani, A. 2001b. The SAMP1/Yit mouse: another step closer to modeling human inflammatory bowel disease. J. Clin. Invest. 107: 667-670.

Tannock, G.W., Munro, K., Harmsen, H.J., Welling, G.W., Smart, J., and Gopal, P.K. 2000. Analysis of the fecal microflora of human subjects consuming a probiotic product containing *Lactobacillus rhamnosus* DR20. Appl. Environ. Microbiol. 66: 2578-2588.

Van der Waaij, L. A., Lindburg, P. C., Mesander, G., and van der Waaji, D. 1996. *In vivo* IgA coating of anaerobic bacteria in human faeces. Gut. 38: 348-354.

Vaughan, E.E., Schut, F., Heilig, H.G., Zoetendal, E.G., de Vos, W.M., and Akkermans, A.D. 2000. A molecular view of the intestinal ecosystem. Curr. Issues Intest. Microbiol. 1: 1-12.

Walter, J., Tannock, G.W., Tilsala-Timisjarvi, A., Rodtong, S., Loach, D.M., Munro, K., and Alatossava, T. 2000. Detection and identification of gastrointestinal *Lactobacillus* species by using denaturing gradient gel electrophoresis and species-specific PCR primers. Appl. Environ. Microbiol. 66: 297-303.

Weiner, H.L. 1997. Oral tolerance: immune mechanisms and treatment of autoimmune diseases. Immunol. Today. 18: 335-343.

Yoshimura, A., Lien, E., Ingalls, R.R., Tuomanen, E., Dziarski, R., and Golenbock, D. 1999. Cutting edge: recognition of Gram-positive bacterial cell wall components by the innate immune system occurs via Toll-like receptor 2. J. Immunol. 163: 1-5.

Zoetendal, E.G., Akkermans, A.D., and de Vos, W.M. 1998. Temperature gradient gel electrophoresis analysis of 16S rRNA from human fecal samples reveals stable and host-specific communities of active bacteria. Appl. Environ. Microbiol. 64: 3854-3859.

Index

A

Adhesion 126
Allergic Rhinoconjunctivitis 241
Allergy 239
    and acquisition of microflora
      248
    and antibiotics 250
    and gut mucosa 248
    and measles 246
    and mycobacteria 247
    and respiratory syncytial virus
      246
    hepatitis A virus 247
Anti-inflammatory activities 186
Asthma 242
    prevalence 243
Atopy 17, 239, 241

B

Bacteroides-Prevotella phylogenetic
    group 6
Barrier effect 125, 192
Bifidobacterium 9, 67, 85
    detection 95
    distribution 97, 98
    identification 92
    isolation 93
    phylogeny 87
    species-specific PCR 90, 92
Bile salt hydrolase 22
Bioinformatics and genome
    annotation 274
Biological freudianism 16
Blood lipids 124

C

Calcium 149
    absorption 155
    bioavailability 159
    measurement 157
Cancer of the colon 13, 121
Clostridium coccoides phylogenetic
    group 6
Clostridium leptum phylogenetic
    group 7
Crohn's disease 13, 176

D

Denaturing gradient gel
    electrophoresis 8, 11, 66
Dietary calcium 154
DNA microarrays 74, 284
    probes 287
    probiotics and microflora 291
DNA probes 43, 51, 115
DNA-RNA hybridisations 8

E

Eczema 241

F

Faecal microflora 12
Fingerprint interpretation 63
Fluorescence *in situ* hybridisation
    automated analysis 7, 41, 45, 115
    protocol 44

G

Gene inactivation 296
Genetic susceptibility and
    inflammatory bowel diseases 199
Genome 263
    sequencing 270
        projects 268
        impact 281
    technology 266
        applications 298
Genomics
    comparative 278
    functional 283
Gradient gel electrophoresis 61

H

Hayfever 241
Homeostasis, 11, 12
Hygiene hypothesis 244
    alternative 254
    inconsistencies 245

I

Immune response 27, 123, 194, 249,
    311
    secretory 313
Immune system
    tolerance 250
Immunoregulation 321
Infection
    sensing 317
Inflammation
    intestinal 321
Inflammatory bowel diseases 14, 175
    and aberrant intestinal microflora
        187
    and antibiotics 201, 204
    and clinical trials 213
    and *Clostridium difficile* 185

and cytomegalovirus 185
and *Escherichia coli* Nissle 1917
    213
and *Lactobacillus* GG 212
and *Listeria monocytogenes* 184
and measles virus 183
and *Mycobacterium
    paratuberculosis* 178
and pathogens 178
and *Saccharomyces boulardii*
    212
and VSL#3 217
Innate defences 315
Intestinal epithelial cells 318

L

Lactobacillus 10, 67

M

Metchnikoff, 2, 13
Microflora associated characteristics
    23, 25, 26
Microbial recognition 316
Microflora and allergy 252
Microflora
    dynamics 53
    variation 47
Models
    animal 15, 112, 188, 209
    colitis, 15
    in vitro 111
Molecular typing 19, 20
Murine characteristics 25, 26
Murine gastrointestinal tract 5

N

Nutrition and microflora 22

P

PCR 8, 87
    competitive 99
    quantitative 43, 87, 99, 100
PCR primers
    Bifidobacterium 90
    DGGE 67
Phylogenetic studies 7
Polysaccharide hydrolysis 135
Prebiotics 107, 109, 153
    and bone health 163
    and calcium 149, 160
    and calcium absorption 165
    as dietary regulators 19
    comparison 110, 116
    effects 151
    manufacture 131
    second generation 118
    stuctures 130
    targeted 127
Probiotic
    and allergy 253
    and inflammatory bowel diseases 175
    genes 280
    genetic tools
    therapy 175, 190, 200, 211
Proteomics 294

Q

Quantitative PCR 43, 87, 99, 100

R

Ribosomal RNA gene, 16S 7, 43, 86, 114

S

Selective culture 42, 113

T

Temperature gradient gel electrophoresis 8, 66
Terminal-restriction fragment length polymorphism 71

U

Ulcerative colitis 14, 176
    and *Escherichia coli* 184

W

Wheezing 242
Woese, 7